Liver Cirrhosis

Frontiers of Gastrointestinal Research

Vol. 8

Series Editor
P. Rozen, Tel Aviv

KARGER

S. Karger · Basel · München · Paris · London · New York · Tokyo · Sydney

Meeting of the Italian Group of Hepatic Cirrhosis,
S. Miniato, January 6–8, 1983

Liver Cirrhosis

Volume Editors
P. Gentilini, Firenze
M.U. Dianzani, Torino

71 figures and 60 tables, 1984

KARGER

S. Karger · Basel · München · Paris · London · New York · Tokyo · Sydney

Frontiers of Gastrointestinal Research

National Library of Medicine, Cataloging in Publication
Liver cirrhosis
Volume editors, P. Gentilini and M.U. Dianzani. - Basel; New York: Karger, 1984.
(Frontiers of gastrointestinal research; v. 8)
1. Liver Cirrhosis I. Gentilini, Paolo II. Dianzani, M.U. (Mario Umberto), 1925 - III. Series
W1 FR946E v. 8 [W1 725 L784]
ISBN 3-8055-3724-7

Drug Dosage

The authors and the publisher have exerted every effort to ensure that drug selection and dosage set forth in this text are in accord with current recommendations and practice at the time of publication. However, in view of ongoing research, changes in government regulations, and the constant flow of information relating to drug therapy and drug reactions, the reader is urged to check the package insert for each drug for any change in indications and dosage and for added warnings and precautions. This is particularly important when the recommended agent is a new and/or infrequently employed drug.

Printed in Switzerland by gdz Genossenschaftsdruckerei Zürich
ISBN 3-8055-3724-7

Contents

Epidemiological Studies

Metabolic Aspects

Complications

Therapeutic Approaches

Foreword

The Italian Liver Cirrhosis Group, supported by the Ministero della Pubblica Istruzione (National Project on 'Liver Cirrhosis') and consisting of 22 University Centres from all over the country, has achieved interesting results in research into chronic liver damage, namely cirrhosis.

The topics are divided into 6 different groups (experimental research on animals, pathogenic, epidemiological, metabolic aspects, complications occurring during cirrhosis and therapeutic approaches). Many of these studies proposed and included in the volume are also original in their methodological formulation.

Cirrhosis itself constitutes a major problem in both European and non-European countries, because of the many factors involved in its aetiopathogenesis. Besides hepatitis B virus infection, which could be combatted with the introduction of vaccine, there are many other elements such as alcohol, viral infections, toxins and drugs, which can attack the liver in different ways, according to individual immunoreactivity.

The aim of this volume is not to thoroughly satisfy all the questions on liver research, but it represents an advance and an up-date in clinical and experimental research into cirrhosis and chronic liver damage.

Firenze, February 25, 1983

P. Gentilini
M.U. Dianzani

Experimental Research

Front. gastrointest. Res., vol. 8, pp. 1–13 (Karger, Basel 1984)

Carbon Tetrachloride-Induced Block of Hepatic Lipoprotein Secretion: Studies of the Pathogenesis Using Isolated Rat Hepatocytes

M.U. Dianzani, G. Poli[1]

Istituto di Patologia Generale dell'Università, Torino, Italia

Introduction

The liver fatty degeneration induced by carbon tetrachloride (CCl_4) poisoning has been shown to be due to the derangement of the lipoprotein secretion pathway produced by the toxic compound [1]. This haloalkane is postulated to be toxic mainly through its homolytical cleavage in the smooth endoplasmic reticulum with the production of the trichloromethyl radical ($CCl_3^{\cdot}$) [2, 3]. Recent experiments have demonstrated rapid conversion of $CCl_3^{\cdot}$, in the presence of O_2, into $CCl_3O_2^{\cdot}$, the trichloromethylperoxy radical [4, 5]. These and other data led to the opinion that the former radical would more probably induce direct change of cell structures by binding to macromolecules and the latter radical, however, would be responsible for the initiation of lipid peroxidation [6].

Whether the CCl_4-induced change in liver lipoprotein secretion is due to its prooxidant effect or to the covalent binding of its metabolites to cell structures is not yet completely clear.

Previous reports from our laboratory produced evidence against a major role of lipid peroxidation in the pathogenesis of this defined cell impairment [7–9]. Those studies mainly concerned the attempt to prevent the block of lipoprotein secretion produced in whole rat hepatocytes by CCl_4 through the increase of their antioxidant potential.

The data reported here further confirm the inefficacy of both a high cell content in glutathione and a cell treatment with the antioxi-

[1] The authors wish to thank the Italian Ministry of Education for financial support.

dant promethazine in preventing the CCl_4-induced damage of liver secretion. In addition, evidence is produced supporting a prominent role of covalent binding of CCl_4 metabolites in the pathogenesis of the cell injury under study.

Materials and Methods

Animals. Male Wistar rats of the Wistar strain (Nossan, Correzzana, Italy) of 200–250 g body weight were used. They were fed on semisynthetic diet, free of any antioxidant (Piccioni, Brescia, Italy) and had free access to water. All experiments started between 09.00 and 10.00 hours.

Reagents. All chemicals were of reagent grade and were obtained from the following sources: collagenase type I, amino acids, ethyleneglycol -*bis*-(β-aminoethyl-ether)-*N,N'*-tetraacetic acid (EGTA), and *N*-2-hydroxyethylpiperazine-*N'*-2-ethane-sulfonic acid (HEPES) from Sigma Chemical Co., St. Louis, Mo.; promethazine-HCl from May and Baker, Dagenham, UK; [U-^{14}C] palmitic acid and [^{14}C] carbon tetrachloride from The Radiochemical Centre, Amersham, UK; other chemicals from BDH Chemicals Ltd., Poole, UK, and Merck, Darmstadt, FRG.

Preparation of Isolated Hepatocytes. Intact rat liver cells were isolated by the collagenase perfusion technique described in preceding papers [1, 10]. Figure 1 shows the procedure used to obtain cells with low or high, nearly physiological, glutathione content. In fact, the addition of EGTA at the first step of the liver perfusion prevents the oxidation of reduced glutathione (GSH) by trace metals (Fe, Cu, Mn) with consequent diffusion of oxidized glutathione (GSSG) into the medium.

CCl_4 Poisoning of Isolated Hepatocytes. The hepatocyte suspension was diluted with a defined salt solution [1] to 5 or 7 × 10^6 cells/ml. Aliquots of 3 ml of the suspension were placed in the main compartment of 50-ml flasks, fitted with a centre well and closed with a screw cap. CCl_4 (7.5 μl/flask) was added in the centre well and allowed to diffuse through the closed system (fig. 2). The CCl_4 final concentration in the cell suspension (129 μ*M*) was checked using [^{14}C]CCl_4 [1]. Flasks were incubated for 30 or 60 min at 37 °C in a shaking water bath. When added to the cell suspension, promethazine reached a final concentration of 10 or 100 μ*M*.

Triglyceride Secretion from Prelabelled Hepatocytes. Cell triglycerides were prelabelled by incubating aliquots of the cell suspension with [^{14}C] sodium palmitate (specific activity 403 mCi/mmol) as elsewhere reported [1]. After 60 min incubation at 37 °C, the cell suspension was diluted, centrifuged, and then resuspended in fresh incubation medium, ready to test CCl_4 action on triglyceride lipoprotein release from the liver cells. At the end of the treatment, cell suspensions were centrifuged and the triglycerides in the supernatants purified and processed for radioactivity measurements as described elsewhere [1].

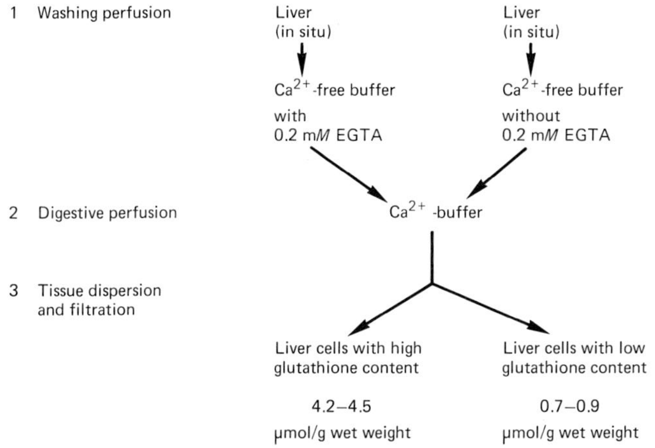

Fig. 1. Flowsheet of the procedure to isolate hepatocytes with high or low glutathione content.

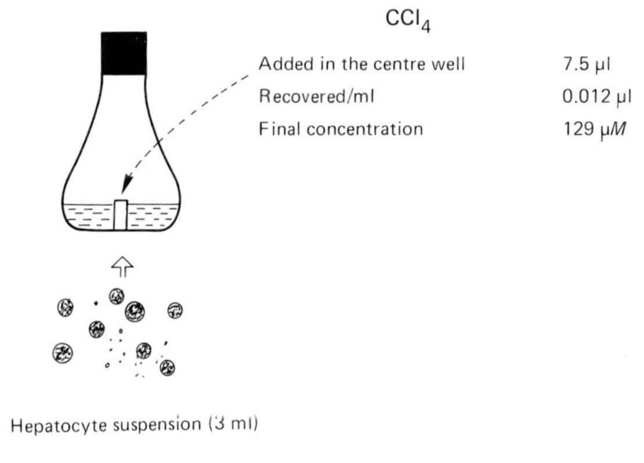

Fig. 2. Isolated liver cell treatment with CCl_4. See also Materials and Methods section.

Separation of Polar and Unpolar Carbonyls Produced during Cell Poisoning with CCl_4. The flowsheet of the procedure for extraction and separation in three main groups the carbonyl products of lipid peroxidation triggered by CCl_4 is reported in table I. In order to determine the total carbonyl groups produced, an aliquot of each sample was withdrawn before thin-layer chromatography (TLC) on Silica Gel 60 and the corresponding maximum absorption at 350 nm measured ($\varepsilon = 26{,}000$).

Table I. Separation of carbonyl compounds produced during CCl_4-induced lipid peroxidation. Only malonaldehyde was not separated with this method

Aldehyde extraction by sample treatment with dinitrophenylhydrazine in *N 1* HCl (2:1) for 2 h in the dark at 25 °C and for 1 h in freezer
Separation of pellet and supernatant by centrifugation at 3,000 rpm for 10 min

Pellet	Supernatant
Extraction of polar and non-polar aldehydes with $CHCl_3$-CH_3OH, mixture (2:1) twice	Extraction of non-polar aldehydes with CH_2Cl_2 twice

Mix the two extracts and keep the mixture in freezer for at least 1 night

Filtration and evaporation of the samples

Thin-layer chromatography in CH_2Cl_2
(samples are resuspended in CH_2Cl_2)

Alkanals, alkenals	fraction III
Acetone	
Osazones	fraction II
Dinitrophenylhydrazine (excess)	
OH-alkanals, OH-alkenals	fraction I
Polar Fraction	

Determination of Binding of CCl_4 Metabolites to Different Cell Suspension Fractions. For these studies at least 1.5–2 × 10^8 cells were necessary. Aliquots of 2 ml of the hepatocyte suspension diluted with the mentioned salt solution to 7 × 10^6 cells/ml were poured in 50-ml flasks and allowed to incubate at 37 °C for 120 min in the presence of unlabelled sodium palmitate. The procedure was essentially that described in a previous report [10]. After incubation, cell aliquots were pooled then centrifuged at 400 *g* × 4 min. The cell pellet was resuspended with fresh incubation medium to 7 × 10^6 cells/ml and aliquots of 6 ml of the suspension were poured in 100-ml flasks without the centre well. Contrary to the usual procedure, [^{14}C] CCl_4 (specific activity 26.9 mCi/mmol) suitably dissolved in acetone to reach a final concentration of about 200 μ*M*, was added directly to the cell suspension. Hepatocytes were then incubated at 37 °C for 30 min in a shaking bath. Assessment of cell viability was performed at the end of the incubation. CCl_4-treated cells did not show significant changes in either Trypan Blue exclusion or lactate dehydrogenase release when compared to untreated hepatocytes incubated under the same conditions in the presence or absence of a corresponding amount of acetone (22 m*M*). After CCl_4 poisoning, cells were separated from the incubation medium and processed separately. (1) Cells: the pellets were weighed and homogenized in 4 volumes (w/v) of 0.25 *M* sucrose using a MSE 150 W ultrasonic disintegrator Mk2 at 20 μ of amplitude for 25 s.

The homogenate was centrifuged at 25,000 *g* 10 min in a Servall SS-1 centrifuge. The resulting pellet which represented cell debris plus nuclei plus mitochondria fractions was retained. The supernatant was centrifuged at 100,000 *g* for 60 min in a Beckman L8-55 centrifuge. The resulting pellet was taken as the microsomal fraction. The fraction of low density lipoproteins was obtained from the cell sap, as hereafter described. Each fraction was then digested with 1 ml of Unisol (Isolab, Akron, Ohio) and mixed with 4 ml of Pico-Fluor 30 (Packard Instruments Co.). The recovered [^{14}C] CCl_4 radioactivity was evaluated with a Packard Prias liquid scintillation counter. (2) Cell incubation medium: the fraction represented by the density lipoproteins was also separated from this and processed for radioactivity measurements as described above.

Preparation of Low Density Lipoprotein Fraction. Because of the small amounts of cells usually employed and the minimal synthesis and secretion of lipoproteins by hepatocytes incubated for a short time, the lipoprotein separation by means of ultracentrifugation procedure was discarded from the beginning of these studies and a procedure of chemical precipitation set up by *Burstein* [11] was adopted. One part of 2% sodium phosphotungstate (pH 7.6) in 1 *M* $MgCl_2$ was added to 9 parts of cell sap or cell incubation in order to precipitate lipoproteins. Rat serum (0.1 ml every 5 ml of the mixture) was added as coprecipitant. The samples were left overnight at 4 °C, then the precipitates were reharvested after centrifugation at 10,000 *g* for 45 min with a Servall SS-1 centrifuge. The selectivity of the precipitation procedure for lipoproteins rich in triglycerides was very recently stressed by *Burstein and Legmann* [12]. By using electrophoresis on polyacrylamide gel we also observed absence of α-lipoproteins and negligible albumin contamination in precipitates obtained without addition of serum and redissolved according to *Burstein* [11]. Further treatment of the supernatant resulting from the centrifugation of precipitates at 10,000 *g* with Na-phosphotungstate and $MgCl_2$ did not produce any further evident material. The finding confirmed the complete precipitation of low density lipoproteins by this method.

Determination of [^{14}C] CCl_4 Radioactivity Covalently Bound to Cell Lipid and Protein. Cell preincubation in the presence of sodium palmitate and cell poisoning with [^{14}] CCl_4 was performed as described above. After separation of defined fractions from cells and incubation medium by ultracentrifugation or chemical precipitation, aliquots of the fractions themselves were processed to measure only the radioactivity covalently bound to the cell lipid or protein. (1) Lipid-bound radioactivity: about one quarter of the total amount of each fraction was extracted 1 h with $CHCl_3$-CH_3OH mixture (2:1 v/v). The chloroform phase was then separated and exsiccated under nitrogen. The extracted lipids were washed 5 times with $CHCl_3$ in order to remove any lipophylic substance including [^{14}C]CCl_4 not covalently bound, and each time they were dried down under nitrogen. The final dried lipid layer was resuspended in 5 ml of Pico-Fluor 30 scintillation cocktail and the radioactivity measured as above. (2) Protein-bound radioactivity: the protein of about one quarter of the single fractions was precipitated with 10% trichloroacetic acid (TCA) and collected by centrifugation at 2,500 *g* for 10 min. The resulting protein pellet was first washed once with 5% TCA, 3 times with a mixture of chloroform-methanol-diethylether (2:2:1 v/v), then once with acetone and again with ether in order to avoid any lipophylic contaminant. Finally, the dried protein layer was solubilized with 0.5 ml of Unisol and mixed with 4.5 ml of scintillation cocktail to measure bound radioactivity.

Incubation of the Isolated Lipoprotein Fraction with [^{14}C]CCl_4. Aliquots of the low density lipoprotein fraction, identical to that processed for determination of covalent bound radioactivity, were dissolved as reported above then suspended in 6 ml of incubation medium and incubated for 30 min at 37 °C in the presence of 200 μ*M* [^{14}C]CCl_4 and in the absence of cells. The determination of lipid and protein-bound radioactivity in the fractions followed the incubation.

Results and Discussion

In the first set of experiments we checked the action of CCl_4 on the triglyceride lipoprotein secretion by hepatocytes whose lipids were prelabelled. The aim was to check whether physiological content of glutathione (4.5 m*M*) was able to significantly reduce the CCl_4-induced block of liver cell secretion obtained using hepatocytes with low glutathione content (about 0.8 m*M*) [1]. Figure 3 shows the percent of inhibition of this cell function provoked by CCl_4. The impairment was practically identical to that obtained after the same incubation time (30 min) with hepatocytes poor in glutathione [1, 7]. Results, reported elsewhere confirmed such a finding already after 10 min of cell treatment with the haloalkane [8]. As expected, the addition of 10 μ*M* promethazine to the cell suspension before its incubation did not prevent CCl_4 damage at all (fig. 3). This strong antioxidant has been found to significantly inhibit not only malonaldehyde production [13] but also the loss of arachidonate [*Slater,* personal communication] and the oxygen uptake [14] in liver microsomes under peroxidative stress. Furthermore, while the above-reported experiments were in progress, we also carried out chromatographic analyses of the complex pattern of carbonyls which are now recognized to originate during lipid peroxidation [15, 16]. In figure 4 preliminary results obtained with intact hepatocytes poisoned with CCl_4 clearly show that promethazine both suppresses malonaldehyde production and during the peroxidative process highly inhibits the formation of several other aldehydes, such as alkanals, alkenals, OH-alkanals, OH-alkenals, and compounds now considered for their potential biological importance [17–19]. Up to now, studies have produced only strong evidence against the primary involvement of lipid peroxidation in CCl_4-induced impairment of liver lipoprotein secretion [7–9, 20]. At present, the leading opinion on the mechanisms by which CCl_4 exerts its toxic effects is summarized in figure 5 [6]. CCl_4 is activated by the cytochrome P-450 enzyme system at the flavoprotein or

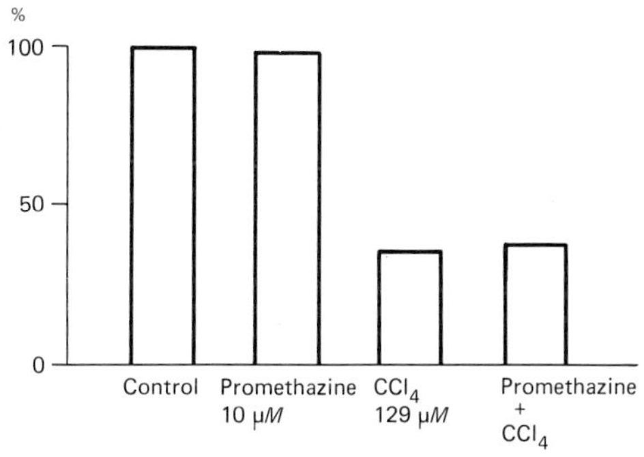

Fig. 3. Lipoprotein secretion by isolated hepatocytes whose lipids were prelabelled. Effect of promethazine on the secretory block exerted by CCl_4. In the typical experiment reported, hepatocytes were incubated for 30 min at 37 °C with or without CCl_4 in the presence or absence of the phenothiazine. The results are expressed as percentage of the control.

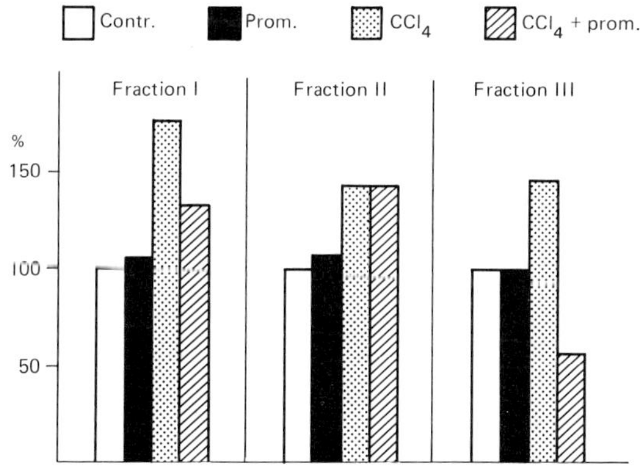

Fig. 4. Effect of promethazine on the CCl_4-induced production of unpolar carbonyls. In the typical experiment reported, cells were incubated for 60 min at 37 °C with or without 129 μM CCl_4 in the presence or absence of 100 μM promethazine. After incubation the unpolar carbonyls produced were extracted and separated in three fractions (see also Materials and Methods section). The results are expressed as percentage of the control.

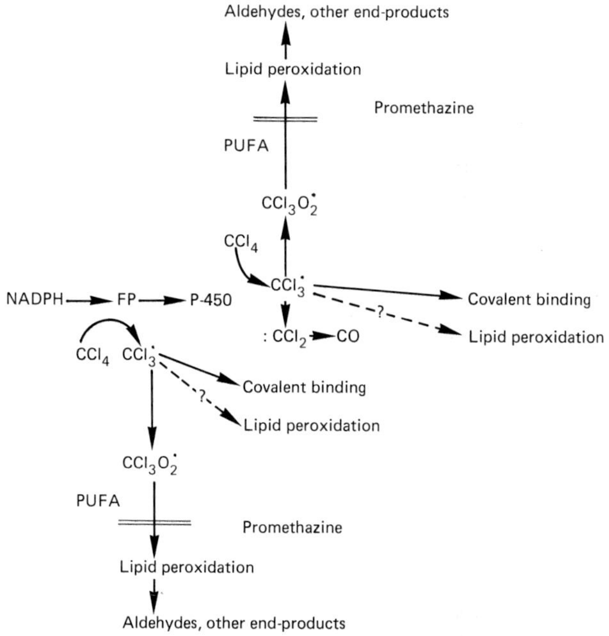

Fig. 5. Scheme of the activation of CCl_4 to the trichloromethyl radical and of the consequent various reactions. The scavenging action exerted by promethazine on the trichloromethylperoxy radical is also reported. NADPH = Nicotinamide-adenine dinucleotide phosphate, reduced; FP = flavoprotein; P 450 = cytochrome P 450; PUFA = polyunsaturated fatty acids.

the haemoprotein level or both, and $CCl_3^{\cdot}$ is originated. The CCl_4-cytochrome P-450 interaction has been postulated to go at least partially through a carbene intermediate [21, 22]. $CCl_3^{\cdot}$ is probably the reactive intermediate mainly responsible for direct cell damage through covalent binding to macromolecules, while the peroxidative breakdown of membrane polyunsaturated fatty acids proceeds much more rapidly with $CCl_3O_2^{\cdot}$ than $CCl_3^{\cdot}$ [23]. Promethazine has only a weak effect on covalent binding of [^{14}C]CCl_4 to microsomal protein [24] and cell protein and lipid [*Poli,* unpublished data] whereas it strongly reduces the CCl_4 stimulation of lipid peroxidation both in the microsomal and in the intact liver cell system [25]. The lack of protection by promethazine against CCl_4 covalent binding suggests that phenothiazine does not effectively trap $CCl_3^{\cdot}$; on the contrary the inhibitory effect on lipid peroxidation supports its main reaction with $CCl_3O_2^{\cdot}$. *Packer* et al. [4] confirmed these hypotheses showing, with pulse radiolysis techniques, that promethazine and $CCl_3^{\cdot}$ react very slowly, whereas the reaction of $CCl_3O_2^{\cdot}$ with promethazine is fast. Since promethazine did not show effectiveness in reducing the impairment of lipoprotein secretion due to CCl_4 in spite of a complete inhibition of the CCl_4 pro-oxidant effect, we postulated that $CCl_3^{\cdot}$-covalent binding could be the main mechanism responsible for this specific CCl_4-induced damage. We then carried out studies on the intracellular distribution of CCl_4 metabolites by incubating intact hepatocytes with radiolabelled CCl_4. One of the aims was to investigate whether CCl_4 metabolites can be found bound to lipoproteins. In table II total radioactivity recovered in different cell suspension fractions is reported. As expected, the largest amount of radioactivity was recovered in the microsomal fraction (100,000 *g* pellet), site of CCl_4 activation. Repeated experiments showed a significant amount of radioisotope bound to the low density lipoprotein fraction. Since it presumably represents both native and mature, in part secreted, lipoproteins, the obtained result could be extremely important in supporting the hypothesis that $CCl_3^{\cdot}$. covalent binding cannot only damage the structure and the secretion of already synthetized lipoproteins but also induce derangement of the lipoprotein assembly itself. As far as microsomal and lipoprotein fractions are concerned, experiments to distinguish between unbound and bound ^{14}C radioactivity were performed and the corresponding results are reported in figure 6. In the two intracellular fractions studied the average amount of bound radioisotope was about 70% of the total, while most of the residual radioisotope was probably dissolved. It is noteworthy that the data concerning the ratio between lipid-bound and protein-bound radioactivity in microsomes are quite similar to those available in the literature and obtained with 'in vivo' experiments. However, the results obtained with lipoprotein fractions cannot, at the present, be compared. For this reason we do not so far have a clear explanation for the difference observed in distribution of bound and unbound radioactivity between intracellular and extracellular lipoproteins. Once obtained, it was necessary to check the reliability of the findings on covalent binding to lipoproteins in other words to exclude or quantify the possibility that unmetabolized CCl_4 and not its metabolites would bind to the lipoproteins. Aliquots of the intracellular and extracellular isolated lipopro-

Table II. Total radioactivity recovered in different fractions of hepatocyte suspensions treated with (^{14}C)-CCl_4 [1]

Cell suspension fractions	dpm	%
Intracellular		
25,000 *G* pellet	694,967 ± 107,667	15
100,000 *G* pellet	2,650,684 ± 366,656	58
Low density lipoproteins	305,598 ± 7,866	7
Residual cell sap	881,104 ± 134,908	20
Extracellular		
Low density lipoproteins	168,422 ± 4,018	13
Residual supernatant	1,110,942 ± 142,971	87

[1] The results are the mean ± SD of three experiments in triplicate. Relative percentages over the total radioactivity respectively recovered within the cells and in the incubation medium are also given.

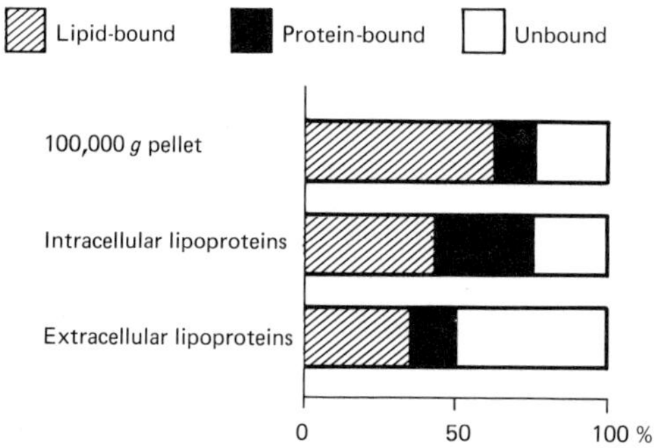

Fig. 6. Recovery of [^{14}C]CCl_4 radioactivity in defined fractions of the hepatocyte suspension. The results are the mean of three experiments in triplicate. Bound and unbound radioactivity are calculated as percentage of the total recovered radioactivity. For details on cell fractionation see the Materials and Methods section.

teins were then incubated under the same conditions as reported above with [^{14}C]CCl_4 but without a source for its activation. In this way it was possible to evaluate how much an activated CCl_4 was bound to cell lipid or protein. As regards lipid binding, the data from two experiments in triplicate showed that only about 0.2 and 0.7% of the total radioactivity added (about 2,300,000 dpm) were bound to intracellular and extracellular lipoproteins, respectively. As regards the binding to cell protein, the recovered radioactivity was in any case less than 0.05–0.1% of the total added. In conclusion, showing CCl_4 metabolites covalently bound to low density lipoproteins, strong evidence has been obtained to support a major role of covalent binding in CCl_4-induced derangement of the lipoprotein secretion pathway. Several observations from different laboratories allow us to believe that $CCl_3^{\cdot}$ is the reactive species involved in the binding.

Summary

CCl_4-induced impairment of the lipoprotein secretion pathway was studied using intact rat hepatocytes. Promethazine addition to the cell suspension before CCl_4 treatment did not protect against the secretory block due to haloalkane although it strongly inhibited its pro-oxidant effect. This was evaluated both in terms of production of malonaldehyde and of several other classes of aldehydes. The distribution analyses of $[^{14}C]CCl_4$ radioactivity in different cell suspension fractions showed a significant amount of the radioisotope covalently bound to the lipid and to the protein components of low density lipoproteins. Suitable experiments demonstrated that the bound radioisotope was represented by CCl_4 metabolites and not by unactivated CCl_4. The reported data strongly support the covalent binding of CCl_4 metabolites to cell structures as the main mechanism by which this compound impairs the lipoprotein secretory pathway of the liver.

References

1 Poli, G.; Gravela, E.; Albano, E.; Dianzani, M.U.: Studies on fatty liver with isolated hepatocytes. II. The action of carbon tetrachloride on lipid peroxidation, protein, and triglyceride synthesis and secretion. Exp. molec. Path. *30:*116–127 (1979).

2 Slater, T.F.: Biochemical studies on liver injury; in Slater, Biochemical mechanisms of liver injury (Academic Press, London 1978).

3 Dianzani, M.U.: Biochemical aspects of fatty liver; in Slater, Biochemical mechanisms of liver injury (Academic Press, London 1978).

4 Packer, J.E.; Slater, T.F.; Willson, R.L.: Reactions of the carbon tetrachloride-related peroxy free radical ($CCl_3O_2^{\cdot}$) with amino acids: pulse radiolysis evidence. Life Sci. *23:*2617–2620 (1978).

5 Packer, J.E.; Willson, R.L.; Bahnemann, D.; Asmus, K.D.: Electron transfer reactions of halogenated aliphatic peroxyl radicals: measurement of absolute rate constants by pulse radiolysis. J. chem. Soc. Perkin Trans. *II:*296–299 (1980).

6 Slater, T.F.: Activation of carbon tetrachloride: chemical principles and biological significance; in McBrien, D.C.H.; Slater, Free radicals, lipid peroxidation and cancer (Academic Press, London 1982).

7 Gravela, E.; Albano, E.; Dianzani, M.U.; Poli, G.; Slater, T.F.: Effects or carbon tetrachloride on isolated rat hepatocytes. Inhibition of protein and lipoprotein secretion. Biochem. J. *178:*509-512 (1979).
8 Dianzani, M.U.; Poli, G.; Gravela, E.; Chiarpotto, E.; Albano, E.: Influence of lipid peroxidation on lipoprotein secretion by isolated heptocytes. Lipids *16:* 823-829 (1981).
9 Poli, G.; Gravela, E.: Lipid peroxidation in isolated hepatocytes; in McBrien, Slater, Free radicals, lipid peroxidation and cancer (Academic Press, London 1982).
10 Gravela, E.; Poli, G.; Albano, E.; Dianzani, M.U.: Studies on fatty liver with isolated hepatocytes. I. The action of Colchicine, Phalloidin, Cytochalasin B, and Cycloheximide on protein and triglyceride synthesis and secretion. Exp. molec. Path. *27:*339-352 (1977).
11 Burstein, M.: Précipitation sélective des lipoprotéines sériques de faible densité par le phosphotungstate de sodium à pH neutre en présence d'une concentration saline élevée. J. Physiol., Paris *54:*647-656 (1962).
12 Burstein, M.; Legmann, P.: Precipitation of serum lipoproteins by anionic polyelectrolytes and bivalent cations; in Lipoprotein precipitation. Monogr. Atheroscler., vol. 11 (Karger, Basel 1982).
13 Slater, T.F.; Sawyer, B.C.: The stimulatory effects of carbon tetrachloride on peroxidative reactions in rat liver fractions in vitro: inhibitory effects of free radical scavengers and other agents. Biochem. J. *123:*823-828 (1971).
14 Slater, T.F.: The inhibitory effects in vitro of phenothiazines and other drugs on lipid peroxidation systems in rat liver microsomes, and their relationship to liver necrosis produced by carbon tetrachloride. Biochem. J. *106:*155-160 (1968).
15 Esterbauer, H.: Aldehydic products of lipid peroxidation; in McBrien, Slater, Free radicals, lipid peroxidation and cancer (Academic Press, London 1982).
16 Esterbauer, H.; Cheeseman, K.H.; Dianzani, M.U.; Poli, G.; Slater, T.F.: Separation and characterization of the aldehydic products of lipid peroxidation stimulated by ADP-Fe^{2+} in rat liver microsomes. Biochem. J. *208:*129-140 (1982).
17 Benedetti, A.; Comporti, M.; Esterbauer, H.: Identification of 4-hydroxynonenal as a cytotoxic product originating from the peroxidation of liver microsomal lipids. Biochem. biophys. Acta *620:*281-296 (1980).
18 Ferrali, M.; Fulceri, R.; Benedetti, A.; Comporti, M.: Effects of carbonyl compounds (4-hydroxyalkenals) originating from the peroxidation of liver microsomal lipids on various microsomal enzyme activities of the liver. Res. Commun. chem. Pathol. Pharmacol. *30:*99-112 (1980).
19 Poli, G.; Chiarpotto, E.; Biasi, F.; Pavia, R.; Albano, E.; Dianzani, M.U.: Enzymatic impairment induced by biological aldehydes in intact rat liver cells. Res. Commun. chem. Pathol. Pharmacol. *38:*71-76 (1982).
20 Pencil, S.D.; Glende, E.A.; Recknagel, R.O.: Loss of calcium sequestration capacity in endoplasmic reticulum of isolated hepatocytes treated with carbon tetrachloride. Res. Commun. chem. Pathol. Pharmacol. *36:*413-428 (1982).
21 Wolf, C.R.; Mansuy, D.; Nastainczyk, W.; Deutschmann, G.; Ulrich, V.: The reduction of polyhalogenated methanes by liver microsomal cytochrome P-450. Molec. Pharmacol. *13:*698-707 (1977).
22 Brault, D.; Bizet, C.; Morliere, P.; Rougee, M.; Land, E.J.; Santus, R.; Swallow,

A.L.: One-electron reduction of ferrideuteroporphyrin IX and reaction of the oxidised and reduced forms with chlorinated methyl radicals. J. Am. chem. Soc. *80:* 1015–1020 (1980).
23 Slater, T.F.: Free radicals as reactive intermediates in tissue injury; in Snyder, Parke, Kocsis, Jollow, Gordon-Gibson, Biological reactive intermediates. 2. Chemical mechanisms and biological effects (Plenum Publishing, New York 1981).
24 Cheeseman, K.H.: The effects of inhibitors of cytochrome P-450-mediated reactions on microsomal lipid peroxidation produced by carbon tetrachloride; in Slater, Garner, Recent advances in lipid peroxidation and tissue injury (Brunel Biochemistry Department, London 1981).
25 Poli, G.; Cheeseman, K.H.; Slater, T.F.; Dianzani, M.U.: The role of lipid peroxidation in CCl_4-induced damage to liver microsomal enzymes: comparative studies in vitro using microsomes and isolated liver cells. Chem.- Biol. Interact. 37: 13–24 (1981).

M.U. Dianzani, MD, Istituto di Patologia Generale dell'Università,
Corso Raffaello 30, I-10125 Torino (Italia)

Front. gastrointest. Res., vol. 8, pp. 14–23 (Karger, Basel 1984)

Acetaminophen Metabolism during Alcoholic Liver Injury

Emanuele Altomare[a], *Gianluigi Vendemiale*[a], *Ottavio Albano*[a], *Charles S. Lieber*[b]

[a] Istituto di Clinica Medica I, Università di Bari, Bari, Italia; [b] Alcohol Research and Treatment Center, Bronx V.A. Medical Center and Mount Sinai School of Medicine (CUNY), New York, N.Y., USA

Introduction

There have been a number of case reports suggesting that alcoholics may have increased susceptibility to the hepatotoxicity of an overdose of acetaminophen [3, 4, 27]. Furthermore, in rats, our group [21, 22] and others [25] have shown that chronic alcohol consumption results in the increased hepatotoxicity of acetaminophen overdosage. The question still remained, however, whether acetaminophen is hepatotoxic in amounts recommended therapeutically in man, when it is associated with chronic alcohol consumption. This question became particularly relevant in view of the observation that unlike chronic pretreatment with ethanol, which exacerbates acetaminophen toxicity [22] the presence of ethanol may in fact have the opposite effect [20]. The present study was undertaken to determine whether an amount of acetaminophen, which results in blood levels therapeutically recommended in man, produces untoward liver effects in sub-human primates when associated with chronic and/or acute ethanol administration.

Materials and Methods

Materials. Acetaminophen (*N*-acetyl-aminophenol) was obtained from Sigma Chemical Co. (St. Louis, Mo.). All other chemicals used represented the best available commercial grades.

Animals. 12 baboons (females) were pair-fed either an ethanol liquid diet containing 50% of total calories or a diet in which ethanol was replaced isocolorically by carbohydrate [12] for a period of 18 months (4 pairs) and 5 years (2 pairs). In addition, 2 chow-fed baboons were studied. The animals and feeding procedures were as described elsewhere [12].

Experimental Designs

(1) Kinetics of acetaminophen metabolism after chronic ethanol consumption. On the eve of each experiment, the baboons were fasted at 4 p.m. to eliminate ethanol. The experiments were started at 8 a.m. the following day preceded by slight ketamone anesthesia during which a urinary catheter and two indwelling intravenous catheters were inserted, one for infusion and the other for blood sampling. During the experiments, a 5% dextrose solution was infused to meet the caloric needs of the animals. A saline solution of acetaminophen (40 mg/kg) was infused after ketamine anesthesia. Blood samples were taken every 30 min for 6 h and urine was collected from 0 to 6 h to evaluate acetaminophen blood levels and its urinary excretion. (2) Effect of an acute dose of ethanol on acetaminophen metabolism. In the second experimental design the same pair-fed baboons were used after an interval of 2–3 months; ethanol was infused 30 min before the acetaminophen injection at a dose of 1.5 g/kg. After acetaminophen injection (40 mg/kg) ethanol infusion was maintained with a Harvard apparatus infusion pump, model 975, at a constant rate of 120–160 mg/kg/h [15]. Remaining procedures were the same as in the first experimental design. Blood samples were taken at 60-min intervals in 0.6 *N* perchloric acid [8] for blood ethanol determination by head space gas chromatography. The same experiments were also carried out in 2 chow-fed baboons, with evaluation of renal function, body temperature, heart rate and blood pressure in addition. (3) Effects of 2 weeks' acetaminophen feeding. In the third experimental design, 8 pair-fed baboons were used. The animals received acetaminophen (85 mg/kg/day) in their respective diets for 2 weeks. Acetaminophen in blood, free acetaminophen and its urinary metabolites were separated by high pressure liquid chromatography (HPLC) according to *Howie* et al. [5] using a Hewlett Packard 1048 B Liquid Chromatograph, with a C18 Bondapak column. The acetaminophen conjugates were identified according to *Buckpitt* et al. [1]. Serum glutamic oxaloacetic (SGOT) activity was measured according to *Karmen* [7]; serum glutamate dehydrogenase (SGDH) was assessed by the method of *Ellis and Goldberg* [2]; serum glutamic pyruvic transaminase (SGPT) was measured according to *Wroblewsky and LaDue* [28].

Statistical Analysis

All measurements were done at least in duplicate and expressed as the means ± standard error of the mean (SEM). Differences were analyzed by Student's test unless otherwise stated [24].

Results

Effect of Chronic and Acute Ethanol Administration on Kinetics of Acetaminophen Metabolism

In baboons fed alcohol chronically, the disappearance of acetaminophen from the plasma was faster than in pair-fed controls (fig. 1).

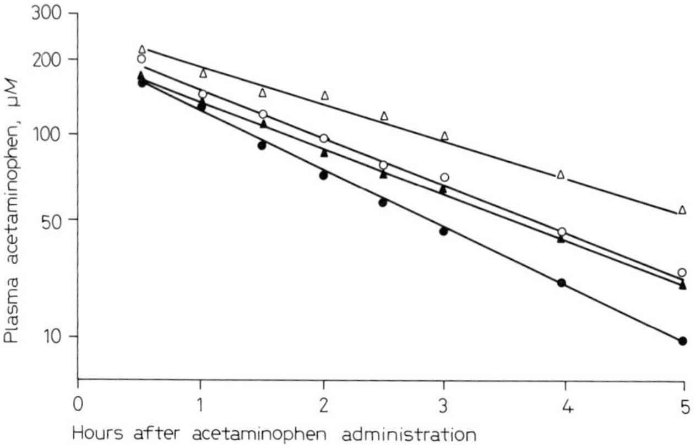

Fig. 1. Plasma elimination of acetaminophen after i.v. injection of 40/mg/kg of the drug in baboons fed alcohol chronically (●) and pair-fed controls (○). In another experiment the same baboons fed alcohol chronically (▲) and pair-fed controls (Δ) received 1.5 g/kg of i.v. ethanol, 30 min before acetaminophen injection (40 mg/kg) and ethanol infusion was continued during the experiment at the rate of 120–160 mg/kg/h. Blood samples were taken at 30-min intervals for 6 h for acetaminophen determination by HPLC. Values are means ± SEM for 6 animals. Slopes were calculated by linear regression.

Acetaminophen half-life was 63.5 ± 4 min in the alcohol-fed group, significantly shorter than in controls (76.8 ± 4 min; $p < 0.01$). When an ethanol infusion was superimposed the difference between the alcohol-fed and control animals persisted, though in both groups the disappearance of acetaminophen was slower. Acetaminophen half-life was 81.7 ± 8.2 min in the alcohol-fed group, versus 98.5 ± 6.4 in the controls ($p < 0.05$). Ethanol infusion significantly increased acetaminophen half-life in both groups: 81.7 ± 8.2 min vs 63.5 ± 4, $p < 0.05$ in baboons fed alcohol chronically and 98.5 ± 6.4 vs 76.8 ± 4.3, $p < 0.01$ in controls (table I). Table I also shows the disappearance rate constants ($K\ h^{-1}$) which were significantly higher in baboons fed alcohol chronically than in pair-fed controls, during ethanol infusion (0.50 ± 0.04 vs 0.42 ± 0.02; $p < 0.05$) and in its absence (0.65 ± 0.04 vs 0.50 ± 0.05, $p < 0.05$). Ethanol infusion resulted in blood ethanol levels of 20–30 m*M*. In the presence of ethanol, plasma disappearance of acetaminophen in baboons fed alcohol chronically was almost the same as in controls

Table 1. Kinetics of acetaminophen disappearance after i.v. administration (40 mg/kg) in baboons fed ethanol chronically with or without acute ethanol administration

	Plasma disappearance, Kh^{-1}		Renal clearance, ml^{-1}		Acetaminophen half-life, min	
	ethanol	saline	ethanol	saline	ethanol	saline
Ethanol-fed baboons (n = 6)	0.50 ± 0.04	0.65 ± 0.04	10.5 ± 3.1	13.0 ± 1.7	81.7 ± 8.2	63.5 ± 4
Pair-fed controls (n = 6)	0.42 ± 0.02	0.50 ± 0.05	7.0 ± 1.2	8.9 ± 1.2	98.5 ± 6.4	76.8 ± 4
p	< 0.05	< 0.05	NS	NS	< 0.05	< 0.01

Blood samples were taken at 30-min intervals for 6 h after acetaminophen injection and urine was collected over the same period. Free acetaminophen in blood and urines was determined by HPLC as described in 'Methods'. The experiment was carried out with saline or ethanol infusion (120–160 mg/kg/h).

which received 40 mg/kg of acetaminophen i.v. without ethanol infusion; the respective half-lives were 81.7 ± 8.2 vs 76.8 ± 4.3 (not significant). In the first experiments, no difference was found in total urinary excretion of acetaminophen. The excretion of free acetaminophen, glucuronide and sulfate conjugate was not significantly different but the excretion of mercapturic acid conjugate was significantly greater in baboons fed alcohol chronically than in controls (1.2 ± 0.2 vs 0.3 ± 0.08% of excreted dose; $p < 0.05$) (fig. 1). When ethanol was infused during the experiment, no statistical differences were found in total acetaminophen urinary excretion, in free acetaminophen excretion or in glucuronide and sulfate conjugates; the urinary excretion of mercapturic acid conjugate in both groups was significantly lower than in the experiment without ethanol infusion, but it remained higher in baboons fed alcohol chronically: 0.2 ± 0.07 vs 1.2 ± 0.2, $p < 0.05$ and in controls it was 0.15 ± 0.04 vs 0.3 ± 0.08, $p < 0.05$ (fig. 2), in the absence or presence of ethanol, respectively.

Effects of 2 Weeks of Acetaminophen Administration

When acetaminophen was given orally at the dose of 85 mg/kg/day for 2 weeks, plasma levels of the drug, measured on the 3rd, 7th and 14th day of the experiment, ranged between 10 and 20 μM (fig. 3),

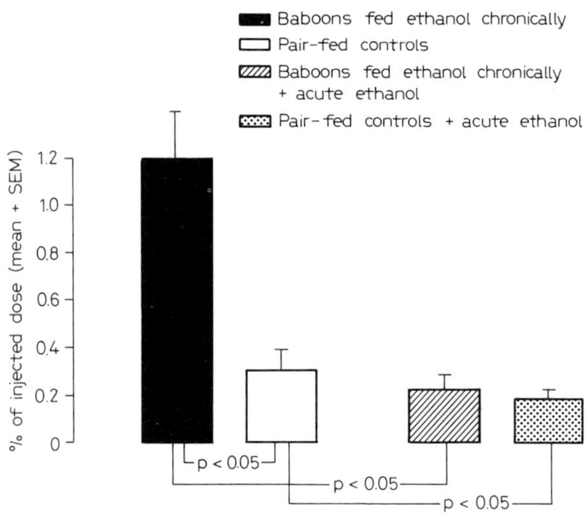

Fig. 2. Effects of chronic ethanol feeding on the urinary excretion of mercapturic acid conjugate of acetaminophen during the 6 h following the injection of acetaminophen (40 mg/kg i.v.) with and without ethanol infusion (120–160 mg/kg/h). Values are means ± SEM for 6 animals.

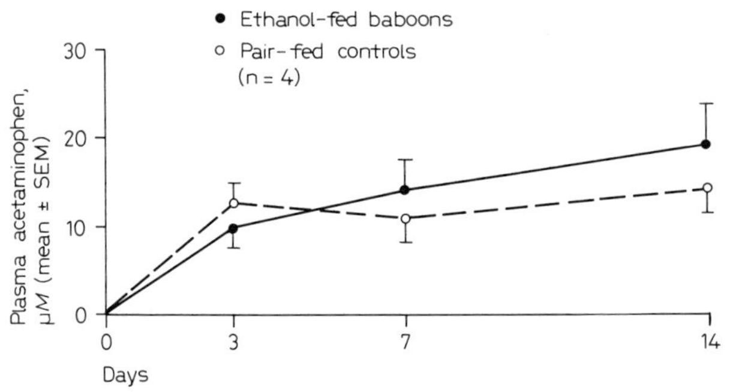

Fig. 3. Blood acetaminophen concentration during chronic intake of the drug (85 mg/kg/day). Acetaminophen was measured in blood by HPLC on the 3rd, 7th and 14th day of the experiment. Values are means ± SEM for 4 animals.

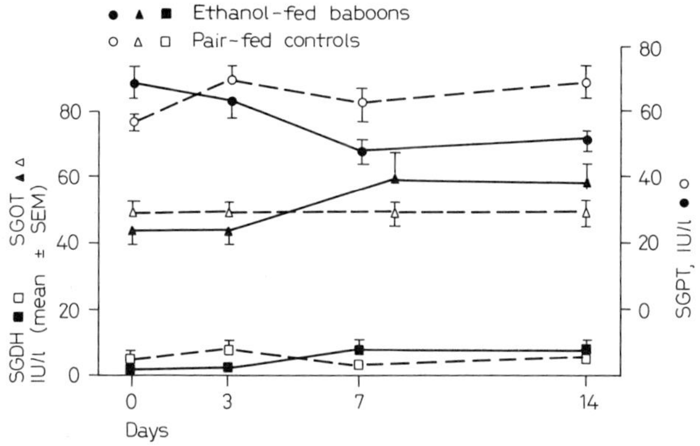

Fig. 4. Time-course of SGOT, SGPT, and SGDH activities during the chronic intake of acetaminophen (85 mg/kg/day for 2 weeks per os). Each point represents the means ± SEM of 4 animals.

concentrations similar to those used therapeutically in man [9b]. SGPT, SGOT, SGDH activities were studied before acetaminophen treatment and on the 3rd, 7th and 14th day of the experiment. No difference was found between the alcohol-fed animals and pair-fed controls during all experimental periods (fig. 4). The urinary excretion of free acetaminophen and its glucuronide and sulfate conjugates, or of mercapturic acid conjugate studied on the 3rd, 7th and 14th day of experiment in 24-hour urines, did not show any difference between baboons fed alcohol chronically and pair-fed controls. The experiments on the 2 chow-fed baboons confirmed the results obtained in baboons fed the control liquid diet; furthermore, they did not show any alteration of renal function, body temperature and blood pressure during all the experimental manipulations.

Discussion

This study shows the interactions between chronic and/or acute ethanol administration and acetaminophen metabolism in the baboons. It was found that chronic intake of acetaminophen in amounts which maintained blood concentrations similar to therapeutic levels used in humans did not produce deleterious effects in baboons drinking alcohol. However, plasma disappearance of i.v. administered acetaminophen was faster in baboons fed alcohol chronically than in pair-fed controls, even during an ethanol infusion. The latter slowed down plasma disappearance of acetaminophen in both groups. It is noteworthy that during ethanol infusion plasma elimination of i.v. administered acetaminophen in baboons fed alcohol chronically approximated the rate observed in pair-fed controls not given ethanol. The urinary excretion of mercapturic acid conjugate of acetaminophen was increased in baboons fed alcohol chronically, but decreased with acute ethanol administration. The interaction of acetaminophen metabolism and ethanol or other drugs has been extensively studied recently. *Wong* et al. [26] have shown in mice that a single, simultaneously administered dose of ethanol protects against acetaminophen-induced hepatotoxicity as assessed by plasma transaminases. It has been demonstrated that chronic ethanol feeding in rats enhances the hepatotoxicity of acetaminophen [22] but that acute ethanol administration in rats prevents acetaminophen-induced hepatotoxicity [23] and that acute ethanol administration in rats fed alcohol chronically decreases acetaminophen-induced hepatotoxicity [*Altomare* et al., submitted]. Other studies have demonstrated that in mice and hamsters *L*-ascorbic acid protects against acetaminophen-induced hepatotoxicity by the inhibition of covalent binding of acetaminophen metabolites to hepatic microsomes [9a], that the deficiency of thiamine in rats enhances the rate of disappearance of acetaminophen from the plasma [19], that propylthiouracil protects acetaminophen hepatotoxicity in rats and mice [16, 29] and that cimetidine protects against acetaminophen hepatotoxicity in rats [13]. The mechanism of these protective effects is not well known. It seems likely that ethanol, cimetidine and propylthiouracil inhibit the production of reactive metabolite(s) of acetaminophen and their covalent binding. It has previously been shown that ethanol itself is metabolized by a microsomal cytochrome P-450-containing enzyme system [10, 11, 14] that ethanol inhibits drug oxidation in vivo and in vitro [17, 18], that chronic alcohol consumption increases covalent binding of reactive metabolite(s) of acetaminophen in rats [22] and that acute ethanol inhibits the production of reactive metabolite(s) of acetaminophen in rats [20]. The interactions of acute and/or chronic ethanol administration with acetaminophen metabolism were investigated in this study in the baboon. When acetaminophen was given i.v. at a dose of 40 mg/kg, the rate of plasma disappearance of the drug was increased in baboons fed alcohol chronically when compared to pair-fed controls. This is most likely due to accelerated metabolism. This finding is supported by an increased excretion of mercapturic acid conjugate in urines. When acute ethanol was perfused during the experiment, the rate of plasma disappearance of acetaminophen decreased in baboons fed alcohol chronically, but was still faster than in pair-fed controls. Also the urinary excretion of mercapturic acid conjugate decreased when ethanol was infused but still remained higher than in pair-fed controls. Since urinary excretion of mercapturic acid conjugate is a commonly used parameter for the production of reactive metabolite(s) of acetaminophen in vivo [6], our data suggest that, in baboons, chronic consumption of alcohol increases the formation of reactive metabolite(s) of acetaminophen, whereas acute ethanol administration inhibits this formation both in baboons fed alcohol chronically and in pair-fed controls.

When acetaminophen was given orally for 2 weeks at a dose of 85 mg/kg/day, the plasma concentrations of the drug ranged between 10 and 20 μM, values similar to human therapeutic plasma concentrations of acetaminophen [9b] and the urinary excretion of mercapturic acid conjugate in baboons drinking alcohol was decreased when compared to pair-fed controls, confirming the results obtained during ethanol infusion. Chronic acetaminophen intake did not produce deleterious effects as assessed by serum enzyme activities at the dosage used both in control baboons and in animals drinking alcohol chronically.

Summary

In baboons fed ethanol chronically, an acute dose of acetaminophen accelerated plasma elimination of the drug and increased urinary acetaminophen mercapturate excretion. With the addition of ethanol infusion, plasma acetaminophen elimination was still stignificantly accelerated in baboons fed ethanol chronically. When the drug was administered chronically for 2 weeks in the diet no significant difference was found in the urinary excretion of acetaminophen mercapturate, in SGOT, SGPT and levels and in plasma acetaminophen levels, between the two groups of animals. These results confirm that chronic ethanol consumption increases the excretion of mercapturic acid conjugate while acute ethanol administration decreases it. A dose of acetaminophen which maintained blood levels similar to those recommended for humans did not produce deleterious effects in baboons drinking alcohol.

References

1 Buckpitt, A.R.; Rollins, D.E.; Nelson, S.D.; Franklin, R.B.; Mitchell, J.R.: Quantitative determination of the glutathione, cysteine and *N*-acetilcysteine conjugates of acetaminophen by high pressure liquid chromatography. Analyt. Biochem. *83:* 168–177 (1977).

2 Ellis, G.; Goldberg, D.M.: Optical conditions for the kinetics assay of serum glutamate dehydrogenase activity at 37 °C. Clin. Chem. *18:*523–527 (1972).

3 Emby, D.J.; Fraser, B.N.: Hepatotoxicity of paracetamol enhanced by ingestion of alcohol: report of two cases. S. Afr. med. J. *87:*299–304 (1977).

4 Goldfinger, R.; Ahmed, K.S.; Pitchumoni, C.S.; Wesley, S.A.: Concomitant alcohol and drug abuse enhancing acetaminophen hepatotoxicity. Am. J. Gastroent., N.Y. *70:*385–388 (1978).

5 Howie, D.; Adriaenssens, P.I.; Prescott, L.E.: Paracetamol metabolism following overdosage: application of high performance liquid chromatography. J. Pharm. Pharmacol. *29:*235–237 (1977).

6 Jollow, D.J.; Thorgeirsson, S.S.; Potter, W.Z.; Davis, D.C.; Gillette, J.R.; Brodie, B.B.: Acetaminophen-induced hepatic necrosis. VI. Metabolic disposition of toxic and non toxic doses of acetaminophen. Pharmacology *12:*251–271 (1974).

7 Karmen, A.A.: Note on the spectrophotometric assay of glutamic-oxaloacetic transaminase in human blood serum. J. clin. Invest. *34:*131–133 (1955).

8 Korsten, M.A.; Matsuzaki, S.; Peinman, L.; Lieber, C.S.: High blood acetaldehyde levels following ethanol administration: differences between alcoholic and non-alcoholic subjects. New Engl. J. Med. *292:*386–389, 1975.

9a Lake, B.G.; Harris, R.A.; Phillips, J.C.; Gangolli, S.D.: Studies on the effect of *l*-ascorbic acid on acetaminophen induced hepatotoxicity. Toxicol. appl. Pharmacol. *60:*229–240 (1981).

9b Licht, H.; Seeff, L.B.; Zimmerman, H.J.: Apparent potentiation of acetaminophen hepatotoxicity by alcohol. Ann. intern. Med. *92:*511 (1980).

10 Lieber, C.S.; De Carli, L.M.: Ethanol oxidation by hepatic microsomes. Adaptive increase after ethanol feeding. Science *162:*917–918 (1968).

11 Lieber, C.S.; De Carli, L.M.: Hepatic microsomal ethanol oxidizing system: in vitro characteristics and adaptive properties in vivo. J. biol. Chem. *245:* 2505–2512 (1970).

12 Lieber, C.S.; De Carli, L.M.: An experimental model of alcohol feeding and liver injury in the baboon. J. med. Primatol. *3:*153–163 (1974).

13 Mitchell, M.C.; Schenker, S.; Avant, G.R.; Speeg, K.V.: Cimetidine protects against acetaminophen hepatotoxicity in rats. Gastroenterology *81:*1052–1060 (1981).

14 Ohnishi, K.; Lieber, C.S.: Reconstitution of the microsomal ethanol oxidizing system (MEOS): qualitative and quantitative changes of cytochrome P-450 after chronic ethanol consumption. J. biol. Chem. *252:*7124–7131 (1977).

15 Pikkarainen, P.H.; Lieber, C.S.: Concentration dependency on ethanol elimination rates in baboons: effect of chronic alcohol consumption. Alcoholism: Clin. exp. Res. *1:*40–43 (1980).

16 Raheja, K.L.; Willem, G.L.; Chaidong, C.; Darbbie, M.: Protective effect of propylthiouracil independent of its hypothyroid effect on acetaminophen toxicity in the rat. J. Pharmac. exp. Ther. *220:*427–432 (1982).

17 Rubin, E.; Gang, H.; Misra, P.S.; Lieber, C.S.: Inhibition of drug metabolism by acute ethanol intoxication. A hepatic microsomal metabolism. Am. J. Med. *49:* 801–806 (1970).
18 Rubin, E.; Lieber, C.S.: Hepatic microsomal enzymes in man and rat; induction and inhibition by ethanol. Science *162:*690–691 (1968).
19 Ruchirawat, M.; Aramphongphan, A.; Tanphaichitr, V.; Bandittanukook, W.: The effect of thiamine deficiency on the metabolism of acetaminophen (paracetamol). Biochem. Pharmacol. *30:*901–1906 (1981).
20 Sato, C.; Lieber, C.S.: Mechanism of the preventive effect of ethanol on acetaminophen-induced hepatotoxicity. J. Pharmac. exp. Ther. *218:*811–815 (1981).
21 Sato, C.; Matsuda, Y.; Lieber, C.S.: Increased hepatotoxicity of acetaminophen after chronic ethanol consumption. Fed. Proc. *38:*916 (1979).
22 Sato, C.; Matsuda, Y.; Lieber, C.S.: Increased hepatotoxicity of acetaminophen after chronic alcohol consumption in the rat. Gastroenterology *80:*140–148 (1981).
23 Sato, C.; Nakano, M.; Lieber, C.S.: Prevention of acetaminophen-induced hepatotoxicity by acute ethanol administration in the rat: comparison with carbon tetrachloride induced hepatotoxicity. J. Pharmac. exp. Ther. *218:*811–815 (1981).
24 Snedecor, G.W.; Cochran, W.G.: Statistical methods; 6th ed. (Iowa State University Press, Ames 1967).
25 Teschke, R.; Stutz, G.; Strohmeyer, G.: Increased paracetamol-induced hepatotoxicity after chronic alcohol consumption. Biochem. biophys. Res. Commun. *91:* 368–373 (1979).
26 Wong, L.T.; Whitehouse, T.W.; Solomonraj, G.; Pual, C.J.: Effect of a single dose of ethanol on the hepatotoxicity and metabolism of acetaminophen in mice. Toxicology *17:*297–309 (1980).
27 Wright, N.; Prescott, L.F.: Potentiation by previous drug therapy of hepatotoxicity following paracetamol overdose. Scott. med. J. *18:*56–58 (1973).
28 Wroblewski, E.; LaDue, J.S.: Serum glutamic pyrivic transaminase in cardiac and hepatic disease. Proc. Soc. exp. Biol. Med. *91:*569–571 (1956).
29 Yamada, T.; Ludwig, S.; Kuhlenkamp, J.; Kaplowitz, N.: Direct protection against acetaminophen hepatoxicity by propylthiouracil. J. clin. Invest. *67:*688–695 (1981).

E. Altomare, MD, Istituto di Clinica Medica I, Università di Bari,
I-70100 Bari (Italia)

Front. gastrointest. Res., vol. 8, pp. 24–45 (Karger, Basel 1984)

Role of Golgi Apparatus on Lipid Metabolism in Ethanol-Induced Liver Injury[1]

U.M. Marinari, A. Casu, M.M. Averame, D. Cottalasso, M.A. Pronzato, G. Nanni[2]

Istituto di Patologia Generale, Università di Genova, Genova, Italy

Introduction

The liver is the main organ capable of oxidizing ethanol to acetaldehyde by utilizing various enzymatic systems which include alcohol/dehydrogenase, the microsomal ethanol-oxidizing system (MEOS) and catalase [see 1 for a review]. Ethanol oxidation produces primary alterations in hepatic lipid metabolism, which modify the synthesis and degradation of lipoproteins or their constituent lipids and proteins [2–4]. Consequently, the production of fatty liver is the first event and the most striking and consistent effect caused by alcohol.

Ethanol-induced changes of liver cells involve ultrastructural and functional alterations of both the mitochondria and the microsomal membranes. In particular, smooth endoplasmic reticulum and Golgi apparatus show a marked proliferation in chronic and acute ethanol intoxication [1, 3, 5–9]. In the latter, the stacked cisternae, vacuoles and vesicles of rat liver Golgi apparatus increase in number and in size and become filled with very low density lipoproteins (VLDLs) [5, 10–12].

The role of rat hepatocyte Golgi apparatus in the synthesis and secretion of lipoproteins is well established [5, 10, 13–15]. During the lipoprotein synthesis cycle, the rate of Golgi membrane turnover appears to be directly proportional to the rate of synthesis and transport of VLDL particles [16]. This functional activity of Golgi apparatus

[1] This investigation was supported in part by the Ministero della Pubblica Istruzione, Roma and by CNR Project Prerent., Rehabilit. Med. – grant n. 82022556.

[2] The valuable technical assistance of *G. Maloberti* and *M. Isola* is acknowledged.

might justify its morphological and metabolic implication in the pathogenesis of various fatty liver degenerations related to an impaired lipoprotein metabolism [5]. Such data suggest that the development of acute ethanol-induced fatty liver could involve an impairment of VLDL synthetic and secretory mechanisms in the hepatocyte Golgi apparatus. Since the pathogenesis of this disease is a crucial problem, as yet undefined [1, 17], the purpose of this study is to investigate the role of ethanol in the impairment of lipoprotein metabolism in rat liver, particularly at the level of the Golgi apparatus which may contribute to the fatty liver by decreasing lipoprotein secretion.

Materials and Methods

Treatment of Animals. Female Sprague-Dawley rats (CD-COBS-Charles River), weighing 200 ± 25 g, were fed on a standard synthetic diet, devoid of antioxidants. Batches of 8–10 animals, fasted for 16 h, received either a single dose of ethanol (6 g/kg body wt as a 50% w/v solution) or isocaloric glucose by oral intubation. As needed, 104 μCi 5% bovine albumin-bound [9, 10-^{3}H] palmitate (Radiochemical Centre, Amersham; spec. act. 500 mCi/mmol) were injected intravenously into each rat. The animals were killed at different times after intubation (1.5, 3, 6, 12 h) and after [^{3}H]palmitic acid injection (5, 10, 20, 30 min).

Preparation of Subcellular Fractions. Immediately after sacrifice the liver was removed, washed in ice-cold 0.25 *M* sucrose, dried and weighed. From the pooled livers the subcellular fractions were isolated in a cold room according to the method previously described [18]. Preparations of Golgi apparatus subfractions were made from total microsomal fraction by discontinuous sucrose density gradient ultracentrifugation according to the method of *Ehrenreich* et al. [11], slightly modified [19]. Three different fractions of purified Golgi membranes (F_1, F_2, F_3) and two residual microsomal fractions (R,S) were isolated.

Isolation of Golgi Apparatus Secretory Content. The pooled (F_1+F_2) and F_3 fractions were resuspended in 0.04 *M* sodium veronal-HCl buffer pH 8.5, in a concentration of about 2 mg protein/ml and mechanically broken by passing them twice through a French-press (LKBX-Press, type X-5) under a pressure approximately 50,000 lb per sq. in. and then centrifuged at 105,000 *g* for 1 h to separate the secretory content (supernate) and the membranes (pellet).

Lipoprotein Preparation. Supernatants obtained in this way were further ultracentrifuged at 165,000 *g* for 17 h in NaCl (1.006 d) to separate F_3-VLDLs and (F_1+F_2)-VLDLs according to *Havel* et al. [20]. Serum VLDLs and high density lipoproteins (HDLs) were isolated with the same method.

Characterization of Golgi Membranes. As a marker enzyme for Golgi membranes, uridine diphosphate (UDP)-galactose: *N*-acetyl-glucosamine galactosyltransferase activity was measured by the method described by *Fleischer* et al. [21]. The specific activity, expressed as nanomoles of galactose transferred to *N*-acetyl-glucosamine/h/mg protein, was 84.36 ± 10.98, 106.40 ± 16.61, 45.83 ± 11.75 for F_1, F_2, F_3 fractions, respectively. An increase of 1,814% was obtained in the enzyme activity of the three combined fractions in relation to the homogenate.

Estimation of Radioactivity. Samples of fractions to be counted were processed by addition of cold 1% phosphotungstic acid (PTA) in 20% trichloroacetic acid (TCA). PTA-TCA precipitates were washed twice with 5% TCA and once with distilled water. Aliquots of precipitates were immediately dissolved in Instagel (Packard) and the specific radioactivity (cpm/mg protein) was determined. Lipids were extracted from other aliquots of PTA-TCA precipitates by the method of *Folch* et al. [22], dried, dissolved in Instagel and the specific radioactivity (cpm/mg triglyceride) was determined.

Lipid Analysis. The triglyceride, total cholesterol and phospholipid-phosphorus determinations were carried out on lipid extracts by the enzymatic method of *Eggstein and Kreutz* [23], *Bowman and Wolf* [24] and *Chen* et al. [25], respectively.

Protein and Apolipoprotein Analysis. Protein was assayed by the method of *Lowry* et al. [26] with bovine serum albumin as a standard. Disc electrophoresis was performed according to *Davis* [27] after delipidation of samples as described in a previous paper [28].

Statistical Evaluation. The standard deviation and the t-value were calculated. The differences between two averages were considered significant when the calculated t-values corresponded to a probability $p < 0.05$.

Chemicals. The reagents used were obtained from BDH, Calbiochem, Fluka, Eastman Kodak and Boehringer-Mannheim.

Results

Triglyceride Content of Liver Subcellular Fractions. In order to evaluate the triglyceride distribution in the various subcellular compartments, the content of this lipid in liver homogenate, cytosol and total microsomal fraction was assayed in normal conditions and during a 36-hour period after ethanol intoxication.

The administration of a single dose of ethanol results in a marked increase of liver triglycerides in homogenate and cytosol. This increase starts in the early stages after intoxication, reaches a maximum within 12 h, maintains a plateau for 12–30 h and then shows an evident decrease 36 h later (table I). The triglyceride levels of the total microsomal fraction significantly increase within 12 h of intoxication and then rapidly decrease (table I). Because of this behaviour, the submicrosomal fractions were studied within 12 h of ethanol administration.

In the Golgi apparatus secretory membranes (F_1, F_2), the triglyceride content significantly increases at 3 h, reaches a peak at 6 h and shows a slight tendency to decrease about 12 h later (table II). On the other hand, in the Golgi apparatus formative membranes (F_3) and in the residual microsome membranes (R,S), the triglyceride content never significantly changes (table II). Taken together, these results suggest an early implication of Golgi secretory membranes in lipoprotein storage during the initial stages of ethanol-induced fatty liver.

Lipid Movements into Liver Subcellular Fractions. A preliminary study was performed in order to make an overall evaluation of fatty liver onset and of its subsequent course until regression. This experiment measured the possible changes of a lipid precursor ([^{3}H] palmitic acid) moving into rat hepatocyte subcellular fractions during a 12-hour

Table I. Triglyceride content of isolated subcellular fractions of rat liver at different times after acute ethanol intoxication (6 g/kg body wt)

Hours	Number of experiments	Homogenate	Cytosol	Total microsome
Control	6	14.29 ± 2.40	4.34 ± 1.35	1.88 ± 0.37
1.5	4	19.31 ± 5.40	5.80 ± 1.64	2.14 ± 1.01
3	4	23.28 ± 5.03[a]	8.21 ± 1.52[a]	2.18 ± 0.17
6	4	31.73 ± 6.21[c]	10.82 ± 2.02[c]	2.52 ± 0.35
12	4	37.37 ± 2.73[c]	15.53 ± 2.85[c]	2.86 ± 0.35[a]
24	4	38.85 ± 5.54[c]	12.13 ± 3.14[b]	2.39 ± 0.43
30	4	39.52 ± 4.37[c]	9.52 ± 3.12	2.07 ± 0.62
36	4	30.87 ± 4.60[c]	7.79 ± 2.03	1.93 ± 0.76

The values are expressed as mg/g wet liver (± SD), compared to control, [a] $p < 0.01$, [b] $p < 0.005$, [c] $p < 0.001$.

Table II. Triglyceride content of isolated subfractions of rat liver Golgi apparatus and residual microsome at different times after acute ethanol intoxication (6 g/kg body wt)

Hours	Number of experiments	Golgi apparatus			Residual microsome	
		F_1	F_2	F_3	R	S
Control	5	0.048 ± 0.009	0.138 ± 0.013	0.362 ± 0.068	0.634 ± 0.050	0.283 ± 0.035
1.5	4	0.057 ± 0.011	0.158 ± 0.045	0.297 ± 0.027	0.599 ± 0.013	0.288 ± 0.060
3	4	0.087 ± 0.017[b]	0.242 ± 0.006[c]	0.389 ± 0.071	0.592 ± 0.080	0.265 ± 0.024
6	4	0.108 ± 0.012[c]	0.270 ± 0.030[c]	0.369 ± 0.040	0.609 ± 0.028	0.299 ± 0.037
12	4	0.080 ± 0.011[b]	0.207 ± 0.019[c]	0.367 ± 0.113	0.707 ± 0.055	0.257 ± 0.030

Conditions as described in table I.

period after ethanol administration. All the animals were sacrificed 10 min after [^{3}H]palmitic acid administration, since the incorporation of the labelled lipid into liver homogenate of normal rats reaches a maximum at this time, according to *Stein and Stein* [10].

The radioactivity of homogenate, cytosol and total microsomal fractions, relative to the protein content, shows a significant increase

Table III. Radioactivity incorporated into liver isolated subcellular fractions 10 min after [^{3}H] palmitic acid intravenous injection in rats at different times after a single ethanol dose administration

Hours	Number of experiments	Homogenate	Cytosol	Total microsome
Control	5	2,963 ± 284	2,049 ± 276	5,828 ± 532
1.5	8	4,450 ± 849^a	3,756 ± 458^c	8,066 ± 749^c
3	4	4,590 ± 749^a	4,318 ± 245^c	8,773 ± 1,081^b
6	4	3,950 ± 323^b	3,405 ± 363^c	6,210 ± 232
12	4	3,285 ± 248	2,808 ± 190^b	4,589 ± 406

The values represent the specific activity of the PTA-TCA precipitates referred to the protein content (cpm/mg protein ± SD),
compared to control, $^a p < 0.01$, $^b p < 0.005$, $^c p < 0.001$.

Table IV. Radioactivity incorporated into liver Golgi apparatus subfractions and residual microsome 10 min after [^{3}H]palmitic acid intravenous injection in rats at different times after a single ethanol dose administration

Hours	Number of experiments	Golgi apparatus			Residual microsome	
		F_1	F_2	F_3	R	S
Control	5	88,805 ± 13,692	42,324 ± 10,345	7,562 ± 1,008	4,606 ± 776	4,002 ± 398
1.5	8	103,654 ± 15,690	48,711 ± 11,590	14,919 ± 3,565^b	6,883 ± 1,176^b	4,631 ± 1,024
3	4	113,519 ± 14,259	47,765 ± 8,012	10,975 ± 2,309	6,612 ± 506^a	4,798 ± 127
6	4	72,428 ± 2,603	39,082 ± 2,440	7,528 ± 332	5,697 ± 604	3,150 ± 314
12	4	55,741 ± 4,397^b	22,587 ± 2,563	5,269 ± 560^a	3,220 ± 315	2,632 ± 236^b

Conditions as described in table III.

starting 1.5 h after ethanol intoxication, reaches a peak 3 h later and successively decreases (table III). Within 12 h, the radioactivity is significantly elevated only in the cytosol, probably due to the presence of a cytoplasmic triglyceride pool. During this period in the total microsomal fraction the specific activity is lower than the normal values, suggesting that the [^{3}H]palmitic acid input in the endoplasmic reticulum decreases. Similar behaviour is also noted in the residual microsome fractions (R, S) (table IV).

The radioactivity of isolated Golgi fractions (F_1, F_2, F_3), relative to the protein content, is very high in the initial period after ethanol treatment. However, the specific activity, after a significant but short-lived increase in F_3 alone, decreases below normal values in all Golgi fractions within 6 h of ethanol intoxication (table IV).

The chloroform/methanol-extracted lipid radioactivity, as compared with the triglyceride content, shows a similar trend in all studied fractions, but is more marked in Golgi membranes (fig. 1). The radioactivity of protein residue after lipid extraction is unchangeably very low. To determine whether the impairment of [^{3}H]palmitic acid incorporation in the various subcellular fractions is due to changes of lipid material input into the Golgi apparatus or of lipoprotein output from the same complex, a time sequence of [^{3}H]palmitic acid distribution in the hepatocyte subcellular fractions was performed in normal and ethanol-intoxicated rats injected with the labelled lipid at different times from 5 to 30 min. In the first 10 min after [^{3}H]palmitic acid injection, its uptake in liver homogenate, cytosol and total microsomal fractions of the ethanol-treated animals is always higher than in normal controls, regardless of the time of ethanol administration (fig. 2a–c). The input reaches a maximum at 1.5 h and gradually decreases at 6 and 12 h after ethanol intoxication. In the case of the 12-hour interval, the incorporation peak occurs later (20–30 min after the labelled lipid injection), showing a marked decrease of [^{3}H]palmitic acid input.

Within 10–30 min, the release of the injected labelled precursor decreases as the time from ethanol intoxication increases, in comparison with the control animals. At 12 h the output of the labelled lipid is almost entirely inhibited and a complete lipid storage is present, particularly in total and residual microsomal fractions. It was observed that the [^{3}H]palmitic acid transport through Golgi apparatus is impaired early (fig. 2d–f). Indeed, in each Golgi fraction (F_1, F_2, F_3) either the labelled lipid input or output becomes significantly lower at 6 h after ethanol intoxication. At this time and especially at 12 h, the lipid accumulation in secretory fractions (F_1, F_2) of Golgi apparatus is significantly higher than in the other studied fractions.

Lipids of Liver Golgi Apparatus Subfractions. From the foregoing results it appears that 6 h after ethanol intoxication the liver Golgi ap-

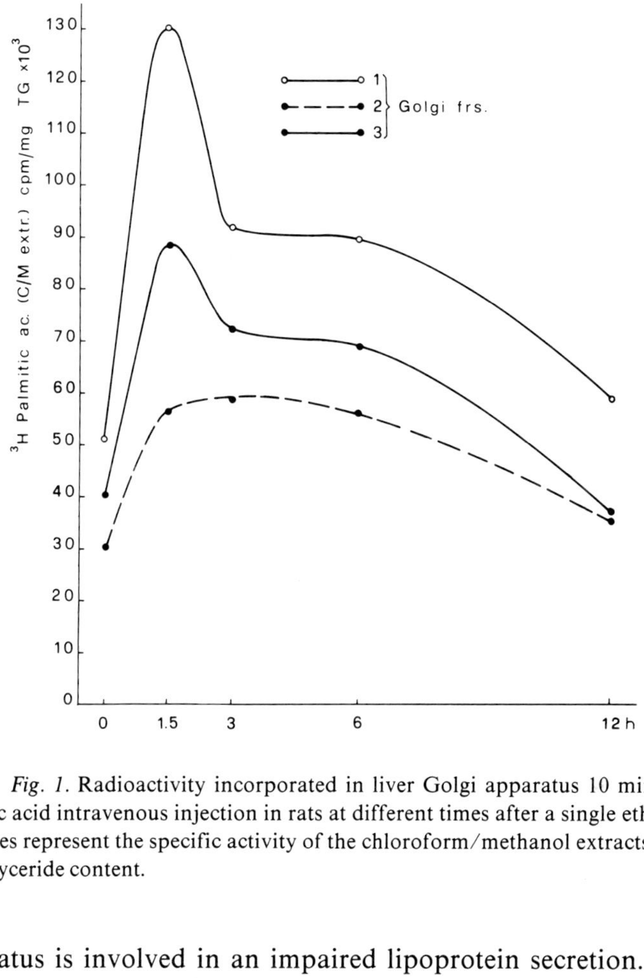

Fig. 1. Radioactivity incorporated in liver Golgi apparatus 10 min after [^{3}H]palmitic acid intravenous injection in rats at different times after a single ethanol dose. The values represent the specific activity of the chloroform/methanol extracts referred to the triglyceride content.

paratus is involved in an impaired lipoprotein secretion. Since these data suggest a specific effect of ethanol on the lipid metabolism of the Golgi apparatus, the lipids extracted from the three subfractions of liver Golgi apparatus membranes and from their secretory contents were examined in normal and in acutely ethanol-intoxicated rats 6 h before sacrifice. Each fraction of the Golgi apparatus isolated from normal rats shows a high lipid concentration. The total lipid-to-protein ratio, in fact, is 2.95 in the secretory fraction F_1, which normally contains the nascent VLDLs; it decreases in the intermediate fraction F_2 (1.57) and in the formative fraction F_3 (1.09). These values never significantly change after ethanol treatment.

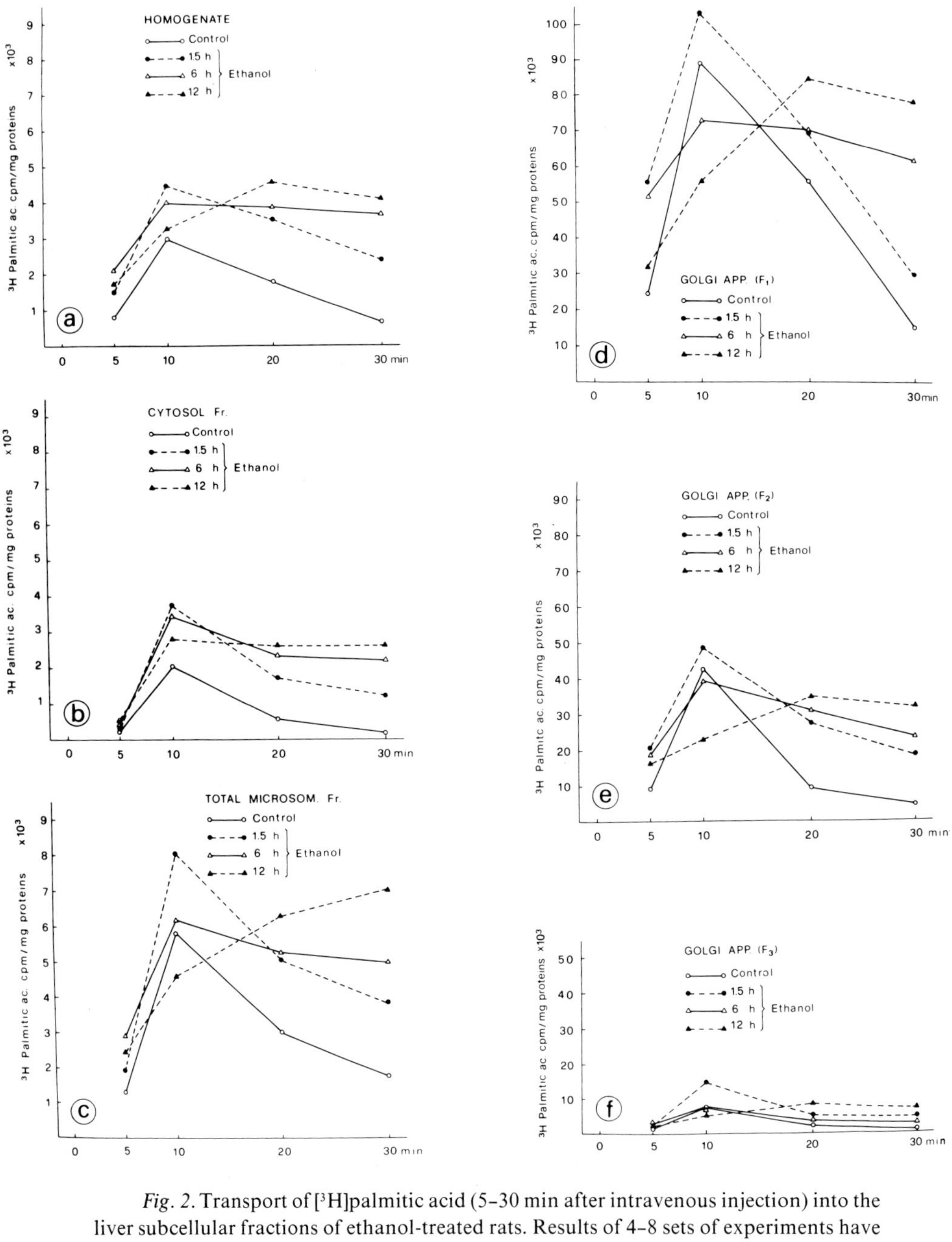

Fig. 2. Transport of [^{3}H]palmitic acid (5–30 min after intravenous injection) into the liver subcellular fractions of ethanol-treated rats. Results of 4–8 sets of experiments have been averaged and represent the specific activity (cpm/mg protein).

The levels of the triglycerides, phospholipids and total cholesterol, expressed as µg/g wet liver, are reported in figure 3. The concentration of these lipids in the whole secretory fractions (F_1, F_2) increases significantly in comparison to the normal 6 h after ethanol intoxication. The major increase is due to the phospholipid and triglyceride levels while the increase of the cholesterol is less marked (fig. 3a, b). The various lipid classes increase in the same manner. The ratio between these, in fact, does not change appreciably, the only exception being the phospholipid-to-neutral lipid ratio which in F_1 is higher after ethanol intoxication. When the combined secretory fractions (F_1+F_2) are disrupted in a French-press, the various lipid classes are concentrated, on average, 75% in the membranes and 25% in the content of the secretory vesicles, without significant changes between the normal values and those obtained after ethanol intoxication. However, the level of each lipid, after ethanol intoxication, significantly increases either in the membranes or in their content where the percentage increase of the triglyceride levels is more evident (fig. 3d). On the other hand, in the Golgi apparatus formative fraction F_3 significant alterations of the lipid concentration are not observed either in the whole or after disruption of the fraction (fig. 3c, e).

Analysis of VLDLs Isolated from Golgi Apparatus. Another step of this study was to investigate, in acute ethanol intoxication, whether the observed changes in lipid metabolism of liver Golgi apparatus secretory region might be correlated to possible alterations of VLDL transport sequences. The disc-electrophoretic patterns of serum and Golgi VLDL-apolipoproteins (apo-VLDLs) are reported in figure 4. One of the major bands of the nascent VLDLs derived from the pooled (F_1+F_2) and F_3 in normal (N) Golgi apparatus is superimposible to the arginine-rich apoprotein (apo-ARP), which is common to VLDLs and HDLs of rat serum. In the same density interval of VLDLs, another apoprotein band, called U, may be separated which shows slower electrophoretic mobility of apo-ARP and does not correspond to any serum apoprotein. After ethanol intoxication (E) only one peak is separable instead of the two major normal bands which comprise the U-protein and apo-ARP. Some polypeptides of normal Golgi-VLDLs migrate to the same region of serum apo-C apoproteins. In this area the F_3-VLDLs show a sharper peak than (F_1+F_2)-VLDLs giving support to a functional differentiation of the Golgi fractions. After ethanol intoxi-

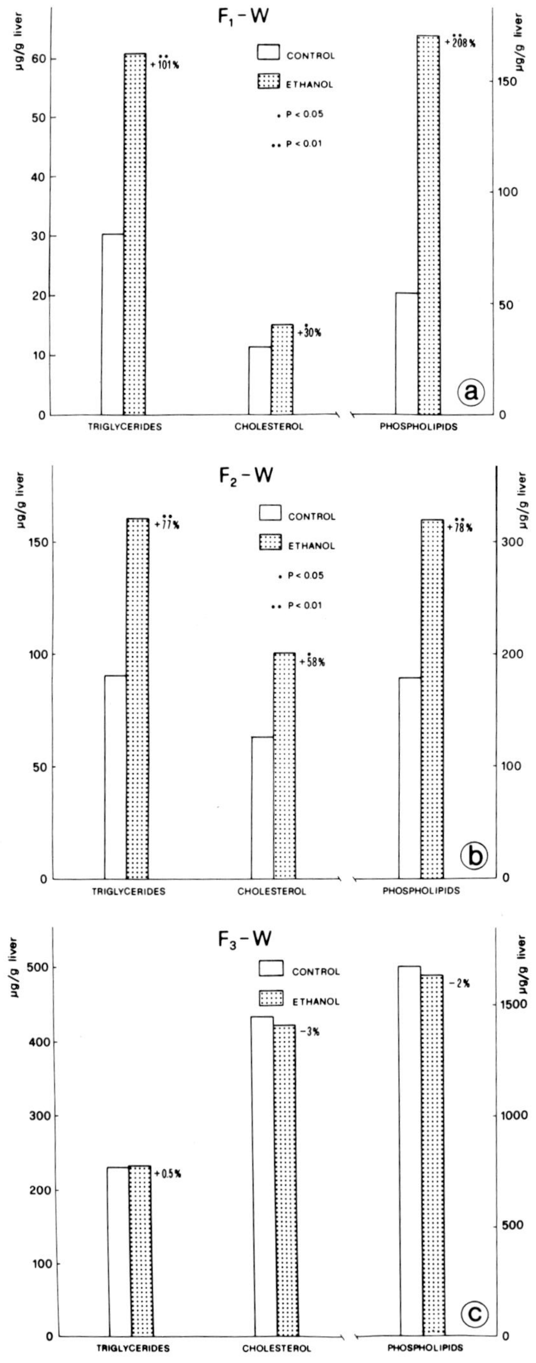

F1 - W
µg/g liver
CONTROL
ETHANOL
• P < 0.05
•• P < 0.01
+101%
+30%
+208%
TRIGLYCERIDES
CHOLESTEROL
PHOSPHOLIPIDS
a
F2 - W
µg/g liver
CONTROL
ETHANOL
• P < 0.05
•• P < 0.01
+77%
+58%
+78%
TRIGLYCERIDES
CHOLESTEROL
PHOSPHOLIPIDS
b
F3 - W
µg/g liver
CONTROL
ETHANOL
+0.5%
-3%
-2%
TRIGLYCERIDES
CHOLESTEROL
PHOSPHOLIPIDS
c

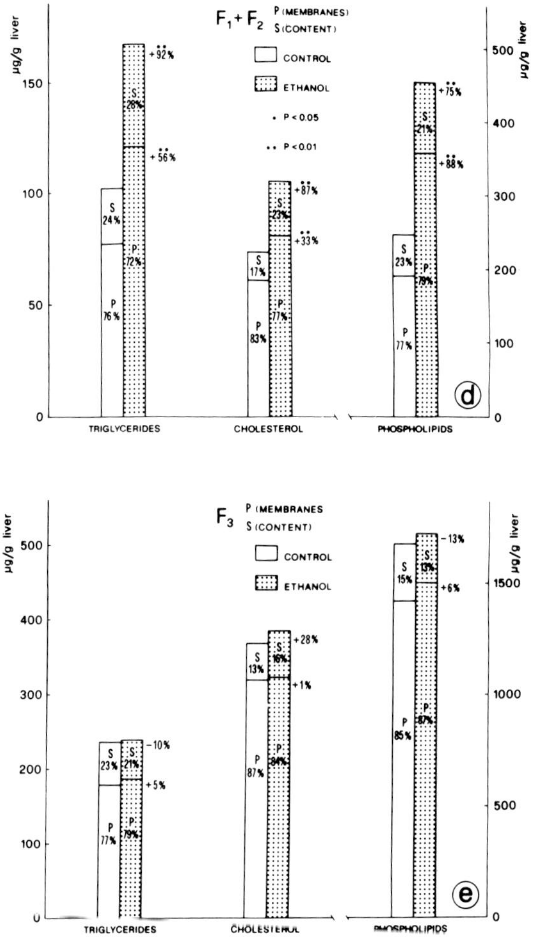

Fig. 3. Lipid composition of Golgi apparatus subfractions from the liver of normal (N) and ethanol-intoxicated (E) rats. Each assay was performed on the whole (W) F_1, F_2, F_3 fractions and on the membranes (P) and their content (S) obtained after disruption of pooled (F_1+F_2) and F_3 fractions. Results, expressed as μg/g wet liver, are the average of 4 to 8 sets of experiments. The percentage increase (%) of E values related to N values was calculated. Values that were significantly different from control are indicated by the given simbols.

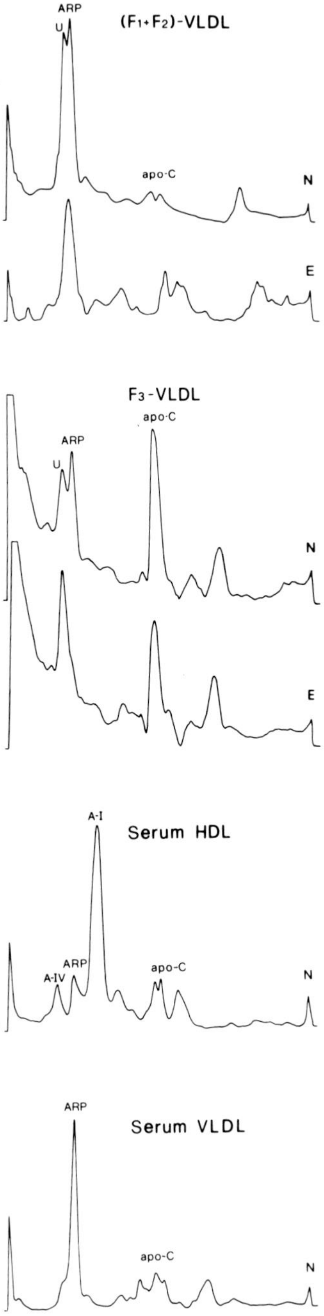
ARP
U
(F1+F2)-VLDL
apo-C
N
E
F3-VLDL
apo-C
ARP
U
N
E
A-I
Serum HDL
ARP
A-IV
apo-C
N
ARP
Serum VLDL
apo-C
N

4

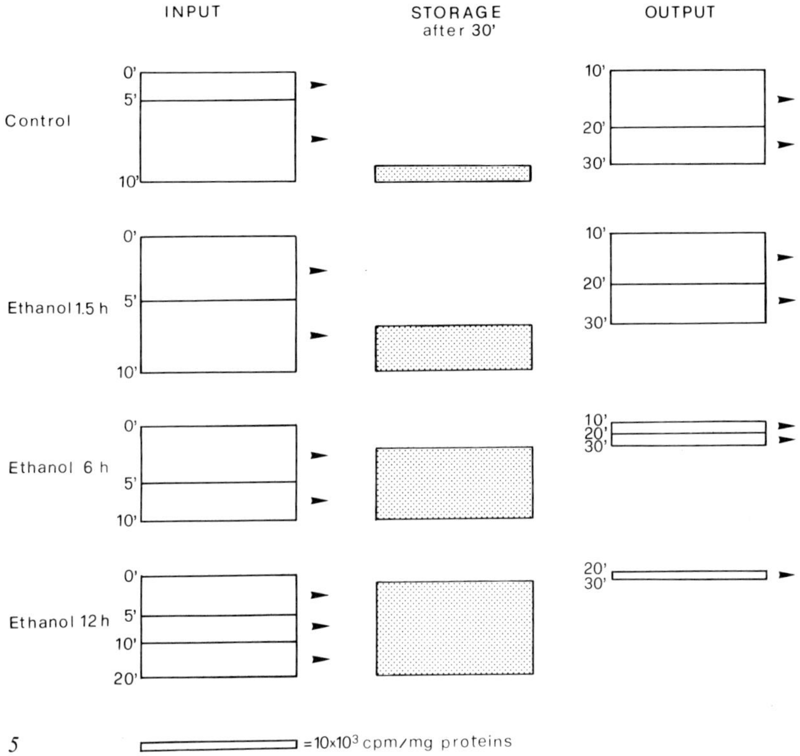

5

Fig. 4. Densitometric scanning of apo-VLDLs isolated from Golgi subfractions and from serum. F_3-VLDL and (F_1+F_2)-VLDL correspond to the VLDL content of the formative and secretory Golgi apparatus, respectively, after French-press disruption and ultracentrifugation. The comparison between serum and Golgi apoproteins was based on the migration coefficent (R_f) of each band. 7.5% SDS-polyacrylamide gel contraining 8 *M* urea was used. Gels were stained with Coomassie blue. N = Normal conditions; E = ethanol intoxication.

Fig. 5. Diagram showing the various steps in the uptake, storage and secretion of [^{3}H]palmitic acid through the combined F_1, F_2, F_3 fractions of Golgi apparatus.

cation apo-C seems to be concentrated in (F_1+F_2) and decreased in F_3 fractions.

Discussion

From the described results it seems that during acute ethanol intoxication there is at least temporary impairment of hepatic secretion of lipoproteins into the blood stream. Since impaired hepatic lipoprotein formation or release has been proposed as the most important event in the pathogenesis of fatty liver produced by various hepatotoxic agents [17], this mechanism, which can account for rapid development of fatty liver better than others, has been claimed also to explain ethanol-induced liver injury. *Schapiro* et al. [29] showed that at high concentrations ethanol causes decreased lipoprotein output from the liver. A decreased serum VLDL level was also reported to occur in rats after acute ethanol intoxication [30]. The defective lipoprotein release into the blood stream must result in a lowering of the serum triglycerides. As to ethanol fatty liver, the existing literature is, at the present controversial: some authors have found an increase [31–33] and others a decrease in serum triglycerides [29, 30, 34]. *Estler* [34] noted an inhibition of lipid secretion from the liver within 2 h of ethanol intoxication and an increase 8 h later. This author thought that the conflicting results reported in literature could be explained partly by the different doses of ethanol used and partly by the different time points of serum lipid assay after ethanol intoxication, which are closely bound up with the biphasic course of the serum triglyceride changes. Unpublished studies from our laboratory on serum triglyceride levels in the course of acute ethanol intoxication broadly agree with *Estler's* [34] results. Other experiments suggest that although elevated serum triglyceride concentrations may occur after acute ethanol administrations they may be inappropriately low in response to the lipid burden presented for secretion. Relative decreases in secretion may contribute to the fatty liver by blocking the release of excess triglycerides [2]. On that account one might infer that the intracellular transport of lipoproteins is impaired. The results reported in this study show that in the secretory membranes of the Golgi apparatus a triglyceride amount, significantly higher than the one found in the untreated rats, has already accumulated 3 h after ethanol intoxication. Therefore, the participation of the secretory membranes of the Golgi apparatus must be considered an early phenomenon in the accumulation of the VLDLs, which contain most of the serum triglyceride [35], during the initial stages of the fatty liver. This behaviour might be referable to an ethanol-induced injury on the transport mechanisms of the lipoproteins which represent a major role evolved by this membrane complex as described by *Chapman* et al. [36]. Such lipoprotein storage could explain the electron microscopic observations of the increasing number of osmiophilic particles in the cisternae and vesicles of Golgi apparatus after acute ethanol treatment. In effect, the patterns of the [^{3}H] palmitic acid incorporation in the different subcellular fractions of the liver clearly demonstrate that the radioactivity increase in the first hours following ethanol intoxication is directly related to the VLDL accumulation in the microsomal fractions and the Golgi apparatus as reported by other authors [10, 11, 37]. The decreased input of the labelled lipid, especially in the Golgi apparatus, where it appears significantly reduced just 6 h after ethanol intoxication, indicates the important implication of this complex in the mechanism of lipid accumulation in the hepatocytes and therefore in the simultaneous decrease of the lipoprotein output. Such an impaired secretory mechanism is more evident, also 12 h after intoxication, in secretory fractions F_1 and F_2 whose behaviour appears to be similar.

The various steps in the transport and storage of [^{3}H]palmitic acid through the three combined fractions of Golgi apparatus are summarized in figure 5. The involvement of Golgi apparatus in ethanol-induced impairment of hepatic lipoprotein secretion may be due to an altered turnover of the membranes responsible for the entire process of liver lipoprotein synthesis, transport and storage [16]. In these membranes the lipid concentration is higher than in other cell organelles [38]. The total lipid-to-protein ratio obtained by us under normal conditions, as well as the concentration of triglycerides and cholesterol, is in agreement with those obtained by *Stremmel and Debuch* [39]. In contrast with these authors, however, the phospholipid content of the whole membranes is increased and the phospholipid-to-neutral lipid ratio is similar to that reported [38]. The secretory vesicles of the Golgi apparatus contain lipoproteins which are precursors of the plasma VLDLs [15, 40, 41]. It is important to observe that the significant increase of the studied lipid classes in the secretory fractions of the Golgi apparatus (F_1+F_2) after ethanol intoxication concerns both the membrane lipids and those present in the secretory vesicles which contain the nascent VLDLs. On the contrary, in the formative Golgi apparatus (F_3), where the VLDL assembly is still incomplete, significant changes of the lipid concentration are not observed. These data might confirm that the secretory region of the Golgi apparatus is selectively implicated in blocking the release of the lipoproteins into the blood stream. The phospholipid-to-neutral lipid ratio is increased especially in the membranes of F_1 fraction but not in its secretory content. This suggests that ethanol acts on the molecular structure of these membranes altering their turnover which could result in possible modifications of the assembly mechanism of the secretory proteins, in particular of the VLDLs which accumulate. Several studies have demonstrated that VLDLs undergo progressive completion during their passage from the formative site (F_3) to the intermediate (F_2) and to the secretory vesicles (F_1) of the Golgi apparatus through the acquisition of peptide components and parallel glycosylation [15, 40–44]. Generally these molecules are somewhat similar to circulating VLDLs in both lipid composition and apoprotein content and may therefore represent a nascent form of VLDLs [36, 40, 45]. Our results indicate that the morphological site of storage and secretion of apo-ARP is in the Golgi apparatus as suggested by *Dolphin and Rubinstein* [44]. Furthermore, the U band might represent either a polymorphous form of apoprotein or a nascent apoprotein. This latter possibility might be due to a charge shift electrophoresis, dependent upon degradation of a possible 'signal' peptide, glycosylation or lipid assembly mechanism, all of which seem to contribute to secretory component maturation. After ethanol intoxication Golgi-VLDLs behave in an interesting way, quite differently from what happens in normal conditions. The overlapping of the U and apo-ARP bands might suggest an alteration of possible polymorphic forms of apo-ARP lipoprotein or an incompleted maturation in the carbohydrates of the nascent apo-ARP. Carbohydrates are known to deviate the electrophoretic mobility of proteins in sodium dodecyl sulphate (SDS) gels [46] and to characterize polymorphic forms of rat apolipoproteins apo-C-III [47] and apo-ARP [48]. Moreover, the behaviour of apo-C, which is concentrated in (F_1+F_2) and decreased in F_3 fractions, implies an impairment both in synthesis and in secretion.

Therefore, the data presented in this report indicate that ethanol treatment modifies the functions accomplished by Golgi subfractions impairing lipoprotein metabolism and, at least temporarily, hepatic lipoprotein secretion. An intriguing question is related to the mechanism of this behaviour. Various authors have demonstrated that ethanol inhibits the synthesis of structural and secretory proteins in different experimental conditions and that it is simultaneously able to inhibit liver protein secretion [49–54]. It is known that the major part of the plasma proteins, including lipoproteins, are secreted by the liver in the form of glycoproteins [15, 44–55]. Some investigators [2, 52, 56] demonstrated that ethanol and its main metabolite, acetaldehyde, are capable of altering the liver metabolism of glycoproteins at the postribosomal level, selectively inhibiting glycoprotein secretion.

Since the glycosylation of the nascent VLDLs takes place in the Golgi apparatus [44, 57–59] it was possible to suggest that ethanol and/or acetaldehyde interfere with the final glycosylation and packaging of the VLDLs into the secretory vesicles of the Golgi apparatus. In fact, the preliminary research findings performed by us have demonstrated that in the course of acute ethanol intoxication, the output of [^{14}C]glucosamine from the secretory fractions of the Golgi apparatus is inhibited early [60]. By comparing the time sequence related to the transit through the Golgi apparatus of the lipid [^{3}H]palmitic acid and of the carbohydrate [^{14}C]glucosamine both precursors of VLDLs, the changes of the uptake and of the release of palmitic acid follow, chronologically, those of the glucosamine.

These results suggest that in ethanol-intoxicated rats the Golgi apparatus secretory membranes are unable to develop their role in the progressive completion of glycoproteins. Consequently, the terminal glycosylation of VLDLs and their secretion might be inhibited.

In conclusion, ethanol seems to selectively damage the explicit function of the Golgi apparatus in the elaboration, glycosylation, transport and secretion of the VLDLs, probably through its main metabolite, acetaldehyde. Further studies are in progress to explain the ethanol-induced changes of glycoprotein metabolism in the Golgi apparatus and their possible association with an impairment of the microtubular system which was recently proposed as an important mechanism responsible for the secretory defect [61].

Summary

The effects of acute ethanol intoxication (6 g/kg body weight) on the lipid metabolism of rat liver subcellular fractions were investigated 'in vivo' at different times after intoxication. Ethanol altered the [^{3}H]palmitic acid transport through the studied fractions. Particularly, in isolated Golgi apparatus, either the lipid input or output were impaired early. The triglyceride, phospholipid and cholesterol levels significantly increased at 6 h after ethanol intoxication either in the secretory Golgi membranes (F_1, F_2), or in their content; at the same time the disc-electrophoretic patterns of nascent VLDLs isolated from the Golgi apparatus showed an altered behaviour of apo-ARP and apo-C apoproteins. These data suggest that, during the first stages of ethanol-induced liver injury, the changes of lipid metabolism might be dependent on the involvement of Golgi secretory membranes in an impaired lipoprotein synthesis, transport and secretion.

References

1 Comporti, M.: Ethanol induced liver injury; in Slater, Biochemical mechanisms of liver injury, pp. 449-516 (Academic Press, New York 1978).

2 Sorrell, M.F.; Tuma, D.J.: Effects of alcohol on hepatic metabolism: selected aspects. Clin. Sci. *57:*481-489 (1979).

3 Baraona, E.; Lieber, C.S.: Effects of ethanol on lipid metabolism. J. Lipid Res. *20:* 289-315 (1979).

4 Robinson, S.F.; Quarfort, S.H.: The effect of ethanol on lipoprotein metabolism. Alcoholism: Clin. Exp. Res. *5:*101-109 (1981).

5 Nanni, G.; Marinari, U.M.: Ruolo dell'apparato di Golgi dell'epatocita nell'elaborazione delle lipoproteine. Fegato *21:*153-198 (1975).

6 Cottalasso, D.; Gambella, G.R.; Pronzato, M.A.; Averame, M.M.; Nanni, G.; Marinari, U.M.: Influence of cycloxilic acid on energy production by hepatocyte mitochondria during acute ethanol intoxication. Arzneimittelforsch. *28:* 1224-1226 (1978).

7 Rubin, E.; Beattie, D.S.; Toth, A.; Lieber, C.S.: Structural and functional effects of ethanol on hepatic mitochondria. Fed. Proc. *31:*131-140 (1972).

8 Rubin, E.; Lieber, C.S.: Early fine structural changes in the human liver induced by alcohol. Gastroenterology *52:*1-13 (1967).

9 Ishii, H.; Joly J.G.; Lieber, C.S.: Effect of ethanol on the amount and enzymes activities of hepatic rough and smooth microsomal membranes. Biochim. biophys. Acta *291:*411-420 (1973).

10 Stein, O.; Stein Y.: Lipid synthesis, intracellular transport, storage and secretion. I. Electron microscopic radioautographic study of liver after injection of tritiated palmitate or glycerol in fasted and ethanol-treated rats. J. Cell Biol. *33:*319-339 (1967).

11 Ehrenreich, J.H.; Bergeron, J.J.M.; Siekevitz, P.; Palade, G.E.: Golgi fractions prepared from rat liver homogenates. I. Isolation procedure and morphological characterization. J. Cell Biol. *59:*45-72 (1973).

12 Franzi, A.T.; Marinari, U.M.; Cadoni, A.; Nanni, G.: Aspetti morfologici e funzionali dell'apparato di Golgi negli epatociti in corso di intossicazione acuta da etanolo. Riv. Istochim. norm. Patol. *21:*238 (1977).

13 Jones, A.L.; Ruderman, N.B.; Herrera, M.G.: Electron microscopic and biochemical study of lipoprotein synthesis in the isolated perfused rat liver. J. Lipid Res. *8:* 429-446 (1967).

14 Hamilton, R.L.; Regen, D.M.; Gray, M.E.; Lequire, V.S.: Lipid transport in liver. I.

Electron microscopic identification of very low density lipoproteins in perfused rat liver. Lab. Invest. *16:*305–319 (1967).
15 Nestruck, A.C.; Rubinstein, D.: The synthesis of apoproteins of very low density lipoproteins isolated from the Golgi apparatus of rat liver. Can. J. Biochem. *54:* 617–628 (1976).
16 Claude, A.: Growth and differentiation of cytoplasmic membranes in the course of lipoprotein granule synthesis in the Golgi complex. J. Cell Biol. *47:*745–766 (1970).
17 Dianzani, M.U.: Patologia generale, vol. I, pp. 1096–1170 (UTET, Torino 1977).
18 Averame, M.M.; Pronzato, M.A.; Cottalasso, D.; Gambella G.R.; Marinari, U.M.; Nanni, G.: Effects of cycloxilic acid on liver subcellular fractions triglyceride content in acute ethanol intoxication. Arzneimittelforsch. *28:*1221–1223 (1978).
19 Nanni, G.; Cottalasso, D.; Gambella, G.R.; Averame, M.M.; Marinari, U.M.: Azione dell'intossicazione acuta con etanolo sull'apparato di Golgi isolato dall'epatocita. Boll. Soc. ital. Biol. sper. *53:*1686–1691 (1977).
20 Havel, R.J.; Eder, H.A.; Bragdon, J.H.: The distribution and chemical composition of ultracentrifugally separated lipoproteins in human serum. J. clin. Invest. *34:* 1345–1353 (1955).
21 Fleischer, B.; Fleischer, S.; Ozawa, H.: Isolation and characterization of Golgi membranes from bovine liver. J. Cell Biol. *43:*59–79 (1969).
22 Folch, J.; Lees, M.; Sloane-Stanley, G.H.: A simple method for the isolation and purification of total lipids from animal tissues. J. biol. Chem. *226:*497–509 (1957).
23 Eggstein, M.; Kreutz, F.H.: Eine neue Bestimmung der Neutralfette im Blutserum und Gewebe. Klin. Wschr. *44:*262–267 (1967).
24 Bowman, R.E.; Wolf, R.C.: A rapid and specific ultramicro method for total serum cholesterol. Clin. Chem. *8:*302–309 (1962).
25 Chen, P.S.; Toribara, T.Y.; Warner, H.: Microdetermination of phosphorus. Analyt. Chem. *28:*1756–1758 (1956).
26 Lowry, O.H.; Rosebrough, N.J.; Farr, A.L.; Randall, R.J.: Protein measurement with the Folin phenol reagent. J. biol. Chem. *193:*265–275 (1951).
27 Davis, B.J.: Disc-electrophoresis. II. Method and application to human serum proteins. Ann. N.Y. Acad. Sci. *121:*404–427 (1964).
28 Casu, A.; Cottalasso, D.; Pronzato, M.A.; Marinari, U.M.; Nanni, G.: Effect of acute ethanol intoxication on rat liver very low density lipoproteins from the Golgi apparatus. IRCS med. Sci. *10:*386 (1982).
29 Schapiro, R.H.; Drummey, G.D.; Shimizu, Y.; Isselbacher, K.J.: Studies on the pathogenesis of the ethanol-induced fatty liver. II. Effect of ethanol on palmitate-1-^{14}C metabolism by the isolated perfused rat liver. J. clin. Invest. *43:* 1338–1347 (1964).
30 Madsen, N.P.: Reduced serum very low density lipoprotein levels after acute ethanol administration. Biochem. Pharmacol. *18:*261–262 (1969).
31 Lieber, C.S.; Jones, D.P.; Mendelson, J.M.; De Carli, L.M.: Fatty liver hyperlipemia and hyperuricemia produced by prolonged alcohol consumption despite adequate dietary intake. Trans. Ass. Am. Physns. *76:*289–300 (1963).
32 Jones, D.P.; Losowsky, M.S.; Davidson, C.S.; Lieber, C.S.: Effects of ethanol on plasma lipids in man. J. clin. Med. *62:*675–682 (1963).
33 Mørland, J.; Øye, I.: Effect of acute and chronic ethanol treatment on hepatic triglyceride release from the rat perfused liver. Eur. J. Pharmacol. *27:*238–244 (1974).

34 Estler, C.J.; Comparison of hepatic triglyceride content and hepatic lipid secretion after various doses of ethanol. Biochem. Pharmacol. *24:*1871–1873 (1975).
35 Lombardi, B.; Ugazio, G.: Serum lipoproteins in rats with carbon tetrachloride-induced fatty liver. J. Lipid Res. *6:*498–505 (1965).
36 Chapman, M.J.; Mills, G.L.; Taylaur, C.E.: Lipoprotein particles from the Golgi apparatus of guinea pig liver. Biochem. J. *128:*779–787 (1972).
37 Bergeron, J.J.M.: Golgi fractions from livers of control and ethanol-intoxicated rats. Enzymic and morphologic properties following rapid isolation. Biochim. biophys. Acta *555:*493–503 (1979).
38 Zambrano, F.; Fleischer, S.; Fleischer, B.: Lipid composition of the Golgi apparatus of rat kidney and liver in comparison with other subcellular organelles. Biochim. biophys. Acta *380:*357–369 (1975).
39 Stremmel, W.; Debuch, H.: The lipids of the Golgi apparatus subfractions from rat liver. Biochim. biophys. Acta *573:*301–307 (1979).
40 Mahley, R.W.; Hamilton, R.L.; Lequire, V.S.: Characterization of lipoprotein particles isolated from the Golgi apparatus of rat liver. J. Lipid Res. *10:* 433–439 (1969).
41 Chapman, M.J.; Mills, G.L.; Taylaur, C.E.: The effect of a lipid-rich diet on the properties and composition of lipoprotein particles from the Golgi apparatus of guinea pig liver. Biochem. J. *131:*177–185 (1973).
42 Bergeron, J.J.M.; Borts, D.; Cruz, J.: Passage of serum-destined proteins through the Golgi apparatus of rat liver. J. Cell Biol. *76:*87–97 (1978).
43 Bouma, M.E.; Amit, N.; Infante, R.: Ultrastructural localization of apo-B and apo-C binding to very low density lipoproteins in rat liver. Virchows Arch. Abt. B Zellpath. *30:*161–180 (1979).
44 Dolphin, P.J.; Rubinstein, D.: Glycosylation of apoproteins of rat very low density lipoproteins during transit through the hepatic Golgi apparatus. Can. J. Biochem. *55:*83–90 (1977).
45 Mahley, R.W.; Bersot, T.P.; Lequire, V.S.: Identity of very low density lipoprotein apoproteins of plasma and liver Gilgi apparatus. Science *168:*380–382 (1970).
46 Banker, G.A.; Cotman, C.W.: Measurement of free electrophoretic mobility and retardation. J. biol. Chem. *247:*5856–5861 (1972).
47 Herbert, P.N.; Windmueller, H.G.; Bersot, T.P.; Shulman, R.S.: Characterization of the rat apolipoprotein. I. The low molecular weight proteins of rat plasma high density lipoproteins. J. biol. Chem. *249:*5718–5724 (1974).
48 Weisgraber, K.H.; Mahley, R.W.; Assmann, G.: The rat arginine rich apoprotein and its redistribution following injection of iodinated lipoproteins into normal and hypercholesterolemic rats. Atherosclerosis *28:*121–140 (1977).
49 Jeejeebhoy, K.N.; Phillips, M.J.; Bruce-Robertson, A.; Ho, J.; Sodtke, U.: The acute effect of ethanol on albumin, fibrinogen and transferrin synthesis in the rat. Biochem. J. *126:*1111–1126 (1972).
50 Perin, A.; Scalabrino, G.; Sessa, A.; Arnaboldi, A.: In vitro inhibition of protein synthesis in rat liver as a consequence of ethanol metabolism. Biochim. biophys. Acta *366:*101–108 (1974).
51 Mørland, J.; Bessesen, A.: Inhibition of protein synthesis by ethanol in isolated rat liver parenchimal cells. Biochim. biophys. Acta *474:*312–320 (1977).
52 Sorrell, M.F.; Tuma, D.J.; Schafer, E.C.; Barak, A.J.: Role of acetaldehyde in the

ethanol-induced impairment of glycoprotein metabolism in rat liver slices. Gastroenterology *73:*137–144 (1977).
53 Baraona, E.; Pikkarainen, P.; Salaspuro, M.; Finkelman, F.; Lieber, C.S.: Acute effects of ethanol on hepatic protein synthesis and secretion in the rat. Gastroenterology *79:*104–111 (1980).
54 Murty, C.N.; Verney, E.; Sidransky, H.: Acute effect of ethanol on membranes of the endoplasmic reticulum and on protein synthesis in rat liver. Alcoholism: Clin. exp. Res. *4:*93–103 (1980).
55 Molnar, J.: A proposed pathway of plasma glycoprotein synthesis. Mol. cell. Biochem. *6:*3–14 (1975).
56 Tuma, D.J.; Sorrell, M.F.: Effects of ethanol on the secretion of lipoproteins by rat liver slices. Gastroenterology *80:*273–278 (1981).
57 Lo, C.H.; Marsh, J.B.: Biosynthesis of plasma lipoproteins: incorporation of (^{14}C)-glucosamine by cells and subcellular fractions of rat liver. J. biol. Chem. *245:*5001–5006 (1970).
58 Wetmore, S.; Mahley, R.W.; Brown, W.V.; Schachter, H.: Incorporation of sialic acid into sialidase-treated apolipoprotein of human, very low density lipoprotein by a pork liver sialyltransferase. Can. J. Biochem. *52:*655–664 (1974).
59 Merritt, W.D.; Morré, D.J.: A glycosyl transferase of high specific activity in secretory vesicles from isolated Golgi apparatus of rat liver. Biochim. biophys. Acta *304:* 397–407 (1973).
60 Nanni, G.; Pronzato, M.A.; Averame, M.M.; Gambella, G.R.; Cottalasso, D.; Marinari, U.M.: Influence of acute ethanol intoxication on rat liver Golgi apparatus glycosylation activities. FEBS Lett. *93:*242–246 (1978).
61 Matsuda, Y.; Baraona, E.; Salaspuro, M.; Lieber, C.S.: Effects of ethanol on liver microtubules and Golgi apparatus. Lab. Invest. *41:*455–463 (1979).

U.M. Marinari, MD, Istituto di Patologia Generale, Università di Genova,
Via L.B. Alberti, 2, I-16132 Genova (Italy)

Front. gastrointest. Res., vol. 8, pp. 46–62 (Karger, Basel 1984)

Reactive Aldehydes (4-Hydroxyalkenals) Originating from the Peroxidation of Liver Microsomal Lipids: Biological Effects and Evidence for their Binding to Microsomal Protein in CCl_4 or $BrCCl_3$ Intoxication[1]

Mario Comporti, Angelo Benedetti, Marco Ferrali, Rosella Fulceri

Istituto di Patologia Generale dell'Università di Siena, Siena, Italy

It is well known that peroxidation of unsaturated lipids in the biological membranes produces deleterious effects on the membrane structure and function [1–3]. Also lipid peroxidation is believed to play an important role in many conditions of cellular damage [4, 5] including carbon tetrachloride (CCl_4)-induced liver injury [6–8]. The mechanism by which lipid peroxidation acts in producing its effects has been the object of several investigations. Molecular alterations of membrane fatty acids are of course important in this respect; it has also been postulated [9] that lipoperoxidation-induced damage to membrane-bound enzymes is a consequence of the loss of membrane structure essential for the enzyme activities. However, in addition to the alterations directly produced by lipid peroxidation on cellular membranes, it must be considered that products evolved from lipid peroxidation can cause damaging effects on biological structures. Supportive evidence for this concept, which implies that the toxic products diffuse from the original locus of peroxidation and act in distant loci, was found in previous studies [10–12]. In our studies [11, 12] in fact, the effects were shown to occur in revealing or target systems, which were separated from the peroxidizing system – liver microsomes, reduced nicotinamide adenine dinucleotide phosphate (NADPH), Fe^{++}, etc. – by a dialysis membrane. The dialysable cytotoxic products originating from lipid peroxi-

[1] This work was supported by a grant from the National Foundation for Cancer Research (USA). Additional funds were derived from a grant of the Consiglio Nazionale delle Ricerche, Roma, Italy.

dation have the capacity of damaging cellular membranes, as shown by the lytic effect on erythrocytes (when the target system, contained in a dialysis tube, consists of blood red cells) [11, 12] and of impairing the activities of membrane-bound enzymes, as shown by the partial inactivation of microsomal glucose 6-phosphatase (G-6-Pase) and cytochrome $P\text{-}_{450}$ (when the target system consists of freshly prepared liver microsomes) [12]. Additional studies [12] showed that these cytotoxic products have a relatively long life span.

In a subsequent series of studies [13, 14] we succeeded in extracting the toxic products from the dialysate using either ethyl ether or ethyl acetate as solvents. The extracts obtained from the dialysate, in fact, retained all the inhibitory activity on microsomal G-6-Pase [13, 14] and the lytic activity on erythrocytes [15].

The products present in the dialysate extract were partially separated (fig. 1) by thin-layer chromatography (TLC) and the individual bands or groups of bands, eluted from the silica gel, were tested for their capacity to inhibit microsomal G-6-Pase [14]. The highest inhibitory activity for microsomal G-6-Pase (fig. 1) was found to occur in a band (area No. 6) which contained most of the carbonyl functional groups present in the unfractionated extract [14]. It appeared therefore that lipid peroxidation products, highly active in inhibiting G-6-Pase, are fatty aldehydes or other carbonyl compounds.

The studies mentioned hitherto show the possibility that these carbonyl compounds originating from the peroxidation of liver microsomal lipids are responsible, at least in part, for some aspects of the liver injury caused by carbon tetrachloride, such as the inhibition of membrane-bound enzymes and the membrane damage. Another very important alteration produced by CCl_4 in the liver cell is the inhibition of protein synthesis; it seemed therefore of interest to investigate whether these carbonyl compounds are capable of inhibiting protein synthesis in a 'cell free' model system, such as the rabbit reticulocyte lysate. It was found [17] that the addition of the carbonyl compounds to the reaction mixture markedly depressed protein synthesis. Consistent inhibition occurred up to concentrations of 0.2–0.4 m*M*. The addition of –SH groups (dithioerythritol or cysteine) to the incubation mixture afforded a marked protection against the inhibition of protein synthesis produced by the carbonyl compounds [17]. Therefore, an interaction of –SH groups with these carbonyl compounds originating from lipid peroxidation can be anticipated.

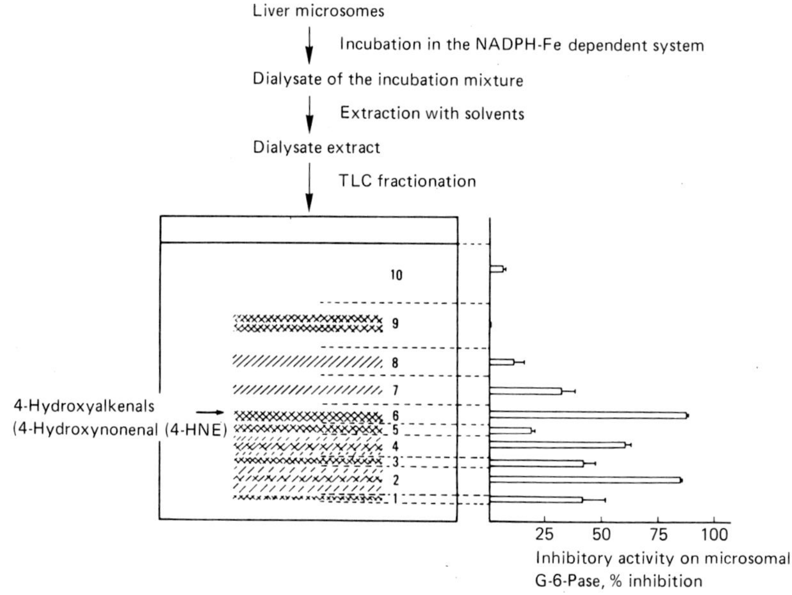

Fig. 1. Diagrammatic representation of TLC fractionation of the products extracted from a dialysate derived from a peroxidizing microsome system and inhibitory activity of the various chromatographic bands on microsomal G-6-Pase. The dialysate from the peroxidizing system was prepared and extracted as reported elsewhere [14]. The developing solvent used was *n*-heptane-ethyl acetate-acetic acid (75:25:1, by vol.). The bands were detected with a *N,N'*-dimethyl-*p*-phenylenediamine spray solution [16]. The plate was divided into 10 parts corresponding to individual bands or groups of bands, as depicted in the figure. The various areas of the chromatoplate (appropriately protected from the spray) were scraped off and the products were eluted, resuspended and tested for the inhibitory activity on the G-6-Pase of freshly prepared liver microsomes (test system), as described elsewhere [14]. The inhibitory activity of the various areas of the chromatoplate was expressed as a percentage of the G-6-Pase activity of the microsomes of the test system incubated in the absence of the products eluted from the chromatoplate. Results are the means ± SEM of 5–8 experiments [modified from ref. 14, with permission].

Since the ultimate effect of CCl_4-intoxication is liver cell necrosis, in further experiments [15, 18] a possible cytotoxic effect of the carbonyl compounds derived from lipid peroxidation on intact hepatocytes was looked for. Table I shows that when isolated hepatocytes are incubated with the carbonyl compounds (0.45 m*M* concentration), 95% of them lose viability, as assessed by the trypan blue exclusion test.

Table I. Loss of viability of isolated hepatocytes exposed to carbonyl compounds originating from the peroxidation of liver microsomal lipids

	Trypan blue stained hepatocytes %	NADH penetration into hepatocytes %	Enzyme release %	
			LDH	GPT
0 time (non-incubated sample)	17.7 ± 1.8	21.7 ± 1.3	12.7 ± 0.8	9.2 ± 0.6
– Carbonyl compounds (control system, + NADPH – microsomes)	17.7	19.1	11.9	7.1
– Carbonyl compounds (control system, + microsomes – NADPH)	17.1 ± 0.7	24.8 ± 1.0	12.1 ± 0.7	8.1 ± 0.6
+ Carbonyl compounds (experimental system, + microsomes + NADPH)	94.9 ± 3.2	90.3 ± 6.2	90.7 ± 8.1	53.6 ± 3.1

The carbonyl compounds (products eluted from the area No. 6 of the chromatoplate) obtained from the dialysate extract derived from the peroxidizing microsome system (plus microsomes, plus NADPH) were resuspended with 1.0 ml of Krebs-Ringer-phosphate solution (pH 7.4) and added to 1.0 ml of a hepatocyte suspension (4×10^6 cells/ml) in the same solution. The concentration of carbonyl functional groups in the final reaction mixture was 0.45 m*M*, as assessed by the UV absorption of a stock solution of the carbonyl compounds in water, assuming a molar absorptivity $\varepsilon = 13{,}600$ [19]. The mixture was incubated aerobically for 60 min at 37 °C. Control samples were performed by eluting the corresponding areas of the chromatoplates to which the dialysate extracts derived from the control systems (plus NADPH, minus microsomes; plus microsomes, minus NADPH) were applied. At the end of the incubation, an aliquot of the reaction mixture was used to determine the percentage of trypan blue-stained cells. Another aliquot was used to determine the penetration of NADH into hepatocytes according to *Högberg* et al. [20]. The penetration was expressed as a percentage of NADH oxidation with lysed cells. An additional aliquot was used to determine the release of enzymes (lactate dehydrogenase = LDH; glutamate-pyruvate transaminase = GPT), as percentage of extracellular activity over total activity determined after cell destruction. The latter was induced by the addition of 0.5% Triton X-100.

These carbonyl compounds thus appear to be highly active in producing liver cell death; they also produce an almost complete permeabilization of the plasma membrane, as measured by the test of permeability to reduced nicotinamide adenine dinucleotide (NADH) [20]. Concomitantly with the membrane permeabilization, a marked release of enzymes (LDH, GPT) from the hepatocyte occurs.

Therefore, some aspects of the CCl_4-induced liver damage are reproduced, at least in these model systems in vitro, by the activity of these carbonyl compounds derived from the peroxidation of liver microsomal lipids. Furthermore, these carbonyl compounds, while inhibiting those microsomal enzymes (such as G-6-Pase, cytochrome P_{-450} and aminopyrine demethylase) which are affected by either lipid peroxidation in vitro or carbon tetrachloride intoxication in vivo, do not inhibit microsomal enzymes, such as NADPH cytochrome *c* reductase, which are not inhibited by lipoperoxidation or CCl_4 poisoning [21].

Studies concerned with the possible mechanisms by which these carbonyl compounds derived from lipid peroxidation act in producing their damaging effects showed [15, 22] that after the carbonyl compounds have been incubated with the liver microsomes, a substantial amount of these products can be recovered bound to the microsomal protein, by allowing the microsomal protein to react with 2,4-dinitrophenylhydrazine (DNPH). Concomitant with the binding, a decrease of total -SH groups of the microsomal protein occurs; such a decrease appears to be accounted for by the amount of carbonyl compounds bound to the microsomal protein. Still in concomitance with the binding, the G-6-Pase activity of the microsomes markedly decreases as usual.

Therefore, a mechanism could be postulated for the binding of 4-hydroxyalkenals to -SH groups of proteins and enzymes similar to that described by *Schauenstein and Esterbauer* [23] (fig. 2). It must be

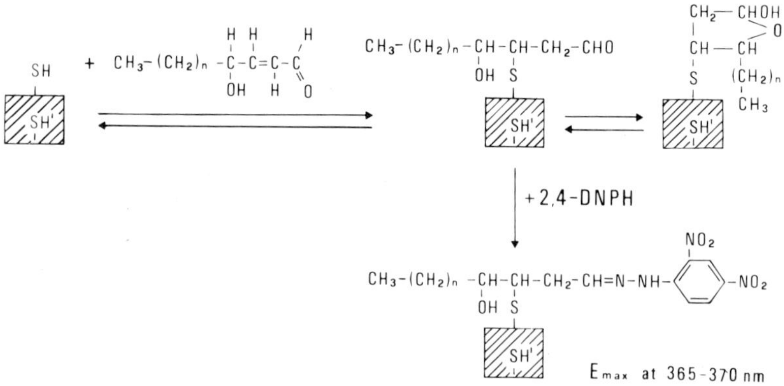

Fig. 2. Scheme of the binding of 4-hydroxyalkenals to -SH groups of proteins and detection of the protein-bound aldehyde by the reaction of DNPH with the aldehyde group [modified from ref. 23].

emphasized in fact, that previous studies by *Schauenstein and Esterbauer* [23] and *Schauenstein* et al. [19] had shown the formation of a particular class of aldehydes - namely 4-hydroxyalkenals - in chemical systems. It was also shown that these aldehydes have a high biological reactivity (they were also used as possible antitumoral agents because of their cytotoxic effects) and easily react with -SH groups of low molecular weight thiols or proteins. Subsequent studies [24] concerned with the molecular characterization of these carbonyl compounds originating from the peroxidation of liver microsomal lipids have, in fact, shown that these products consist of 4-hydroxyalkenals, almost entirely (more than 95%) of 4-hydroxy-*trans*-2,3-nonenal

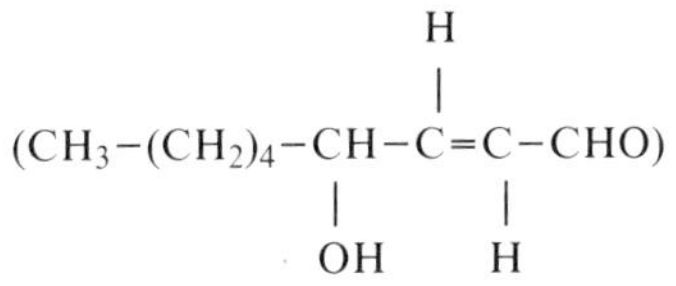

with small amounts of 4-hydroxyoctenal, 4-hydroxydecenal and 4-hydroxyundecenal. The structure was ascertained by means of high pressure liquid chromatography (HPLC), ultraviolet, infrared and mass spectrometry.

4-hydroxynonenal (4-HNE), prepared by chemical synthesis [25], has been found to reproduce, qualitatively and quantitatively, all the cytopathological effects brought about by the biogenic carbonyl compounds - namely the inhibition of microsomal G-6-Pase [24], cytochrome $P\text{-}_{450}$ [26], aminopyrine demethylase [21], the results with NADPH cytochrome *c* reductase [21], the inhibition of protein synthesis in a reticulocyte lysate [17], the cytotoxic effect on isolated hepatocytes [26] and the lytic activity on erythrocytes [24].

The next step in the problem was the following: is there evidence of alkenals (or any other carbonyl compounds) being bound to the microsomal protein in in vivo conditions - such as CCl_4 or bromotrichloromethane ($BrCCl_3$) intoxication - in which peroxidation of lipids of hepatic endoplasmic reticulum has been demonstrated?

As previously mentioned, a number of studies [19, 23] have demonstrated that reactive aldehydes such as alkenals or 4-hydroxy-2-alkenals react with -SH groups of low molecular weight thiols (glutathione, cysteine, etc.), protein and enzymes and that, in the latter case, the reaction is responsible for the enzyme inactivation. This reaction leads to the formation of a modified protein, in which the aldehyde moieties are bound covalently by thioether linkages to the cysteine residues of the polypeptide chain [19, 23]. Since the aldehyde grouping remains free during this reaction, the number of aldehyde molecules attached to the protein can be estimated quantitatively from the ultraviolet-visible spectrum of the DNPH-treated protein [19, 23]. These protein-bound dinitrophenylhydrazones show an absorption spectrum which, under the test conditions used, has a maximum in the 365–370 nm range. Therefore, the appearance of such a spectrum in the microsomal protein allowed to react with DNPH should be indicative of aldehyde binding. Furthermore, the evaluation of the spectrum should allow a quantitative estimation of the amount of alkenals or other carbonyls bound to the microsomal protein [27].

Liver microsomes were prepared from control, CCl_4- and $BrCCl_3$-poisoned rats (the animals were starved overnight before being used). Microsomal membranes were solubilized with Triton X-100 and then allowed to react with a DNPH reagent [28]. The lipids and the excess reagent were then removed by repeated extraction with chloroform:methanol (2:1, v/v). Finally, the DNPH-treated non-lipoidal residue was dissolved in 7 *M* guanidine. The absorption spectrum was then recorded. It must be noted that butylated hydroxytoluene (BHT) and ethylenediaminetetracetic acid (EDTA) were present in the homogenization, solubilization and extraction medium to prevent artifactual lipid oxidation.

Figure 3a shows the ultraviolet-visible spectra of DNPH-treated non-lipoidal residues of liver microsomes (microsomal protein) from control, CCl_4- and $BrCCl_3$-treated rats (2,600 μmol/100 g body wt, per os). The dotted lines (fig. 3a) show the absorption spectra of the blank samples, i.e. the samples treated with a 'reagent' in which DNPH was omitted. Because of this intrinsic absorption of the microsomal protein, the proper hydrazone spectra (fig. 3b) were obtained by recording the absorption of the DNPH-treated samples (from control, CCl_4 and $BrCCl_3$ groups) against their respective blank samples. These spectra seem to demonstrate that relatively high amounts of DNPH-reacting groups are present even in microsomal protein of non-intoxicated rats. A somewhat similar spectrum with an absorption maximum in the 355 nm region has been obtained [*Nöhammer and Poli*, personal communication] in cytospectrophotometric examination of control liver cells, in studies concerned with lipid peroxidation of isolated hepatocytes. Intoxication with CCl_4 or $BrCCl_3$ leads to 1.5- and 1.8-fold increase in the

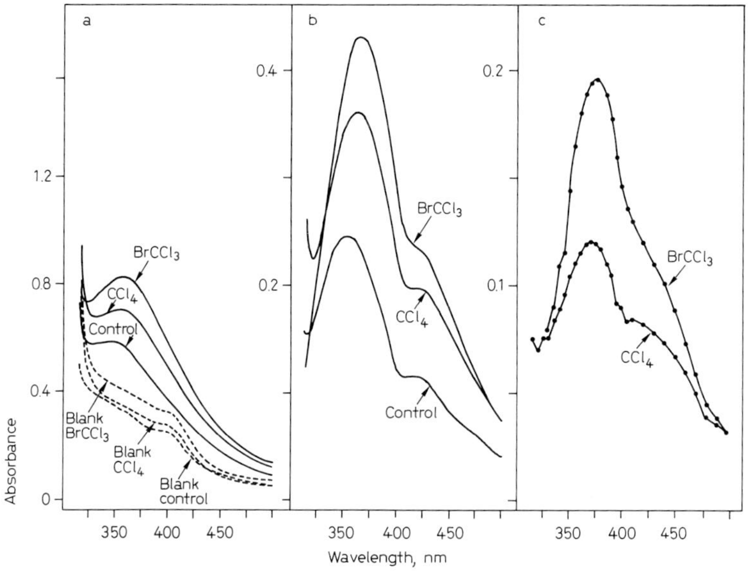

Fig. 3. *a* Solid lines: absorption spectra of DNPH-treated non-lipoidal residues of liver microsomes derived from control, CCl_4-poisoned and $BrCCl_3$-poisoned rats. Dotted lines: absorption spectra of blank samples. The rats were killed 2 h after poisoning. A typical experiment is reported. The protein concentration of the sample dissolved in 7 *M* guanidine was 8 mg/ml. The spectra were recorded against 7 *M* guanidine in the reference cuvette. *b* Spectra for control, CCl_4-poisoned and $BrCCl_3$-poisoned rats. The spectra for DNPH-treated samples were recorded against the corresponding blank samples in the reference cuvette. *c* Difference spectra calculated from the spectra of figure 3b by subtracting the spectrum of the control sample from the spectra of the samples of the intoxicated rats [from ref. 27, with permission].

absorption intensity and concomitantly to a shift of the maximum to longer wavelengths (265–275 nm). The 'intoxicated minus control' difference spectra (fig. 3c) (calculated by subtracting the spectrum of the control from the spectra of the intoxicated rats) show a maximum in the range of 370–375 nm. This could be indicative for the dinitrophenylhydrazone derivative of a saturated aldehyde, which is expected to be formed according to the scheme shown above (fig. 2).

Results similar to those of the in vivo experiments, although magnified from a quantitative point of view, were also obtained in a number of in vitro studies. First of all, incubation of liver microsomes with

4-HNE results in a significant binding of the aldehyde to the microsomal protein, as shown by the hydrazone spectra reported in figure 4. Secondly, a high content of protein-bound carbonyls, as shown by the hydrazone spectra reported in figure 5, was found in liver microsomes peroxidized for 20 min in the NADPH-Fe-dependent system (60 μM Fe^{++}). Therefore, these in vitro studies strongly suggest that the appearance of the spectrum characteristic of dinitrophenylhydrazones in the microsomal protein is due to the binding of aldehydes derived from lipid peroxidation.

Figure 6 shows the results obtained in a time-course study of microsomal lipid peroxidation induced by NADPH and two different concentrations of Fe^{++} (60 and 6 μM, fig. 6a and b, respectively). As can be seen in both instances the amounts of DNPH-reacting groups in the non-lipoidal residue of liver microsomes increases with the incubation time and is correlated ($r = 0.999$ and 0.900 in fig. 6a and b, respectively) to the amount of malonic dialdehyde (MDA) formed in the incubation mixture. Therefore, the appearance of DNPH-reacting groups in the microsomal protein can be assumed as a good index of lipid peroxidation.

The results (absorption or difference absorption values at 375 nm) of the in vivo studies are reported in full in table II. As can be seen, in both the intoxications (2,600 μmol/100 g body wt, per os) the amount of DNPH-reacting groups (bound carbonyls) in the non-lipoidal residue of liver microsomes increases from 15 min up to 2 h of intoxication. At both times the effect is more pronounced with $BrCCl_3$ than with CCl_4, a finding which is in agreement with the different level of microsomal lipid peroxidation and toxicity found after the in vivo intoxication [29] or the in vitro treatment [30] with these two toxins. When a lower dose (260 μmol/100 g body wt) of CCl_4 was used (table II) the amount of protein-bound carbonyls was approximately halved, compared to that found with the standard dose (2,600 μmol/100 g body wt).

Since it has been demonstrated that 4-hydroxyalkenals and other lipid peroxidation products exhibit a marked chemotactic activity [31] and induce an increased vascular permeability [32, 33], their continuous production could be a contributing factor in the pathogenesis of the events leading to the development of the liver cirrhosis which is known to result from chronic CCl_4 treatment. It seemed therefore of interest to investigate whether aldehydes bound to the liver microsomal

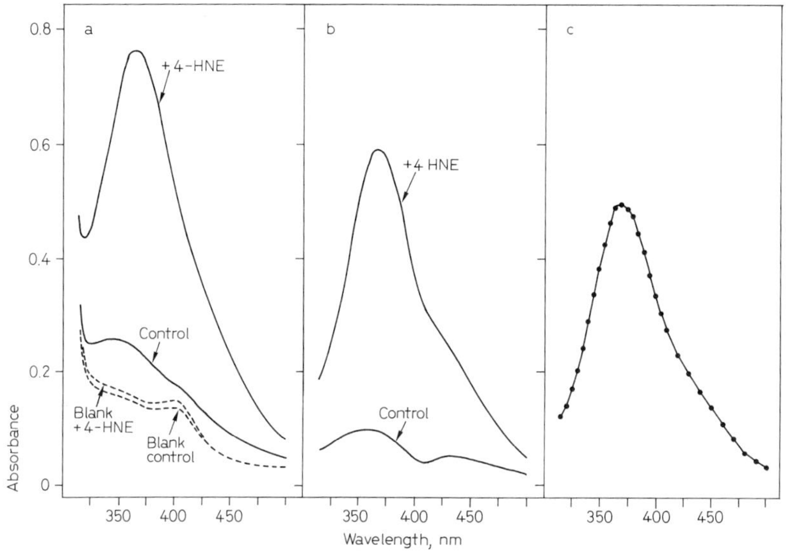

Fig. 4. a Solid lines: absorption spectra of DNPH-treated non-lipoidal residues of either liver microsomes allowed to react with synthetic 4-HNE (at the concentration of 0.45 m*M*) or control microsomes. Dotted lines: absorption spectra of blank reacted samples. A typical experiment is reported. The protein concentration of the sample dissolved in 7 *M* guanidine was 4 mg/ml. The spectra were recorded against 7 *M* guanidine in the reference cuvette. The means (2 experiments) of the absorptions at 375 nm of the DNPH-reacted samples were 0.270 and 0.907 for control microsomes and microsomes incubated with 4-HNE, respectively. The means of the absorptions at 375 nm of the blank-reacted samples were 0.151 and 0.156 for control microsomes and microsomes incubated with 4-HNE, respectively. *b* Spectra for liver microsomes allowed to react with synthetic 4-HNE and control microsomes. The spectra for DNPH-treated samples were recorded against the corresponding blank samples in the reference cuvette. The means of the absorptions at 375 nm were 0.119 and 0.751 for control microsomes and microsomes incubated with 4-HNE, respectively. *c* Difference spectrum 'microsomes reacted with 4-HNE minus control microsomes' calculated from spectra of figure 4b by subtracting the spectrum of the control sample from the spectra of the sample for microsomes reacted with 4-HNE [from ref. 27, with permission].

protein could still be detected (and if so to what extent) after a chronic CCl_4 treatment leading to liver cirrhosis. Liver cirrhosis was rapidly produced in rats according to the method described by *McLean* et al. [34], which consists in the administration of phenobarbital in the drinking water and of CCl_4 by inhalation. CCl_4 was given twice weekly for 5–7 weeks. The animals did not receive CCl_4 3 days in advance and

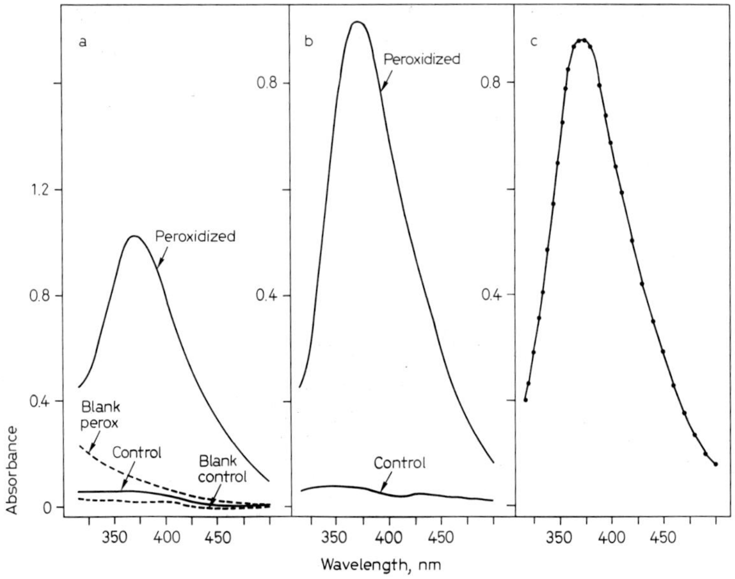

Fig. 5. a Solid lines: absorption spectra of DNPH-treated non-lipoidal residues of either liver microsomes peroxidized in the NADPH-Fe-dependent system or non-peroxidized microsomes. Dotted lines: absorption spectra of blank reacted samples. The concentration of $FeSO_4$ was 60 μM. The incubation time was 20 min. A typical experiment is reported. The protein concentration of the sample dissolved in 7 *M* guanidine was 1 mg/ml. The spectra were recorded against 7 *M* guanidine in the reference cuvette. The means (2 experiments) of the absorptions at 375 nm of the DNPH-reacted samples were 0.062 and 0.970 for non-peroxidized and peroxidized microsomes, respectively. The means of the absorptions at 375 nm of the blank reacted samples were 0.030 and 0.130 for non-peroxidized and peroxidized microsomes, respectively. *b* Spectra for either liver microsomes peroxidized in the NADPH-Fe-dependent system or non-peroxidized microsomes. The spectra for DNPH-treated samples were recorded against the corresponding blank samples in the reference cuvette. The means of the 'DNPH minus blank' difference absorptions at 375 nm were 0.032 and 0.840 for non-peroxidized and peroxidized microsomes, respectively. *c* Difference spectrum 'peroxidized microsomes minus non-peroxidized microsomes' calculated from spectra of figure 5b by subtracting the spectrum of the sample for non-peroxidized microsomes from the spectrum of the sample for peroxidized microsomes [from ref. 27, with permission].

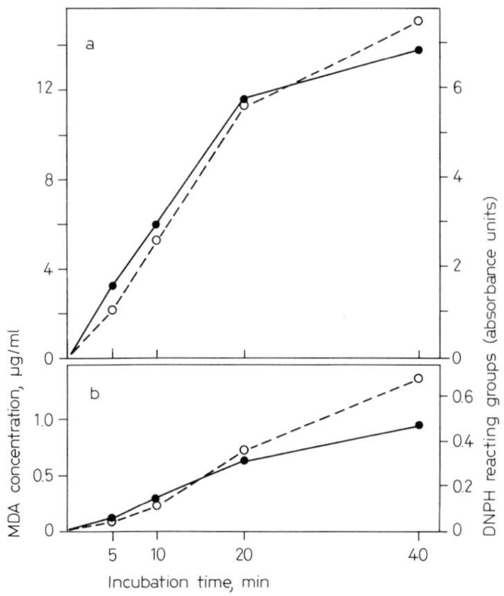

Fig. 6. Time-course appearance of DNPH-reacting groups in non-lipoidal residues of liver microsomes peroxidizing in the NADPH-dependent system and concomitant evolution of MDA. DNPH-reacting groups are expressed as absorbance units for a protein concentration of 8 mg/ml. The absorbance units were calculated from the values at 375 nm of the difference spectra 'peroxidized microsomes minus non-peroxidized microsomes'. *a* 60 μM $FeSO_4$. *b* 6 μM $FeSO_4$. ● = MDA concentration; ○ = DNPH-reacting groups [from ref. 27, with permission].

were starved overnight before sacrifice. Cirrhosis was ascertained by histological examination in all the livers used. The livers were processed as in the acute experiments. As shown in table III, the amount of carbonyls bound to the liver microsomal protein seems to increase with the length of the chronic treatment. It seems therefore that the production of reactive aldehydes continues during the chronic treatment with CCl_4. It must also be noted that some protein-bound carbonyls are present in liver microsomes of the rats treated with phenobarbital alone for a relatively long period of time.

From the difference spectra obtained (see the results reported in legends to fig. 4 and 5 and the results of tables II, III, and fig. 6) and the molar extinction coefficient ($\varepsilon = 21{,}000$) determined [19, 28] for the dinitrophenylhydrazone of 4-hydroxypentenal previously bound to vari-

Table II. Data for spectrophotometric analyses of 2,4-dinitrophenylhydrazine-treated or blank-reacted non-lipoidal residues of liver microsomes from control, CCl_4-poisoned and $BrCCl_3$-poisoned rats

	Dose of toxins µmol/100 g body wt	Time after poisoning	2,4-Dinitrophenylhydrazine-treated samples E_{375}	Blank-reacted samples E_{375}	2,4-Dinitrophenylhydrazine – blank ΔE_{375}	Intoxicated – control ΔE_{375}
Control (2)	–	15 min	0.504	0.277	0.227	–
CCl_4 (2)	2,600	15 min	0.542	0.274	0.268	0.041
$BrCCl_3$ (2)	2,600	15 min	0.631	0.301	0.330	0.103
Control (11)	–	2 h	0.481 ± 0.017	0.270 ± 0.015	0.211 ± 0.004	–
CCl_4 (1)	260	2 h	0.575	0.310	0.265	0.054
CCl_4 (5)	2,600	2 h	0.667 ± 0.020	0.330 ± 0.027	0.337 ± 0.024	0.126
$BrCCl_3$ (9)	2,600	2 h	0.763 ± 0.040	0.363 ± 0.025	0.400 ± 0.017	0.189

Figures represent the means (± SEM) of the absorption or difference absorption values at 375 nm. The number of experiments is given in parentheses. For each experiment 3 pooled livers were used. The protein concentration of the sample dissolved in 7 *M* guanidine was 8 mg/ml [from ref. 27, with permission].

Table III. Data for spectrophotometric analyses of 2,4-dinitrophenylhydrazine-treated or blank-reacted non-lipoidal residues of liver microsomes from rats chronically treated with phenobarbital and CCl_4, and the respective controls

	2,4-Dinitrophenylhydrazine-treated samples E_{375}	Blank-reacted samples E_{375}	2,4-Dinitrophenylhydrazine – blank ΔE_{375}	Intoxicated – control ΔE_{375}
Control (2)	0.628	0.325	0.303	–
Phenobarbital (2)	0.605	0.282	0.323	0.020
Phenobarbital + CCl_4 (9 exposures) (2)	0.856	0.465	0.391	0.088
Control (2)	0.635	0.478	0.157	–
Phenobarbital (2)	0.494	0.258	0.236	0.079
Phenobarbital + CCl_4 (14 exposures) (2)	1.144	0.763	0.381	0.224

Figures represent the absorption or difference absorption values at 375 nm. The number of rats is reported in parentheses. The protein concentration of the sample dissolved in 7 *M* guanidine was 8 mg/ml. All the livers from the rats chronically treated with phenobarbital and CCl_4 showed moderate to severe cirrhosis on histological examination.

Table IV. Protein-bound aldehydes calculated from the hydrazone spectra of the 2,4-dinitrophenylhydrazine-treated samples assuming a molar extinction coefficient of 21,000 [19, 28]

	nmol/mg microsomal protein
Microsomes peroxidized in vitro for 20 min (NADPH - 60 μM Fe^{++})	38.0
Microsomes peroxidized in vitro for 20 min (NADPH - 6 μM Fe^{++})	2.5
Microsomes allowed to react with 4-HNE (0.45 mM)	6.7
CCl_4 intoxication (2,600 µmol/100 g body wt) - 2 h	0.8
$BrCCl_3$ intoxication (2,600 µmol/100 g body wt) - 2 h	1.1
Chronic treatment with phenobarbital and CCl_4 (9 exposures)	0.5
Chronic treatment with phenobarbital and CCl_4 (14 exposures)	1.3

[from ref. 27, with permission, with additional data].

ous proteins, the figures from table IV could be calculated for protein-bound aldehydes in the case of liver microsomes peroxidized in the presence of 60 or 6 μM Fe^{++}, in the case of liver microsomes allowed to react with synthetic 4-HNE and in the case of CCl_4 or $BrCCl_3$ intoxication (including the chronic intoxication with CCl_4). As can be seen, the amount of protein-bound carbonyls found in the in vivo intoxications is not so very different from the amount of protein-bound carbonyls found in the in vitro condition in which lipid peroxidation is induced by the lower amount (6 μM) of Fe^{++}. It should be added that a significant 30% decrease of microsomal G-6-Pase activity was observed in this in vitro condition. Furthermore, the possibility must be considered that sufficient amounts of toxic aldehydes are formed in the in vivo situation at discrete sites in the membrane to block relatively few, but essential, functional groups, such as active -SH groups of enzymes.

Summary

Peroxidation of liver microsomal lipids leads to the formation of dialysable products with cytopathological activities. The most active products, isolated by thin-layer chromatography of a dialysate extract and bearing most of the carbonyl functional groups present in the extract, exhibited a number of biological effects such as inhibition of membrane-bound enzymes (microsomal glucose 6-phosphatase, cytochrome $P\text{-}_{450}$ and aminopyrine demethylase), hemolytic activity, inhibition of protein synthesis in a cell-free model system and cytotoxic activity on isolated hepatocytes. These biologically active carbonyl compounds were successively identified as a particular class of aldehydes, namely 4-hydroxyalkenals (the majority of them being represented by 4-hydroxynonenal). Since 4-hydroxyalkenals react easily with -SH groups of low molecular weight thiols, proteins and enzymes, evidence was searched for aldehydes bound to microsomal protein in in vivo conditions (CCl_4 and $BrCCl_3$ intoxications) in which peroxidation of lipids of hepatic endoplasmic reticulum has been demonstrated previously. The spectrophotometric analysis of 2,4-dinitrophenylhydrazine-treated proteins of liver microsomes from the intoxicated rats showed absorption spectra similar to those formed in the reaction of alkenals with -SH groups of proteins or low molecular weight thiols, indicating the presence of protein-bound aldehydes. These were also detected in the case of chronic treatment with CCl_4 leading to liver cirrhosis. The pathological significance of the binding of reactive aldehydes to membrane proteins is discussed.

References

1 Tappel, A.L.; Zalkin, H.: Lipid peroxidation in isolated mitochondria. Archs. Biochem. Biophys. *80:*326–332 (1959).

2 Desai, I.D.; Sawant, P.L.; Tappel, A.L.: Peroxidative and radiation damage to isolated lysosomes. Biochim. biophys. Acta *86:*277–285 (1964).

3 McKnight, R.C.; Hunter, F.E., Jr.; Oehlert, W.H.: Mitochondrial membrane ghosts produced by lipid peroxidation induced by ferrous ion. I. Production and general morphology. J. biol. Chem. *240:*3439–3446 (1965).

4 Menzel, D.B.: Toxicity of ozone, oxygen, and radiation. A. Rev. Pharmacol. *10:* 379–394 (1970).

5 Plaa, G.L.; Witscki, H.: Chemicals, drugs, and lipid peroxidation. Annu. Rev. Pharmacol. Toxicol. *16:*125–141 (1976).

6 Comporti, M.; Saccocci, C.; Dianzani, M.U.: Effect of CCl_4 in vitro and in vivo on lipid peroxidation of rat liver homogenates and subcellular fractions. Enzymologia *29:*185–204 (1965).

7 Recknagel, R.O.; Ghoshal, A.K.: Lipoperoxidation as a vector in carbon tetrachloride hepatotoxicity. Lab. Invest. *15:*132–146 (1966).

8 Slater, T.F.: Necrogenic action of carbon tetrachloride in the rat: a speculative mechanism based on activation. Nature, Lond. *209:*36–40 (1966).

9 Wills, E.D.: Effects of lipid peroxidation on membrane-bound enzymes of the endoplasmic reticulum. Biochem. J. *123:*983–991 (1971).

10 Roders, M.K.; Glende, E.A., Jr.; Recknagel, R.O.: Prelytic damage of red cells in filtrates from peroxidizing microsomes. Science *196:*1221–1222 (1977).

11 Benedetti, A.; Casini, A.F.; Ferrali, M.: Red cell lysis coupled to the peroxidation of liver microsomal lipids. Compartmentalization of the hemolytic system. Res. Commun. chem. Pathol. Pharmacol. *17:*519–528 (1977).

12 Benedetti, A.; Casini, A.F.; Ferrali, M.; Comporti, M.: Effects of diffusible products of peroxidation of rat liver microsomal lipids. Biochem. J. *180:*303–312 (1979).

13 Casini, A.F.; Benedetti, A.; Ferrali, M.; Fulceri, R.; Comporti, M.: Estrazione di prodotti tossici dializzabili originatisi dalla perossidazione dei lipidi microsomali epatici. Boll. Soc. ital. Biol. sper. *54:*893–899 (1978).
14 Benedetti, A.; Casini, A.F.; Ferrali, M.; Comporti, M.: Extraction and partial characterization of dialysable products originating from the peroxidation of liver microsomal lipids and inhibiting microsomal glucose 6-phosphatase activity. Biochem. Pharmacol. *28:*2909–2918 (1979).
15 Benedetti, A.; Casini, A.F.; Ferrali, M.; Fulceri, R.; Comporti, M.: Cytotoxic effects of carbonyl compounds (4-hydroxyalkenals) originating from the peroxidation of microsomal lipids; in Slater, Garner, Recent advances in lipid peroxidation and tissue injury, pp. 56–85 (Brunel Printing Services, Uxbridge 1981).
16 Kirchner, J.G.: Thin-layer chromatography, p. 159 (Wiley & Sons, New York 1967).
17 Benedetti, A.; Barbieri, L.; Ferrali, M.; Casini, A.F.; Fulceri, R.; Comporti, M.: Inhibition of protein synthesis by carbonyl compounds (4-hydroxyalkenals) originating from the peroxidation of liver microsomal lipids. Chem.-Biol. Interact. *35:* 331–340 (1981).
18 Casini, A.F.; Benedetti, A.; Ferrali, M.; Fulceri, R.; Comporti, M.: Effetto citotossico di composti carbonilici (4-idrossialchenali) derivati dalla perossidazione dei lipidi microsomali epatici sugli epatociti isolati. Boll. Soc. ital. Biol. sper. *57:* 1470–1476 (1981).
19 Schauenstein, E.; Esterbauer, H.; Zollner, H.: Aldehydes in biological systems, p. 25 (Pion, London 1977).
20 Högberg, J.; Moldéus, P.; Arborgh, B.; O'Brien, P.J.; Orrenius, S.: The consequences of lipid peroxidation in isolated hepatocytes. Eur. J. Biochem. *59:* 457–462 (1975).
21 Ferrali, M.; Fulceri, R.; Benedetti, A.; Comporti, M.: Effects of carbonyl compounds (4-hydroxyalkenals) originating from the peroxidation of liver microsomal lipids on various microsomal enzyme activities of the liver. Res. Commun. chem. Pathol. Pharmacol. *30:*99–112 (1980).
22 Ferrali, M.; Benedetti, A.; Casini, A.F.; Fulceri, R.; Comporti, M.: Sul meccanismo di azione dei composti carbonilici (4-idrossialchenali) derivati dalla perossidazione dei lipidi microsomali epatici. Boll. Soc. ital. Biol. sper. *57:*1443–1449 (1981).
23 Schauenstein, E.; Esterbauer, H.: Formation and properties of reactive aldehydes; in Submolecular biology and cancer, Ciba Fdn Ser., No. 67, pp. 225–244 (Excerpta Medica, Amsterdam 1979).
24 Benedetti, A.; Comporti, M.; Esterbauer, H.: identification of 4-hydroxynonenal as cytotoxic product originating from the peroxidation of liver microsomal lipids. Biochim. biophys. Acta *620:*281–296 (1980).
25 Esterbauer, H.; Weger, W.: Über die Wirkung von Aldehydes auf gesunde und maligne Zellen. 3. Mitt.: Synthese von homologen 4-Hydroxy-2-alkenalen. II. Mh. Chem. *98:*1994–2000 (1967).
26 Benedetti, A.; Casini, A.F.; Ferrali, M.; Fulceri, R.; Comporti, M.: Unpublished results.
27 Benedetti, A.; Esterbauer, H.; Ferrali, M.; Fulceri, R.; Comporti, M.: Evidence for aldehydes bound to liver microsomal protein following CCl_4 or $BrCCl_3$ poisoning. Biochim. biophys. Acta *711:*345–356 (1982).

28 Schauenstein, E.; Taufer, M.; Esterbauer, H.; Kylianek, A.; Seelich, T.: Über die Reaktion von Protein-SH-Gruppen mit 4-Hydroxypentenal. Mh. Chem. *102:* 517–529 (1971).
29 Burdino, E.; Gravela, E.; Ugazio, G.; Vannini, V.; Calligaro, A.: Initiation of free radical reactions and hepatotoxicity in rats poisoned with carbon tetrachloride or bromotrichloromethane. Agents Actions *4:*244–253 (1973).
30 Slater, T.F.; Sawyer, B.C.: The stimulatory effects of carbon tetrachloride and other halogenoalkanes on peroxidative reactions in rat liver fractions in vitro. General features of the systems used. Biochem. J. *123:*805–814 (1971).
31 Torrielli, M.V.; Giroud, J.P.; Curzio, M.; Negro, F.: Lipoperossidazione ed attività fagocitaria nelle reazioni infiammatorie; in Atti XVI Congresso Nazionale della Società Italiana di Patologia, pp. 337–354 (Copisteria Scientifica Universitaria, Torino 1981).
32 Ugazio, G.; Torrielli, M.V.; Burdino, E.; Sawyer, B.C.; Slater, T.F.: Long-range effects of products of carbon tetrachloride-stimulated lipid peroxidation. Biochem. Soc. Trans. *4:*353–356 (1976).
33 Benedetti, A.; Ferrali, M.; Casini, A.F.; Pieri, S.; Comporti, M.: Foot-edema induced by carbonyl compounds originating from the peroxidation of liver microsomal lipids. Biochem. Pharmacol. *29:*121–124 (1980).
34 McLean, E.K.; McLean, A.E.M.; Sutton, P.M.: Instant cirrhosis. An improved method for producing cirrhosis of the liver in rats by simultaneous administration of carbon tetrachloride and phenobarbitone. Br. J. exp. Path. *50:*502–506 (1969).

M. Comporti, MD, Istituto di Patologia Generale dell'Università di Siena,
Via del Laterino 8, I-53100 Siena (Italy)

Front. gastrointest. Res., vol. 8, pp. 63–74 (Karger, Basel 1984)

Post-Ischemic Recovery and Acute Phase Reaction in the Liver Cells

Common Patterns of Protein Synthesis

Lidia Bardella, G. Cairo, Luisa Schiaffonati, A. Bernelli-Zazzera

Istituto di Patologia Generale dell'Università degli Studi di Milano, Centro di Studio per la Patologia Cellulare del CNR, Milano, Italy

The interruption of the blood supply to the liver cells causes definite biochemical alterations in RNA and protein synthesis [7, 22, 26]. If the blood supply is restored before the 'point of no return' of the tissue, the capacity for RNA synthesis recovers with a temporary rebound well over normal levels. At variance with regenerating liver, the increase in RNA synthesis in post-ischemic livers is not followed by thymidine incorporation into DNA and there is no cell replication; for this reason the process of recovery has been defined as 'intracellular repair' [26]. The activation of RNA synthesis occurring during post-ischemic intracellular repair is sustained by the enhanced activity of the engaged (endogenous template-dependent) polymerases, while the free forms do not change in activity; the increase in number of the transcribing polymerase molecules plays the main role in this enhancement [5]. The resumption of the capacity to synthesize nascent peptide chains is associated with the reconstitution of the polysomal pattern, but there is no indication of overshooting above the normal level of protein synthesis [9]. This lack of compensatory activation does not rule out the possibility of a shift in the proportions of the different proteins synthesized by post-ischemic liver, which in turn could depend on transcriptional or post-transcriptional events. Evidence in favor of this possibility has recently been given [4, 25]. Changes in the pattern of plasma protein due to reprogramming of liver protein synthesis are known to occur during the inflammatory response and constitute the so-called 'acute phase reaction' [27]. In the present study we completed the description of the changes in protein composition and in the pattern of functional messenger RNAs (mRNAs) of rat liver during post-ischemic repair and we measured the same parameters in the liver during the acute phase reaction. The comparison between the results obtained under these two conditions shows a certain number of common features, together with some quantitative differences.

Materials and Methods

Animals Male albino rats (Wistar strain) weighing approximately 200 g were housed, fed and subjected to liver ischemia as previously described [26]. Inflammation was induced by a single subcutaneous injection of 0.5 ml of sterile turpentine/100 g body weight, into the back.

Chemicals. L-[^{35}S]-Methionine (spec. radioactivity > 800 Ci/mmol), methyl-[^{14}C] molecular weight standards, and the Translation kit used for in vitro protein synthesis were all from N.E.N. Chemicals G.m.b.h., Dreieich, FRG. Oligo(dT)-cellulose (type 3) was obtained from Collaborative Research, Waltham, Mass., USA. The scintillation fluid contained 5.5 g of Permablend III (Packard Instrument Co.)/liter in toluene. Solutions for RNA extraction and assay were autoclaved before use. All glassware was heated to 150 °C for 5 h.

In vivo Protein Synthesis. Rats were injected intravenously with 250 μCi of ^{35}S-methionine and killed after 12 min. Livers were homogenized in 3 volumes of 0.25 *M*-sucrose at 0 °C and adjusted to 1% Triton X-100 and 1% Na-deoxycholate. The homogenate was centrifuged for 10 min at 2 °C and 10,000 *g*. Incorporation into total protein was followed by spotting 5-μl aliquots of the supernatant fraction on Whatman 3 MM paper followed by washing through the hot trichloroacetic acid (TCA) procedure of *Mans and Novelli* [18]; 95% of all the acid insoluble radioactivity in the whole homogenate was recovered in the detergent-treated supernatant. Samples were taken, diluted in 125 m*M* Tris, pH 6.8, 2% SDS (sodium dodecyl sulfate), 1% glycerol and 5% β-mercaptoethanol and subjected to electrophoresis. The relative proportion of albumin synthesis was determined by precipitation of labelled albumin with specific antibody. Reactions were performed with several different ratios of liver postmitochondrial supernatant to antialbumin to ensure that immunoprecipitation was quantitiative. Quantification of the incorporation of ^{35}S-methionine into proteins, other than albumin, relative to the incorporation into total protein, was done by weighing the area under the peaks of the densitometric tracing of the fluorograms of SDS-polyacrylamide gel electrophoresis (PAGE). Care was taken to remain within the linear range of the film. The molecular weights of the proteins were determined by their relative mobilities (R_F = protein migration distance/dye migration distance) on SDS-PAGE compared to the standard molecular weight markers run under identical conditions. A standard curve was constructed by plotting the logarithm of molecular weight of standard proteins against their R_F values.

Separations of Poly-(A)$^+$ and Poly-(A)$^-$RNAs. Total cellular RNA, isolated from unfractionated liver homogenates by phenol-chloroform extraction as described by *Keller* *and Taylor* [14], was dissolved in binding buffer (0.5 *M* NaCl, 10 m*M* Tris-HCl, pH 7.4, 0.5% *N*-lauroyl sarcosine) and fractionated on an oligo(dT)-cellulose column involving two binding steps. The polyadenylated fraction was eluted with 10 m*M* Tris-HCl, pH 7.4, and precipitated twice with 2 volumes of ethanol in the presence of 0.2 *M* NaCl [4]. The unbound material was then subjected to two additional cycles of affinity chromatography. At each cycle the small proportion of RNA retained by the column was eluted and translated in vitro. The final poly-$(A)^-$ fraction was then precipitated with 2 volumes of ethanol in the presence of 0.2 *M* NaCL.

In vitro Translation of RNA. Cell free protein synthesis was carried out in a total volume of 25 μl containing 40 μCi of ^{35}S-methionine and various amounts of RNA with a reticulocyte lysate translation assay system. Aliquots were removed for determination of radioactivity into total protein or for electrophoretic analysis as described for in vivo synthesis.

Immunoprecipitation. Immunological quantitation of albumin by precipitation with anti-rat serum albumin antibodies was performed as previously described [4].

Gel Electrophoresis. Radioactive proteins synthesized both in vivo and in vitro as well as molecular weight markers were subjected to SDS-PAGE as described by *Laemmli* [16] on 10% acrylamide slab gels. Gels were then processed for fluorography [17], exposed to Kodak X-Omat AR films and scanned with a LKB microdensitometer.

Determination of Radioactivity. Acid-insoluble radioactivity in total protein and immunoprecipitates was determined by liquid scintillation spectrometry in a Tri-Carb A-300 (Packard Instruments Co. Inc.) with an efficiency of 75% and a background of 15 cpm; the standard error was 5% or better.

Results and Discussion

Figure 1 shows the pattern of liver proteins labelled in vivo during a 12-min pulse of ^{35}S-methionine; the latter amino acid was used instead of ^{3}H-leucine, as done before [4], because of a higher specific activity which allows the identification of a larger number of proteins with a better resolution. The short labelling period ensures that practically all of the labelled plasma proteins are retained within the cells [20]. In addition, the calculation of radioactivity in any single protein species, relative to that in total protein, normalizes possible differences in specific radioactivities among different animals as well as minor variations in the absorption of the amino acid by the liver [14]. Equal amounts of labelled peptides were applied to each of the gel slots; in practice this also corresponds to equal amounts of protein, as revealed

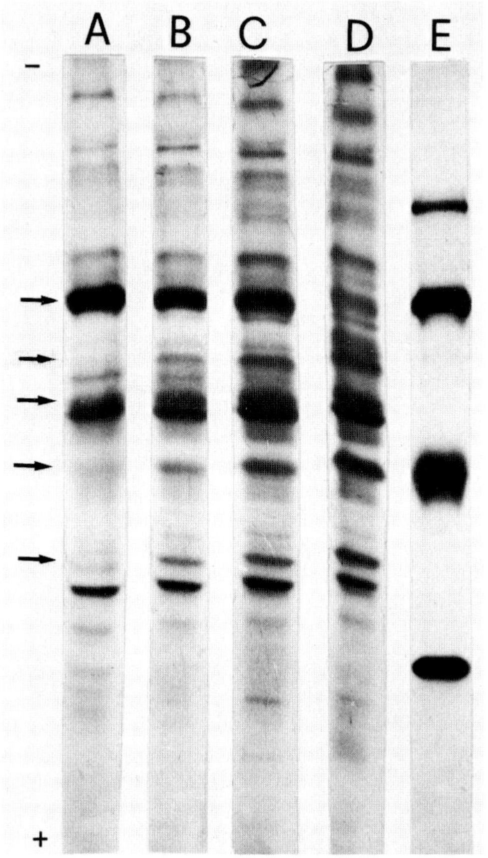

Fig. 1. Fluorograms of SDS-PAGE of total protein from rat liver post-mitochondrial supernatant labelled in vivo. Total proteins were prepared as described in the Materials and Methods section. Gel electrophoresis and fluorography were performed as described in the text. Arrows on the left of the gel indicate the migration of the proteins considered in the present study. Equal amounts of labelled peptides were applied to each of the gel slots. Lanes: A = control rat liver; B = sham-operated rat liver (24 h); C = post-ischemic rat liver (24 h); D = turpentine-treated rat liver (24 h); E = ^{14}C-labelled marker proteins [phosphorylase B (MW 97,000), bovine serum albumin (MW 69,000), ovalbumin (MW 46,000), carbonic anhydrase (MW 30,000)], from the top to the bottom.

by Coomassie-blue staining, in tracks A, B and C, while track D is stained less intensely. This confirms that overall protein synthesis is not affected either in sham-operated or post-ischemic livers but increases during the acute phase reaction [27]. The samples of sham-operated, post-ischemic and turpentine-treated rats show variations of at least five bands. The relative mobilities of these bands, compared to molecular weight (MW) markers, allow their identification as albumin (65,000 MW) [28] Bβ- and γ-fibrinogen chains (58,000 and 51,000 MW, respectively) [19], α_1-acid glycoprotein (44,000 MW) [23] and haptoglobin (35,000 MW) [12]. These latter proteins are known to increase both in plasma and liver cells during the inflammatory response [27]. Quantitation of the changes by measurements made on densitometric tracings shows a substantial invariance of albumin and modest increases (+30%) of Bβ- and γ-fibrinogen, α_1-acid glycoprotein and haptoglobin in sham-operated rats. Post-ischemic livers show a clear decrease in albumin (which is reduced to 55%) and increases of 65% in γ-fibrinogen and α_1-acid glycoprotein, the other changes being analogous to those of sham-operated rats. The pattern of turpentine-treated rats deviates still more from the normal, with a bare 35% of albumin left, and increases of 95% in Bβ-fibrinogen and α_1-acid glycoprotein. Precipitation of labelled post-mitochondrial supernatants with anti-albumin specific antibodies confirmed the data obtained by densitometric scanning. In conclusion, surgical trauma seems to produce a modest change in liver protein composition; this change is amplified when the liver recovers from a direct injury and becomes even more pronounced when the presence of an inflammatory process in the organism triggers the acute phase response in the liver cells. Increase in acute phase reactants is accompanied by a simultaneous decrease of albumin synthesis. In this way increase in the demand of a specific set of proteins could be accomplished in the presence of a quantitatively normal (post-ischemic) or slightly increased (turpentine-treated) protein synthesis.

The pattern of protein synthesized by eukaryotic cells is regulated by a variety of mechanisms which are indicated, comprehensively, as 'gene control' and operate at different levels; the most important single regulatory factor seems to be the amount of functional mRNA available in the cytoplasm of the cells [8]. To clarify the mechanism underlying the changes in protein pattern described above we studied the population of mRNA coding for specific rat liver proteins both during the acute phase response and the post-ischemic recovery. To this purpose, poly-(A)$^+$-RNAs were isolated and tested in vitro in a rabbit reticulocyte lysate system (fig. 2).

Characterization of labelled translation products by SDS-PAGE followed by autoradiography showed that the changes observed in turpentine-treated and post-ischemic rats occur in the hepatic concentration of 4 mRNA species. The products of these mRNA were found to have MW = 65,000, 57,600, 38,000, 23,000. On the basis of the immunochemical reactivity, the protein of MW 65,000 was identified as albumin, and it was found to decrease by 45–50% both in post-ischemic recovery and in the response to acute inflammation. The molecular weight of the proteins which increase in acutely inflamed rats correspond to fibrinogen Bβ-chain [19], haptoglobin [12] and α_1-acid-glycoprotein [23] from the higher to lower molecular weights, respectively. Proteins of the same molecular weight also increase among the translational product of poly-(A)$^+$ mRNA from post-ischemic liver. The percentage of increase is the same in both livers (150% of the control for fibrinogen and haptoglobin, and 250% of the control for α_1-acid glycoprotein). The changes observed in the lysate system are in reasonably good agreement with those seen in vivo. It is appropriate to point out that additional metabolic controls, active in the liver in vivo, do not operate in the reconstructed system in vitro; but on the whole, it seems that under the present circumstances, alterations in the amount of functional mRNA may account for most of the charges observed in liver protein in vivo. By means of hybridization with specific cDNA probes, it has recently been demonstrated that the increased synthesis of fibrinogen [21] and α_1-acid glycoprotein [23] during the inflammatory response depends on transcriptional mechanisms. Translational assays alone cannot ascertain whether transcriptional control is responsible for the same kinds of variation in protein synthesis in post-ischemic liver, because they do not reveal changes resulting in the accumulation of non-functional mRNA sequences or changes in specific mRNA stabilities. Though most mRNAs are polyadenylated at their 3′-end, products capable of directing protein synthesis are also found among the poly-(A)$^-$ RNA population [13]. We tried to characterize the proteins synthesized in vitro by poly-(A)$^-$ RNAs having messenger activity (fig. 3). All experiments were carried out in parallel with corresponding poly-(A)$^+$ preparations. On paired samples it is easier to see how products that increase in the lysate with poly-(A)$^+$ RNA from

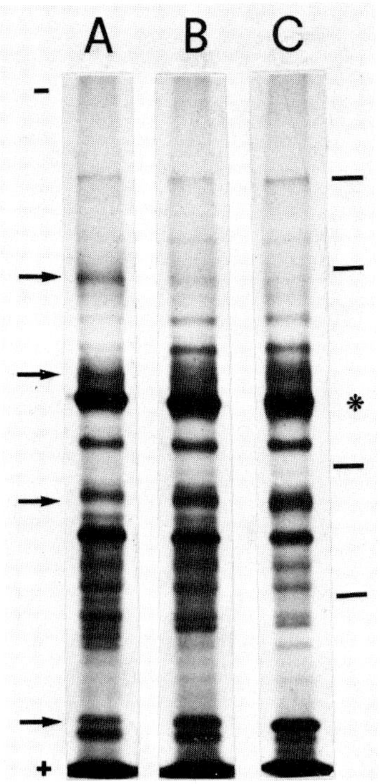

Fig. 2. Fluorograms of SDS-PAGE of total translational products of total cellular poly-(A)$^+$ RNA. Total proteins were synthesized in a rabbit cell-free reticulocyte lysate system primed with poly-(A)$^+$ RNA and prepared for electrophoresis as described in the Materials and Methods section. Slab gel electrophoresis and fluorography were performed as described in the text. Bars on the right of the gel represent the migration position of the marker proteins run in a parallel slot of the gel; asterisk represents an endogenous product of the reticulocyte lysate (not shown). Arrows on the left of the gel represent the migration of the proteins considered in the present study. Equal amounts of peptides labelled in vitro were applied to each of the gel slots. Lanes: A = control rat liver; B = post-ischemic rat liver (24 h); C = turpentine-treated rat liver (24 h).

treated rats show analogous increases with the corresponding poly-(A)$^-$ fractions. The observation indicates that the mRNAs for these proteins exist in both poly-(A)$^+$ and poly-(A)$^-$ forms, which respond in a similar and proportionate way to the treatments applied in the present research. The accuracy of the separation procedure ensures that little, if any, contamination by poly-(A)$^+$ RNA occurs in the poly-(A)$^-$

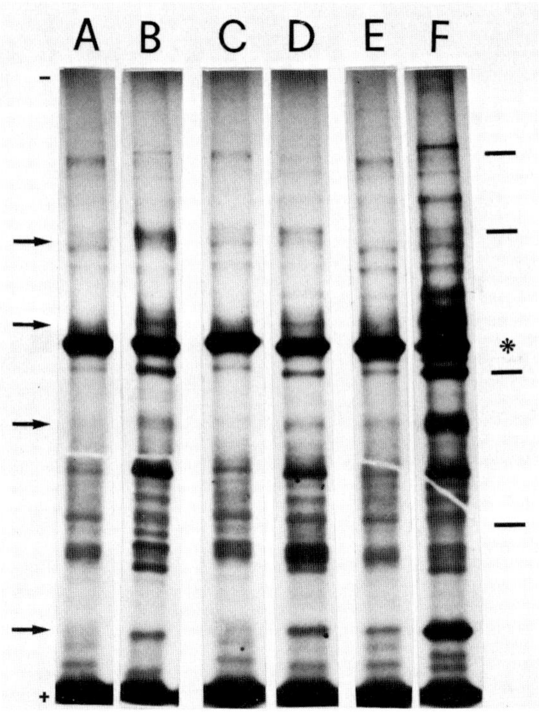

Fig. 3. Fluorograms of SDS-PAGE of total translational products of total cellular poly-$(A)^+$ and poly-$(A)^-$ RNA. Total proteins were synthesized in a rabbit reticulocyte lysate system primed with poly-$(A)^+$ (0.5 μg) and poly-$(A)^-$ (5 μg) RNA and prepared and analysed by slab gel electrophoresis and fluorography as described in the text. Details as in figure 2. Lanes: translational products of poly-$(A)^-$ RNA from control rat liver (A); poly-$(A)^+$ RNA from control rat liver (B); poly-$(A)^-$ RNA from post-ischemic rat liver (24 h) (C); poly-$(A)^+$ RNA from post-ischemic rat liver (24 h) (D); poly-$(A)^-$ RNA from turpentine-treated rat liver (24 h) (E); poly-$(A)^+$ RNA from turpentine-treated rat liver (24 h) (F).

fractions. This is further confirmed by the results of the translational control, performed according to *Kioussis* et al. [15] and by the absence in the poly-$(A)^-$ lanes of bands strongly positive in their poly-$(A)^+$ counterparts, while faint comigrating bands are present in both of them. In turpentine-treated rats, the increase of acute phase reactants is correlated to the process of inflammation and a protein fraction synthesized by activated polymorphonuclear leukocytes has been impli-

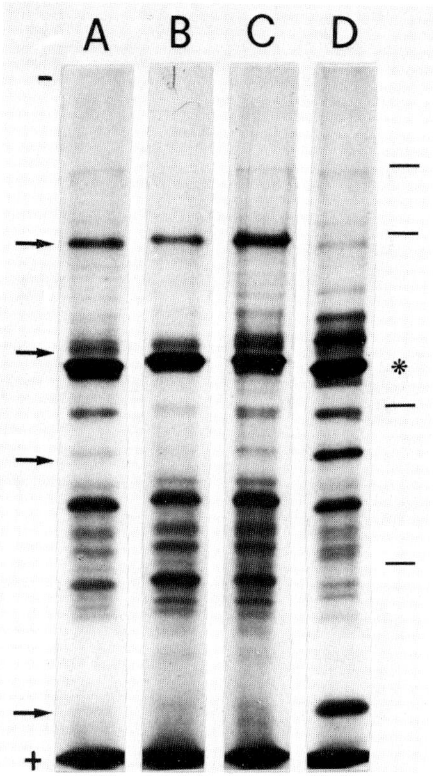

Fig. 4. Fluorograms of SDS-PAGE of total translational products of poly-(A)$^+$ RNA. Details of preparation and analysis as in fig. 2. Lanes: A = control rat liver; B = ischemic rat liver (60 min); C = post-ischemic rat liver (2 h); D = post-ischemic rat liver (24 h).

cated as a mediator in the acute response [24]. However, the meaning and the trigger of the increase of the same type of protein in post-ischemic liver could be different. The interruption of blood supply causes severe alterations in many metabolic pathways of the liver cell: i.e. reduced oxygen uptake [3], decreased adenosine triphosphate (ATP) concentration [10] inhibition of protein [2] and RNA synthesis [26] with a prompt recovery if ischemia does not last more than 60 min.

Recently, *Hammond* et al. [11] found an apparent connection between energy metabolism and stress-induced proteins in ischemic heart. Moreover, the effect of the heat shock on *Drosophila* larvae can be reproduced by hypoxia [1]. In order to investigate if the reduction in oxygen supply could be responsible for the alterations in the protein pattern found in post-ischemic liver, we studied the population of mRNA in ischemic liver at the end of a period of 60-min ischemia and at an early time (2 h) after the restoration of the blood supply (fig. 4). The population of mRNA of the liver cell is well preserved at the end of 60-min ischemia, in keeping with the fact that this time-period is within the limits of reversibility of the injury, and protein synthesis, though temporarily inhibited, can resume very soon even under conditions preventing the synthesis of new mRNA [6]. During the first 2 h post-ischemia, that correspond to the reconstitution of the polysomal pattern and to the restoration of a normal capacity for protein synthesis, there are no appreciable variations in the population of functional mRNAs; the albumin mRNA is still well represented and active. Only after this time-point are the mechanisms set in motion that will lead, later on, to the pattern of mRNA and protein synthesis similar to the acute phase reaction. Products released from the injured liver cells, or possibly from white blood cells accumulating in post-ischemic livers, could trigger the changes in mRNAs, and finally in protein composition, occurring in post-ischemic liver cells: besides their possible role in inflammation, these changes could be instrumental to the process of intracellular repair. The proteins investigated in the present study are export proteins that will be finally secreted from the cell; the analysis of structural liver cell proteins of post-ischemic and turpentine-treated rats will provide further information on the meaning of the shift in protein synthesis during post-ischemic recovery.

Summary

The changes in the pattern of liver protein composition during post-ischemic recovery are largely superimposable to those occurring during the acute phase reaction; however, decrease in albumin and increase in α_1-acid glycoprotein are less pronounced in post-ischemic livers. In both conditions, changes in protein composition are associated with corresponding rearrangements in the relative amounts of total functional mRNAs – both poly-(A)$^+$ and poly-(A)$^-$ RNAs – suggesting that transcriptional mechanisms play the main role in these shifts in protein synthesis. No changes in the relative amounts of functional mRNAs occur in 60-min ischemic livers, and during the early phases (2 h) of recovery from ischemia. The latency of the post-ischemic response is similar to the one observed in the acute phase reaction.

References

1 Ashburner, M.; Bonner, J.J.: The induction of gene activity in *Drosophila* by heat shock. Cell *17:*241–254 (1979).

2 Bernelli-Zazzera, A.; Cajone, F.; Simonetta, M.; Schiaffonati, L.: Further studies on ribosomal damage in liver ischemia. Exp. molec. Path. *17:*121–131 (1972).

3 Bernelli-Zazzera, A.; Gaja, G.: Some aspects of glycogen metabolism following reversible or irreversible liver ischemia. Exp. molec. Path. *3:*351–368 (1964).

4 Cairo, G.; Schiaffonati, L.; Aletti, M.G.; Bernelli-Zazzera, A.: Effect of post-ischaemic recovery on albumin synthesis and relative amount of translatable albumin messenger RNA in rat liver. Biochem. J. *204:*197–202 (1982).

5 Cairo, G.; Schiaffonati, L.; Bernelli-Zazzera, A.: Cell repair after liver injury. Stimulation of RNA synthesis, engaged polymerases, number of RNA transcribing molecules and elongation rate in post-ischemic liver nuclei. Exp. molec. Path. *35:* 231–241 (1981).

6 Cajone, F.; Ragnotti, G.; Bernelli-Zazzera, A.; Bassi, M.: State and function of liver polysomes during recovery from ischemia. Exp. molec. Path. *14:*392–403 (1971).

7 Cajone, F.; Schiaffonati, L.; Bernelli-Zazzera, A.: Protein synthesis in liver injury. Soluble factors of protein synthesis in the cytosol from ischemic rat liver. Lab. Invest. *34:*387–393 (1976).

8 Darnell, J.E.: Variety in the level of gene control in eukaryotic cells. Nature, Lond. *297:*365–371 (1982).

9 Ferrero, M.E.; Orsi, R.; Bernelli-Zazzera, A.: Cell repair after liver injury. Membranes of the endoplasmic reticulum, ribosomes and amino acids in post-ischemic livers. Exp. molec. Path. *32:*32–42 (1980).

10 Gaja, G.; Ferrero, M.E.; Piccoletti, R.; Bernelli-Zazzera, A.: Phosphorylation and redox states in ischemic livers. Exp. molec. Path. *19:*248–265 (1973).

11 Hammond, G.L.; Lay, Y.K.; Markert, C.L.: Diverse forms of stress lead to new patterns of gene expression through a common and essential metabolic pathway. Proc. nat. Acad. Sci. USA *79:*3485–3488 (1982).

12 Haugen, T.H.; Hanley, J.M.; Heath, E.C.: Haptoglobin. J. biol. Chem. *256:* 1055–1057 (1981).

13 Katinakis, P.K.; Slater, A.; Burdon, R.H.: Non-polyadenylated mRNAs from eukaryotes. FEBS Lett. *116:*1–7 (1980).

14 Keller, G.H.; Taylor, J.M.: Effect of hypophysectomy on the synthesis of rat liver albumin. J. biol. Chem. *251:*3766–3773 (1976).

15 Kioussis, J.; Cato, A.C.B.; Slater, A.; Burdon, R.H.: Polypeptides encoded by polyadenylated and non-polyadenylated messenger RNAs from normal and heat shocked HeLa cells. Nucl. Acids Res. *9:*5203–5214 (1981).

16 Laemmli, U.K.: Cleavage of structural proteins during the assembly of the head of bacteriophage T4. Nature, Lond. *227:*680–685 (1970).

17 Laskey, R.A.; Mills, A.D.: Quantitative film detection of ^{3}H and ^{14}C in polyacrylamide gels by fluorography. Eur. J. Biochem. *56:*335–341 (1975).

18 Mans, R.J.; Novelli, G.D.: Measurement of the incorporation of radioactive amino acids into protein by a filter paper disk method. Archs. Biochem. Biophys. *94:* 48–53 (1961).

19 Nickerson, J.M.; Fuller, G.M.: In vitro synthesis of rat fibrinogen: identification of pre Aα, pre Bβ, and pre γ polypeptides. Proc. natn. Acad. Sci. USA *78:* 303–307 (1981).

20 Peters, T., Jr.: The biosynthesis of rat serum albumin. J. biol. Chem. *237:* 1186–1189 (1962).

21 Princen, J.M.G.; Nieuwenhuizen, W.; Mol-Back, G.P.B.M.; Yap, S.H.: Direct evidence of transcriptional control of fibrinogen and albumin synthesis in rat liver during the acute phase response. Biochem. biophys. Res. Commun. *102:* 717–723 (1981).

22 Ragnotti, G.; Cajone, F.; Bernelli-Zazzera, A.: Structural and functional changes in polysomes from ischemic livers. Exp. molec. Path. *13:*295–306 (1970).

23 Ricca, G.A.; Hamilton, R.W.; McLean, J.W.; Conn, A.; Kalinyak, J.E.; Taylor, J.M.: Rat α_1-acid glycoprotein mRNA. J. biol. Chem. *256:* 10362–10368 (1981).

24 Rupp, R.G.; Fuller, G.M.: The effects of leucocytic and serum factors on fibrinogen biosynthesis in cultured hepatocytes. Expl. Cell Res. *118:*23–30 (1979).

25 Schiaffonati, L.; Cairo, G.; Bardella, L.: Total protein synthesis and albumin synthesis in livers of post-ischemic rats. IRCS med. Sci. *10:* 190 (1982).

26 Schiaffonati, L.; Cairo, G.; Bernelli-Zazzera, A.: RNA synthesis by nuclei and nucleoli from ischemic liver cells. J. cell. Physiol. *97:*487–496 (1978).

27 Schreiber, G.; Howlett, G.; Nagashima, M.; Millership, A.; Martin, H.; Urban, J.; Kotler, L.: The acute phase response of plasma protein synthesis during experimental inflammation. J. biol. Chem. *257:* 10271–10277 (1982).

28 Taylor, J.M.: Schimke, R.T.: Synthesis of rat liver albumin in a rabbit reticulocyte cell-free protein synthesizing system. J. biol. Chem. *248:*7661–7668 (1973).

Dr. A. Bernelli-Zazzera, Istituto di Patologia Generale,
Via Mangiagalli 31, I-20133 Milano (Italy)

Front. gastrointest. Res., vol. 8, pp. 75–84 (Karger, Basel 1984)

Changes in Renal Function Following Temporary Occlusion of the Thoracic Duct

A. Satta[a], *R. Menghi*[b], *R. Faedda*[a], *G.F. Branca*[a], *A. Uzzau*[b], *L. Chiandussi*[a], *F. Bresadola*[b], *E. Bartoli*[a]

[a] Istituto di Patologia Medica, Cattedra di Semeiotica, Scuola di Specializzazione in Nefrologia; [b] Istituto di Patologia Chirurgica, University of Sassari, Sassari, Italy

Introduction

The mechanisms causing Na retention in liver cirrhosis are controversial. According to the traditional explanation, the increased sinusoidal resistance traps a larger amount of plasma within the splanchnic area, leading to a drop in effective plasma volume, that is sensed by volume receptors located in the thorax [1]. The kidney responds appropriately to stimuli elicited by the effective plasma volume only, and retains Na [2]. This interpretation is favored by the diuretic response to volume expansion, whether selective, by increasing the effective plasma volume during water immersion [3, 4], or unselective, such as that occurring during infusions of isotonic fluids [5, 6], during reinfusion of ascitic fluid [7, 8], and during LeVeen shunt [9]. With these maneuvers cardiac output rises, renal vascular resistances fall, renal Na reabsorption is reduced, partly because of renin and aldosterone suppression, and Na excretion increases [9]. According to the overflow theory [10], increased renal Na reabsorption is the primary event, leading to volume expansion, which will preferentially occur in areas of high vascular compliance, namely the splanchnic area. The increased sinusoidal resistances will favor splanchnic trapping of the retained fluid. In favor of this interpretation are the experiments by *Levy* [11–13] and *Levy and Allofey* [14] as well as the measurements of an increased plasma volume in cirrhosis [10] and the occurrence of Na retention with normal glomerular filtration rate (GFR) [15, 16] even before apparent ascites formation [14]. Through both theories require an increased sinusoidal resistance, this would lead to Na retention under normal circumstances only according to the traditional explanation. We decided to investigate this problem in dogs with an experimental protocol whereby sinusoidal resistances alone could be reversibly increased.

Material and Methods

The experiments were performed on 7 mongrel dogs, weighing 10–20 kg. The animals were allowed only water to drink during the 12 h preceding the study. They were anesthetized with Na thiopental (Farmitalia, Milan, Italy), 50 mg/kg given slowly i.v., repeated if necessary. They were placed on a surgical table, intubated and connected to a volume respirator and ventilated at a tidal volume of 20 ml/kg, delivered at a rate of 24/min. The femoral artery was cannulated with a polyethylene tube and connected to a pressure transducer for continuous recording of blood pressure (Soxil Strain gauge P23 ID, Gould) connected with a Soxil registrator. The femoral vein was cannulated with a polyethylene tube and connected with a peristaltic pump to deliver the solution containing the glomerular marker and the normal saline used to replace the surgical losses and as maintenance infusion. The left kidney was exposed through a flank incision, freed from its surrounding fat, and its ureter cannulated with a polyethylene tube of appropriate size. The right kidney was drained through a bladder catheter: no attempt was made to obtain complete collections for clearance measurements from this kidney. A small polyethylene catheter was inserted into the left renal vein through a short collateral that was tied off proximally. This catheter was connected to a syringe for renal vein sampling. The thoracic lymph duct (TD) was carefully isolated and dissected via a left subclavicular incision, its insertion into the left subclavian vein ligated, and the duct cannulated with a PVC catheter, which was securely fastened with silk ties. The catheter was connected to a bottle containing normal saline, mounted on a vertical ruler with centimeter divisions and the zero point carefully aligned at the same horizontal level of the upper surface of the liver. Pressure within the duct could be increased by raising the saline container along the vertical ruler. Finally, in 3 dogs interstitial hepatic pressure measurements were taken directly through a 21-gauge needle, connected to a pressure transducer through a rigid line filled with saline. After all surgical procedures had been completed, the animals were infused with normal saline solution containing inulin (Carlo Erba, Milan, Italy) 10 $g \cdot l^{-1}$ at a rate of 7 $mg \cdot min^{-1}$ and with normal saline as a maintenance solution at a rate of 1 $ml \cdot min^{-1}$. The surgical losses were estimated and replaced with 300–500 ml of an isotonic solution of composition identical to dog plasma, lacking proteins, Mg and urea, and containing Ca^{++}, Cl^-, HCO^-_3, Na^+, K^+ and glucose. The experiment was started 30–60 min later, when a steady state for urine flow rate, plasma inulin and blood pressure had been reached. The experimental design was as follows: baseline data were obtained by renal clearance measurements when the TD was kept at a pressure of 4 cm H_2O, then repeated 20 min after the TD pressure had been elevated to 40 cm H_2O (experimental high TD pressure) and finally repeated during post-control conditions, 20 min after return of the TD pressure to control values. For each experimental condition (baseline, high TD pressure, post-control) at least two clearance measurements were performed. For each clearance, a complete urine collection of 15–20 min duration was performed from the catheter inserted into the left ureter, and a blood sample was collected at the midpoint of urine collection from the femoral artery and the renal vein. Blood pressure was monitored continuously, while interstitial liver pressure was measured at the beginning and at the end of each period; the average value was used in subsequent calculations. When the TD was connected with an increased pressure, a variable volume of fluid flowed inside the lymph duct to distend it: no measurement was started until this flow had come to an end and the pressure inside the system stabilized. In each urine and plasma sample the following measurements were performed: Na and K (flame photometer, Radiometer, Copenhagen), osmolality (Osm, advanced Fiske Osmometer), inulin (anthrone method) [17]. The urine volume was measured in a calibrated cylinder. For each clearance period the following calculations were made: urine flow rate ($\dot{V}$) in $ml \cdot min^{-1}$, GFR $ml \cdot min^{-1}$ (as $U_{In}/P_{In} \cdot \dot{V}$), renal plasma flow (RPF $ml \cdot min^{-1}$, as GFR/In-extraction), extraction of inulin by the kidney (In-extraction, as $1 - RV_{In}/P_{In}$, where RV is the renal vein, P plasma, In is Inulin concentration), filtration fraction (FF = GFR/RPF), urine (U) and plasma (P) concentrations for Na, K and osmolality (U_{Na}, U_K, U_{osm}), urinary excretion per minute of Na and K ($U_{Na} \cdot \dot{V}$, $U_K \cdot \dot{V}$). For each animal, the results obtained during each experimental condition (baseline, high TD pressure, post-control) were averaged and these values used to calculate the overall means. For each of the above-mentioned variables, means and standard error of the mean were calculated, and differences from the experimental condition (high TD pressure) with respect to either baseline or post-control were assessed by t test and, when appropriate, by variance analysis [18].

Results

The results obtained are reported in table I and figure 1. During baseline conditions, when the TD pressure was 4 cm H_2O, GFR averaged 15.4 ± 1.8, and RPF 60.6 ± 9.7 $ml \cdot min^{-1}$ in the left kidney of our dogs. After raising the TD pressure to 40 cm H_2O, the interstitial hepatic pressure rose from 9.5 ± 0.2 to 21.3 ± 2.5 cm H_2O ($p < 0.01$): GFR fell significantly to 10.2 ± 1.5 $ml \cdot min^{-1}$, with similar changes in RPF, so that the FF remained at 0.31 ± 0.04, not significantly different from the control value of 0.30 ± 0.03. Concomitantly, $U_{Na} \cdot \dot{V}$ fell from 18.8 ± 3.0 to 9.4 ± 2.8 $\mu Eq \cdot min^{-1}$. K excretion did not change significantly, and U_{osm} rose insignificantly from 709 ± 188 to 812 ± 123 $mosm \cdot kg^{-1}$. Thus, thoracic duct obstruction and high pressure were transmitted to the liver interstitium with a fall in renal hemodynamics and Na excretion. When the TD pressure was released in post-control conditions, the hemodynamic changes were fully reversible: GFR and RPF returned to control values (13.8 ± 2.8 and 58.5 ± 7.5 $ml \cdot min^{-1}$,

Table I. Means ± SEM are reported for GFR, RPF in $ml \cdot min^{-1}$, U_{osm} in $mosm \cdot kg^{-1}$ as well as electrolyte concentrations (U_{Na} and U_K in $mEq \cdot l^{-1}$) and excretion ($U_{Na} \cdot \dot{V}$ and $U_K \cdot \dot{V}$ in $\mu Eq \cdot min^{-1}$). The results of significance testing is reported between the data measured during baseline and experimental (high TD pressure) periods, as well as between the latter and post-control. NS = Not significant; p = statistic probability, with respect to control

	GFR	RPF	U_{osm}	$U_{Na} \cdot \dot{V}$	$U_K \cdot \dot{V}$	U_{Na}	U_K
Baseline	15.4 ±	60.6 ±	709 ±	18.8 ±	7.9 ±	87.1 ±	48.6 ±
	1.8	9.7	188	3.0	1.8	22.1	18.6
p	< 0.02	< 0.05	NS	< 0.01	NS	NS	NS
High TD pressure	10.2 ±	40.9 ±	812 ±	9.4 ±	5.0 ±	89.5 ±	53.8 ±
	1.5	7.6	123	2.8	0.3	23.9	5.1
p	NS	NS	NS	< 0.05	NS	NS	NS
Post-control	13.8 ±	58.5 ±	765 ±	5.1 ±	4.1 ±	59.7 ±	31.8 ±
	2.8	7.5	159	0.1	0.3	13.4	3.8

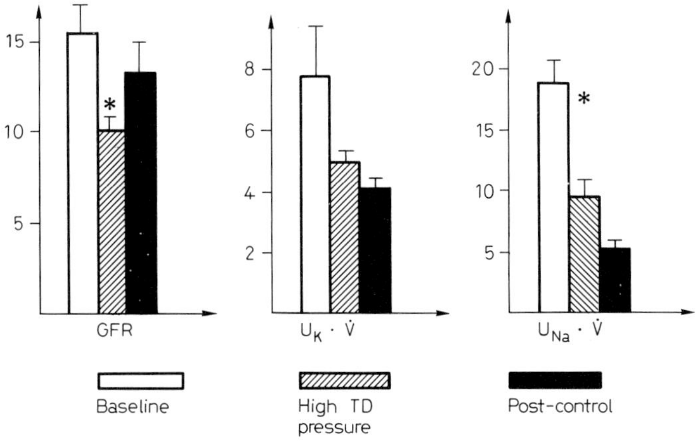

Fig. 1. GFR in $ml \cdot min^{-1}$ and the excretion of Na ($U_{Na} \cdot \dot{V}$ and K ($U_K \cdot \dot{V}$) in $\mu Eq \cdot min^{-1}$ are reported for baseline, high TD pressure and post-control measurements. The asterisk indicates statistical significance with respect to baseline.

respectively), and U_{osm} fell from 812 ± 123 to 765 ± 159 quite close to the control value of 709 ± 188 $mosm \cdot kg^{-1}$. In contrast, Na excretion remained remarkably unchanged at 5.1 ± 0.1 $\mu Eq \cdot min^{-1}$. There was a slight fall in post-control K excretion, from 5.0 ± 0.3 to 4.1 ± 0.3 $\mu Eq \cdot min^{-1}$ ($p > 0.05$). Thus, during the post-control period, when TD pressure was brought back to zero and the liver interstitial pressure fell towards control levels, there was a dissociation between the hemodynamic changes and those in electrolyte excretions. While the former were fully reversible, the latter remained at the same level as in the experimental period. It should be remembered that renal tubular reabsorption of Na actually rose during the post-control period, since the unchanged excretion occurred in the face of a large rise in GFR. Actually, Na reabsorption was increased even with respect to the control period (baseline).

There was no significant and consistent change throughout the experiment in Na (140 ± 3 $mEq \cdot l^{-1}$), K (3.0 ± 0.3 $mEq \cdot l^{-1}$) and osmolar (P_{osm} 301 ± 5 $mosm \cdot kg^{-1}$) concentrations in plasma. Plasma aldosterone was measured in 2 animals: it averaged 32 during control, 25 during the experimental period, and rose to 80 $ng \cdot ml^{-1}$ in the post-control period.

Discussion

The present study was undertaken to clarify the mechanisms of Na retention in liver cirrhosis. In the human disease there is an increase in sinusoidal resistances, due to the histologic changes in the liver. Consequently, blood is trapped within the splanchnic area and escapes from it through porta-caval shunts in close proportion to the rise in sinusoidal resistances [19]. According to the traditional theory of Na retention, splanchnic volume trapping is the primary event leading to increased renal Na reclamation through a fall in effective plasma volume [20], whereas in the overflow theory [10–14] only the perpetuating mechanism for Na retention is attributable to the rise in sinusoidal resistance, which forces the fluid primarily retained by the kidney into the splanchnic area and peritoneal cavity. In order to distinguish between these two possibilities, we decided to acutely increase sinusoidal resistances to portal flow in an experimental setting where the kidney was not retaining Na and the plasma volume was not expanded. This was accomplished by applying an increased pressure to the liver interstitium through the TD: the interstitial pressure was expected to squeeze the hepatic sinusoids, reducing their size and raising their resistances. In addition, the experimental design allowed for a prompt reversal of this situation in each experiment and this was expected to permit an assessment of the reversibility of the observed changes. The data show that raising the pressure applied to the TD caused a significant rise in the interstitial pressure directly measured by a needle. Associated with these changes there was a fall in GFR and Na excretion. The latter could be attributed to the drop in filtered Na and does not mean an increased tubular reclamation of filtered Na, since the absolute amount absorbed was actually reduced owing to the presence of the phenomenon of glomerulotubular balance [21] that protects against anuria under these circumstances. The rise in U_{osm} demonstrates that the fall in renal hemodynamics is uniform and causes a pattern of nephron hypoperfusion [22]: it was not attributable to anti-diuretic hormone (ADH) secretion since P_{osm} was unchanged and the rise in U_{osm} was reversible upon discontinuation of the high TD pressure. The most likely explanation is that water reabsorption along the distal tubule and collecting ducts takes place from a lower volume escaping proximal resorption [23], an explanation similar to that accounting for the limited diluting power of patients with liver cirrhosis [24]. In addition, a concomitant decrease in medullary blood flow, possibly parallel to that in overall RPF, might contribute to the higher U_{osm}, because of more efficient solute trapping [25] within the medullary interstitium. The reasons causing the hemodynamic derangement of the kidney during the period of high interstitial pressure can only be inferred from the present studies: although nervous and humoral factors cannot be excluded, it is likely that the rise in sinusoidal resistances produced the trapping and increased volume within the splanchnic vasculature, with a fall in effective plasma volume. This is followed, directly because of circulating volume contraction, and indirectly because of the associated adrenergic discharge and renin-angiotensin activation, by a fall in renal hemodynamics. When sinusoidal resistances return to their normal level, the sequestrated volume is slowly returned to the systemic circulation. Furthermore, effective plasma volume is restored and the mechanisms leading to reduced renal perfusion are suppressed. If the drop in Na excretion were a mere consequence of that in GFR, it would be expected to revert to normal when GFR returns to control levels. Our data instead demonstrate that Na excretion remains low at almost negligeable levels in spite of the large rise in filtered Na that occurs in the post-control period. At the same time even K excretion, which had not significantly changed during the experimental period, fell slightly and insignificantly. This could indicate that less Na was available to the distal exchange sites during the post-control period, resulting in a fall in K distal secretion and, eventually, excretion. If this interpretation is correct, it should be taken as indirect evidence that proximal tubular reabsorption had risen during the post-control period. Therefore, the observed rise in plasma aldosterone, which occurred only during the late post-control period, was not an important factor in Na retention. Thus, our data indicate that the mechanisms leading to Na retention are different from those causing the fall in renal hemodynamics, although they may be triggered by identical stimuli. The immediate effect reducing Na excretion may also be due to the fall in filtered Na, but it must be slowly replaced by some longer-acting mechanism, not promptly reversible with volume reexpansion, and presumably acting on the proximal tubule, independent of renal, systemic and splanchnic hemodynamics. Since in human liver cirrhosis there is a significant rise in proximal tubular reabsorption [6, 24, 26], one is tempted to speculate that this is somehow related to the increased interstitial pressure of the liver, present both in liver cirrhosis [27] as well as in our studies during the TD obstruction. The importance of lymph drainage from the liver has been considered only for the mechanism of ascites formation [28], which apparently either stops or falls when the drainage of the TD is improved by clearance and subsequent reimplantation in the subclavian vein [29, 30]. It is not inconceivable that interstitial liver edema may exert different effects, presently unknown. Although it is tempting to speculate that the factor leading to late, presumably hemodynamic-independent, enhancement in proximal transport is humoral, a similar conclusion is unwarranted and not substantiated by the present data.

In conclusion, our experiments of reversible interstitial hypertension in the liver of dogs demonstrate that GFR and renal hemodynamics fall with the development of liver edema and return to normal upon its release, while Na reabsorption rises. Presumably, this occurs in the proximal tubule, as a late event independent of renal hemodynamics. However, it could be triggered by hemodynamic events during the initial stage of TD obstruction. These data demonstrate that the mechanisms leading to Na retention in cirrhosis may be secondary to the rise in interstitial hepatic pressure and/or increased sinusoidal resistances, and that renal Na retention occurs secondarily, presumably in response to the fall in effective plasma volume.

Summary

The mechanism of Na retention and hemodynamic impairment of the kidney during liver disease was investigated in dogs. The animals were studied before, during and following release of increased liver interstitial pressure obtained by applying an average hydrostatic pressure of 40 cm H_2O to the thoracic lymph duct. While liver interstitial pressure rose by 11.5 cm H_2O, GFR fell from 15.4 ± 1.8 to 10.2 ± 1.5 ($p < 0.02$) and rose to 13.8 ± 2.8 $ml \cdot min^{-1}$ when the hydrostatic pressure applied to the thoracic duct was released. Changes in the same direction occurred in RPF, while U_{osm} rose from 709 ± 188 to 812 ± 123 $mosm \cdot kg^{-1}$ and Na excretion fell from 19 ± 3 to 9 ± 2 $\mu Eq \cdot min^{-1}$ ($p < 0.05$). These latter changes were not reversible upon reopening of the thoracic duct. Thus, interstitial liver hypertension can impair renal hemodynamics, reduce GFR and lead to Na retention. Since the latter effect is not reversible, it is possible that the stimulus leading to more avid Na reclamation by the renal tubule, though triggered by reversible hemodynamic changes, will persist in their absence and maintain an increased proximal reabsorption. This is indicated by the constant K excretion, suggesting that the increased Na reclamation had taken place prior to distal exchange sites. The fall in renal hemodynamics may be attributed to trapping of blood within the splanchnic area by the augmented sinusoid resistance.

References

1 Tristani, F.; Cohn, J.: Systemic and renal hemodynamics in oliguric hepatic failure: effect of volume expansion. J. clin. Invest. *46:* 1894–1906 (1967).
2 Papper, S.; Vaamonde, C.: The kidney in liver disease; in Diseases of the kidney, pp. 1139–1154 (Strauss, Welt, Boston 1971).
3 Epstein, M.: Cardiovascular and renal effects of head-out water immersion in man. Application of the model in the assessment of volume homeostasis. Circulation Res. *39:* 619–628 (1976).
4 Epstein, M.; Pins, D.; Arrington, R.; DeNunzio, A.; Engstrom, R.: Comparison of water immersion and saline infusion as a means of inducing volume expansion in man. J. appl. Physiol. *39:* 66–70 (1975).
5 Vlahcevic, Z.; Adam, N.; Jick, H.; Moore, E.; Chalmers, T.: Renal effects of acute expansion of plasma volume in cirrhosis. New Engl. J. Med. *272:* 387–391 (1965).
6 Chiandussi, L.; Bartoli, E.; Arras, S.: Reabsorption of sodium in the proximal renal tubule in cirrhosis of the liver. Gut *19:* 497–503 (1978).
7 Yamahiro, H.; Reynolds, T.: Effects of ascitic fluid infusion on sodium excretion, blood volume, and creatinine clearance in cirrhosis. Gastroenterology *40:* 497–503 (1961).

8 Borghetti, A.; Novarini, A.: Fisiopatologia idrosalina e reinfusione di liquido ascitico concentrato; in Incontro medico-chirurgico sui problemi terapeutici dell'ascite nel cirrotico, pp. 23-49 (Fiaccadori, Parma 1975).
9 Blendis, L.; Greig, P.; Langer, B.; Baigrie, R.; Ruse, J.; Taylor, B.: The renal and hemodynamic effects of the peritoneovenous shunt for intractable hepatic ascites. Gastroenterology *77:*250-257 (1979).
10 Lieberman, F.; Reynolds, T.: Plasma volume in cirrhosis of the liver: its relation to portal hypertension, ascites and renal failure. J. clin. Invest. *46:*1297-1308 (1967).
11 Levy, M.: Renal function in dogs with acute selective hepatic venous outflow block. Am. J. Physiol. *227:*1074-1083 (1974).
12 Levy, M.: Effects of acute volume expansion and altered hemodynamics on renal tubular function in chronic caval dogs. J. clin. Invest. *51:*922-938 (1972).
13 Levy, M.: Renal function during graded elevation of portal venous pressure. Am. J. Physiol. *227:*1084-1087 (1974).
14 Levy, M.; Allofey, J.: Temporal relationships between urinary salt retention and altered systemic hemodynamics in dogs with experimental cirrhosis. J. Lab. clin. Med. *92:*560-566 (1978).
15 Klinger, E.; Vaamonde, C.; Vaamonde, L.; Lancestremere, R.; Morosi, H.; Frisch, E.; Papper, S.: Renal function changes in cirrhosis of the liver. Archs. intern. Med. *125:*1010-1015 (1970).
16 Chaimovitz, C.; Szylman, P.; Alroy, G.; Better, O.: Mechanism of increased renal tubular sodium reabsorption in cirrhosis. Am. J. Med. *52:*198-202 (1972).
17 Führ, J.; Kaczmarczyk, J.; Krüttgen, C.: Eine einfache colorimetrische Methode zur Insulinbestimmung für Nieren-Clearance-Untersuchungen bei Stoffwechselgesunden und Diabetikern. Klin. Wschr. *33:*729-730 (1955).
18 Snedecor, G.; Cochran, W.: Statistical methods (Ames 1972).
19 Groszmann, R.; Wotelanski, R.; Cohn, J.: Quantitation of portal systemic shunting from the splenic and mesenteric beds in alcoholic liver diseases. Am. J. Med. *53:* 715-722 (1972).
20 Epstein, M.: Renal sodium handling in cirrhosis; in The kidney in liver disease, pp. 35-53 (Epstein, New York 1978).
21 Bartoli, E.: Intraluminal control of glomerulotubular balance (GTB). Adv. Physiol. Sci. *11:*269-278 (1981).
22 Van Giesen, G.; Reese, M.; Kiil, E.; Rector, F.; Seldin, D.: The characteristics of renal hypoperfusion in dogs with acute and chronic reduction in glomerular filtration rate as disclosed by the pattern of water and solute excretion after hypotonic saline infusion. J. clin. Invest. *43:*416-424 (1964).
23 Berliner, R.; Davidson, D.: Production of hypertonic urine in the absence of pituitary antidiuretic hormone. J. clin. Invest. *36:*1416-1427 (1957).
24 Bartoli, E.; Satta, A.; Faedda, R.; Olmeo, N.; Soggia, G.; Branca, G.: A furosemide test in the functional evaluation of the human nephron in vivo. J. clin. Pharmacol. *23:*56-64 (1983).
25 Ang, G.; Landahi, H.; Bartoli, E.: Transient and steady state simulation of the renal countercurrent mechanism. Comput. Biol. Med. *7:*87-111 (1977).
26 Bartoli, E.; Satta, A.; Faedda, R.; Olmeo, N.; Soggia, G.; Branca, G.: Reabsorption by different segments of the human nephron in the hepato-renal syndrome (HRS). Hepatology *2:*172 (1982).

27 Lebrec, D.; Paynard, T.; Hillon, P.; BenHamou, J.: Propranolol for prevention of recurrent gastrointestinal bleeding patients with cirrhosis. A controlled study. New Engl. J. Med. *305:*1371–1374 (1981).

28 Witte, C.; Witte, M.; Dumont, A.: Lymph imbalance in the genesis and perpetuation of the ascites syndrome in hepatic cirrhosis. Gastroenterology *78:* 1059–1068 (1980).

29 Serenyi, P.; Magyar, Z.; Szabo, G.: Cervical lymphato-venous shunt in treatment of ascites in caval constricted dogs and in patients with hepatic cirrhosis. Lymphology *9:*53–54 (1976).

30 Bresadola, F.; Mannella, P.; Sortini, A.; Buccoliero, F.; Pilinzi, V.: Il drenaggio interno del dotto toracico nella terapia chirurgica dell'ascite cirrotica. Minerva chir. *31:*1051–1054 (1976).

Prof. E. Bartoli, Istituto di Patologia Medica, Viale San Pietro 12,
I-07100 Sassari (Italy)

Front. gastrointest. Res., vol. 8, pp. 85–90 (Karger, Basel 1984)

Quantitative Analysis of the Perisinusoidal Cells in Human Liver: The Lipocytes

A.M. Jezequel, G. Novelli, C. Venturini, F. Orlandi

Istituto di Patologia Sperimentale (Cattedra di Farmacologia Cellulare); Istituto di Semeiologia, Diagnostica e Terapia Strumentale (Cattedra di Gastroenterologia); Facoltà di Medicina e Chirurgia, Università di Ancona, Italy

Among the numerous questions still unanswered regarding the development and evolution of cirrhosis in man, the nature of the cell responsible for the formation of hepatic fibrosis, the initial site of deposition of fibers and the reason for the irreversibility of the phenomenon are the most puzzling. Although a number of morphological data have already been accumulated, based either on histological or ultrastructural studies, much remains to be done to elucidate these particular aspects. There is no doubt that the problems are made even more difficult due to the lack of a suitable animal model of liver cirrhosis which would duplicate the human disease. The purpose of our research in this area has been to characterize the various cell types present in the liver tissue, paying special attention to a population of subendothelial cells, the lipocytes (or fat-storing cells, Ito cells) which have been considered as potential fibroblasts. Typical fibroblasts are rare in the normal liver. They may be occasionally seen in portal areas, but they are not easily identified under the light microscope, and no quantitative data (number, volume density, pathological variations, etc.) are so far available. More attention has been given to the lipocytes which are situated in the space of Disse, between the endothelium and the hepatocytes [1, 3, 7, 15]. In ultrathin sections, they appear triangular or elongated in shape and often extend in recesses between hepatocytes. They usually have a large, pale nucleus and the cytoplasm contains one or more lipid droplets and prominent cisternae of rough endoplasmic reticulum (fig. 1, 2). As a rule, no evidence of phagocytic activity is present. Under the light microscope, these cells escape observation in pa-

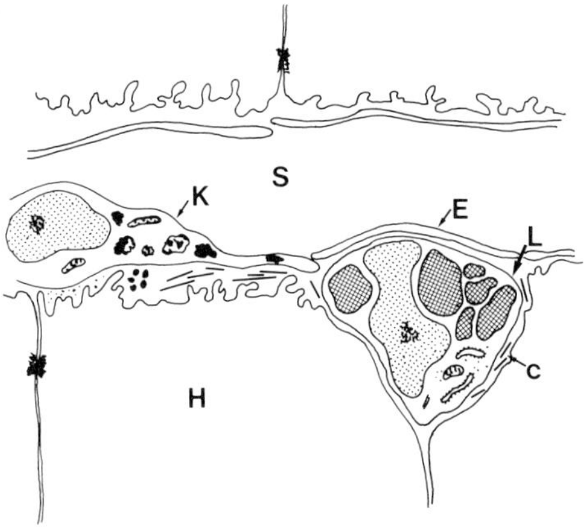

Fig. 1. Diagram illustrating the situation of the lipocytes in the hepatic parenchyma. The lipocytes (L) are situated in the space of Disse between the endothelium (E) and the hepatocytes (H). Large lipid droplets (hatched areas) are present in the cytoplasm, often indenting the nucleus. Small clumps of collagen fibrils (c) are seen next to the lipocytes. S = Sinusoid; K = Kupffer cell.

raffin-embedded material, due to the dissolution of the lipid droplets during processing of the material. However, in tissues fixed with osmium tetroxide and embedded in resin mixture (Epon-Araldite for instance) the lipid droplets are well preserved throughout, making it possible to identify the lipocytes even under the light microscope (fig. 3). This is the technique that we have been using to conduct a quantitative analysis of lipocytes in normal human liver, as a preliminary study for the analysis of these cells in pathological situations.

Material and Methods

The material consists of 1-μm sections of needle liver biopsies fixed in 1% osmium tetroxide buffered at pH 7.3 in phosphate buffer. The morphometric analysis is conducted with the point-counting technique [16] as used in other stereological studies on human liver [10, 11]. The volume density Vv of a component = Pi/Pt, where Pi is the

---

Fig. 2. This electron micrograph shows a large lipocyte (L), with lipid droplets (ld) filling the cytoplasm and deforming the nucleus (N). A group of collagen fibrils (c) is seen between the lipocyte and the hepatocytes (H). Lipid droplets (*) are also present in the cytoplasm of hepatocytes. S = Sinusoid; E = endothelium. Osmium-fixed, Epon/Araldite-embedded human liver. × 9,000.

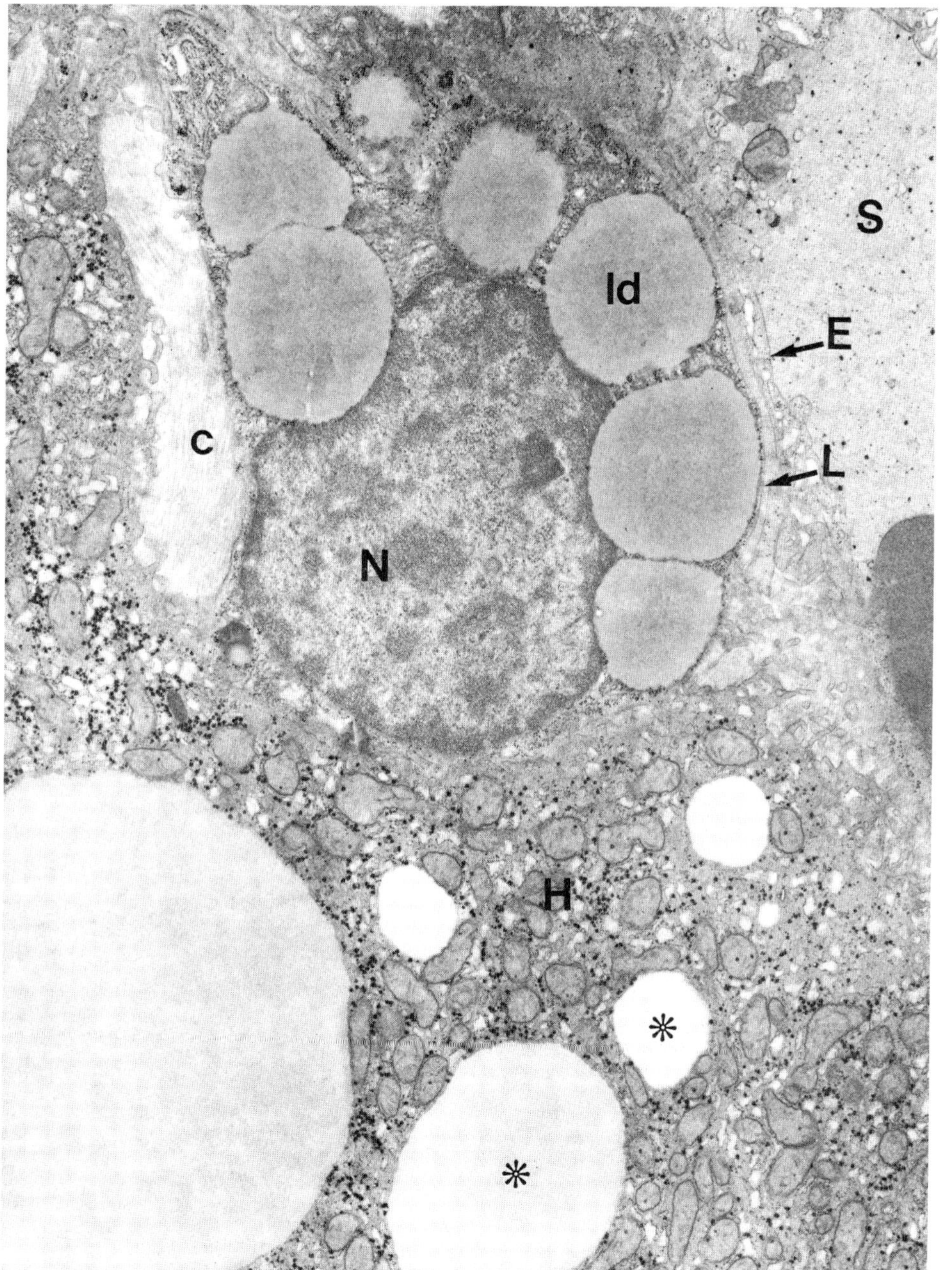
S
ld
E
L
c
N
H
*
*

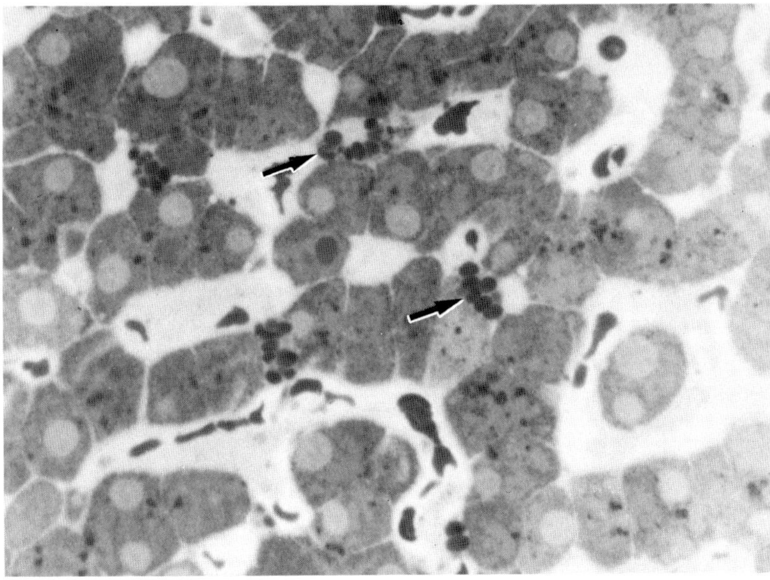

Fig. 3. Light micrograph of a 1-μm section of osmium-fixed, Epon/Araldite-embedded human liver stained with basic fuchsin. The arrows point to lipocytes bulging into the lumen of the sinusoids and filled with numerous dense lipid droplets. ×425.

number of points of a graticule falling upon the structure considered and Pt is the total number of points falling upon the element selected for reference (hepatocytes, parenchyma, etc.). A Leitz graticule is used, graduated at 10 intervals/1 cm and inserted into the 8× eyepiece of a Leitz orthoplan microscope. The counting is performed using a 40× objective and 5 sections from 5 different blocks are examined for each biopsy. At least 2 and up to 5 non-overlapping fields can be examined in each 1-μm section. In addition, the lobular distribution of the lipocytes may be estimated by selecting areas centered on the portal spaces or on the central veins so that a layer of parenchyma 3-6 hepatocytes wide is selected in the frame of the graticule.

Also the Ito cell index [5] or number of lipocytes per 1,000 hepatocytes may be calculated by counting the number of hepatocytes within the frame of the graticule. The technical details have been extensively reported elsewhere [3, 10, 11].

Results

The results which we have obtained in a first group of 8 biopsies show that in the normal human liver the volume density of lipocytes is 1.88 ± 0.25% (mean ± 1SD) of the parenchymal volume and the Ito cell index is 63.45 ± 19.18 showing no relationship with the volume density of parenchymal steatosis.

The lipocytes were present in all the sections examined and were more numerous in centrolobular than in periportal areas (2.49 ± 0.51 versus 0.43 ± 0.58%, respectively, $p < 0.001$). Indeed, the periportal areas were often free of lipocytes.

Discussion

An increase of lipocytes has been said to occur after administration of carbon tetrachloride to rats [12] and in a variety of situations in man, such as chronic hypervitaminosis A together with perisinusoidal fibrosis [6] and in the initial stages of vinylchloride injury [13]. An increase in size and number of lipocytes has also been noted after therapy with methotrexate [4], a drug which has been implicated in the development of chronic hepatitis. Evidence has also been presented that the lipocytes may be involved in the synthesis of type III collagen [9]. It is of interest that other perisinusoidal cells are occasionally observed in the hepatic parenchyma. Under the electron microscope these cells are similar to the lipocytes, but they lack the characteristic lipid droplets, appearing as young fibroblasts or sometimes as undifferentiated mesenchymal cells identical to the pericytes which accompany small vessels. The suggestion has been made that primitive mesenchymal cells could differentiate into 'myofibroblasts' or 'contractile fibroblasts' such as observed in chronic hypervitaminosis A [6] in alcoholic cirrhosis [14] or in focal nodular hyperplasia [2]. 'Modulation' of one cell type to another is also suggested by the presence of transitional aspects between lipocytes and fibroblasts in the fibrous septa following carbon tetrachloride injury [12] or in chronic active hepatitis [8] where deposition of collagen is enhanced. The behavior of fibroblasts and lipocytes appears also similar in response to the action of antimicrotubular agents [8].

Thus, indirect evidence suggests a relationship between the deposition of intralobular collagen and the presence of perisinusoidal cells which may appear as typical fibroblasts, more often as fat-laden cells (lipocytes). These are probably various aspects of undifferentiated mesenchymal cells occasionally seen in the space of Disse. The quantitative data that we have obtained in normal liver may serve as a baseline in further studies conducted on pathological material.

References

1 Bronfenmajer, S.; Schaffner, F.; Popper, H.: Fat-storing cells (lipocytes) in human liver. Archs Path. *82:*447–453 (1966).

2 Callea, F.; Mebis, J.; Desmet, V.J.: Myofibroblasts in focal nodular hyperplasia of the liver. Virchows Arch. Abb. A Path. Anat. Histol. *396:*155–166 (1982).

3 Giampieri, M.P.; Jezequel, A.M.; Orlandi, F.: The lipocytes in normal human liver. Digestion *22:*165–169 (1981).

4 Hopwood, D.; Nyfors, A.: Effect of methotrexate therapy in psoriatics on the Ito cells in liver biopsies assessed by point counting. J. clin. Path. *29:*698–703 (1976).

5 Horvath, E.; Kovacks, K.; Ross, R.C.: Liver ultrastructure in methotrexate treatment of psoriasis. Archs Derm. *108:*427–428 (1973).

6 Hruban, Z.; Russel, R.M.; Boyer, J.L.; Glagov, S.; Bagheri, S.A.: Ultrastructural changes in livers of two patients with hypervitaminosis A. Am. J. Path. *76:*451–468 (1974).

7 Ito, T.; Nemoto, M.: Über die Kupfferschen Sternzellen und die 'Fettspeicherungszellen' ('Fat-storing cells') in der Blutkapillarenwand der menschlichen Leber. Okajimas Folia anat. jap. *24:*243–258 (1952).

8 Jezequel, A.M.; Koch, M.M.; Orlandi, F.: Hepatic fibrosis. Role of perisinusoidal cells? Ital. J. Gastroent. *4:*560–565 (1980).

9 Kent, G.; Gay, S.; Inouye, T.; Bahu, R.; Minick, D.T.; Popper, H.: Vitamin A containing lipocytes and formation of type III collagen in liver injury. Proc. natn. Acad. Sci. USA *73:*3719–3722 (1976).

10 Koch, M.M.; Freddara, U.; Lorenzini, I.; Giampieri, M.P.; Jezequel, A.M.; Orlandi, F.: A stereological and biochemical study of the human liver in uncomplicated cholelithiasis. Digestion *18:*162–177 (1978).

11 Koch, M.M.; Giampieri, M.P.; Lorenzini, I.; Jezequel, A.M.; Orlandi, F.: Effect of chenodeoxycholic acid on liver structure and function in man: a stereological and biochemical study. Digestion *20:*8–21 (1980).

12 McGee, J.O.; Patrick, R.: The role of perisinusoidal cells in hepatic fibrogenesis. An electron microscopic study of acute carbon tetrachloride injury. Lab. Invest. *26:* 429–440 (1972).

13 Popper, H.; Thomas, L.B.; Schaffner, F.; Maltoni, C.; Selikoff, I.J.: Interaction between sinusoidal cells and hepatocytes in human and experimental angiosarcoma induced by environmental factors; in Wisse, Knook, Kupffer cells and other liver sinusoidal cells, pp. 173–181 (Elsevier/North-Holland, Amsterdam 1977).

14 Rudolph, R.; McClure, W.J.; Woodward, M.: Contractile fibroblasts in chronic alcoholic cirrhosis. Gastroenterology *76:*704–709 (1979).

15 Wake, K.: 'Sternzellen' in the liver: perisinusoidal cells with special reference to storage of vitamin A. Am. J. Anat. *132:*429–462 (1977).

16 Weibel, E.R.; Stäubli, W.; Gnägi, H.R.; Hess, F.A.: Correlated morphometric and biochemical studies on the liver cell. J. Cell Biol. *42:*68–91 (1969).

A.M. Jezequel, MD, University of Ancona, CP 538, I-60100 Ancona (Italy)

Front. gastrointest. Res., vol. 8, pp. 91–105 (Karger, Basel 1984)

Humoral and Cell-Mediated Autoimmunity in Chronic Liver Disease: Hepatocyte Membrane-Bound Immunoglobulins, Autologous Cytotoxicity and Liver-Specific Autoantibodies

R. Meliconi, M. Baraldini, G.F. Stefanini, Maura Mazzetti, F. Miglio, G. Gasbarrini

III Patologia Medica, Università di Bologna, Bologna, Italy

Introduction

Since they were first detected in liver patients by *Hopf* et al. [1] in 1975, hepatocyte membrane-bound immunoglobulins have been the evidence, demonstrable in vivo, of (auto)immune phenomena directed against the membrane of liver cells. Indeed, the detection of IgG fixed to hepatocytes isolated from patients' biopsies supported the hypothesis of antibodies reacting with membrane antigens in vivo, in patients with chronic active hepatitis (CAH). These membrane IgG were detected with two different distributions on the membrane: linear or granular according to the aetiology of CAH, autoimmune or viral – hepatitis B virus (HBV)-induced [1–3]. In the present investigation, the prevalence and surface distribution of membrane IgG and the amount of IgG-positive cells have been evaluated in different forms of chronic liver disease. We then studied the possible relationship between these findings and: (a) T and non-T cytotoxicity to autologous hepatocytes; (b) circulating liver-specific autoantibodies, in order to evaluate the possible involvement of membrane-fixed IgG in the postulated immune mechanism of ongoing liver injury in CAH and to obtain some insight into the composition and characterization of these cytophilic antibodies.

Materials and Methods

Study A (Membrane IgG and Autologous Cytotoxicity)

Patients. 45 patients were studied. 31 presented CAH, 13 of whom had underlying cirrhosis: 6 cases of autoimmune CAH, 11 $HBsAg^+$, 14 $HBsAg^-$/anti-HBc^+, 5 cases with inactive alcoholic cirrhosis and 9 subjects who underwent gallbladder surgery with surgical liver biopsy presenting normal liver or minimal non-specific changes. 5 patients with CAH were on immunosuppressive treatment. Diagnosis of CAH was made, following the criteria of an international committee [4].

Preparation of Hepatocytes from Biopsies

A fragment of the biopsy was placed in a Petri dish with RPMI 1640 (Gibco, UK) at 37°C and gently broken up mechanically using two 12-gauge needles. The cells were filtered through a 1-ml syringe filled with nylon wool (Leucopak, Fenwal Laboratories, USA) and washed twice in RPMI 1640. Hepatocyte morphology was checked by phase-contrast microscopy and hepatocyte viability by trypan blue exclusion test. The range of viable cells was: (a) 82–91% immediately after the isolation, (b) 55–64% after 24 h culture, 46–58% after 48 h culture. The percentages of viable cells were similar to the percentages of adherent hepatocytes still present in control wells after 24 and 48 h culture. Liver cells were counted and the concentration adjusted to 1×10^4 cells/ml in RPMI 1640+10% fetal calf serum (FCS) (Gibco, UK). The cells were then seeded in microculture plates (Falcon 3034, USA) (80 hepatocytes/well). The plates were incubated for 18–24 h at 37°C in 95% O_2 + 5% CO_2 atmosphere.

Immunofluorescence Test for Membrane-Fixed IgG

50 µl of hepatocytes (1–2 × 10^3 cells) isolated from the biopsy as described above were incubated with 100 µl of $F(ab')_2$ goat anti-human IgG antiserum labelled with fluorescein isothiocyanate (Nordic, Tillburg, The Netherlands) for 30 min at 37°C. The cells were then washed twice with RPMI 1640 with addition of 2.5% bovine serum albumin (BSA), mounted on a slide and examined under a Zeiss microscope equipped with epi-illumination. The results were expressed as the percentage of membrane-bound IgG-positive cells. Only cells with intact membrane, as checked by phase contrast microscopy, were counted.

Lymphocyte Preparation

40 ml of peripheral blood from the patient was incubated with carbonyl iron powder (2 mg/ml) at 37°C for 30 min, in order to remove the macrophages [5]. Lymphocytes were separated after Ficoll-Hypaque, washed and adjusted to a concentration of 9–10 × 10^6/ml in RPMI 1640 + 10% FCS and the cells were then incubated with sheep red blood cells (SRBC) (1:1 v/v) previously activated by neuraminidase (Behring, Mannheim, Cod RKD 05) [6]. After centrifugation at 100 *g* for 10 min the pellet was incubated in an ice bath for 1 h. The pellet was gently resuspended and this suspension was stratified on a Ficoll-Hypaque density gradient (d = 1.077) and centrifuged at 500 *g* for 15 min.

E rosette-forming cells were found in the pellet while the non-rosetting lymphocytes were present at the interface of the gradient. The latter lymphocytes (non-T) were collected and washed twice, while T lymphocytes were obtained after osmotic lysis of

SRBC-forming E rosette cells (2 min incubation with 0.15 *M* NH_4Cl-$KHCO_3$ at 4°C) [7, 8].

Control experiments showed that more than 90% (82–95%) of E rosette forming cells were present in T-enriched lymphocyte subpopulations. T-depleted subpopulations presented less than 4% of E rosette-forming cells and 50–60% of cells positive for surface membrane immunoglobulins detected by immunofluorescence using F $(ab')_2$ fragments of anti-human $F(ab')_2$ conjugated with TRITC (Nordic, Tillburg, The Netherlands). Lymphocyte viability was 95–100% as determined by fluorescein diacetate test [9], monocytes were less than 3% as determined by morphology and peroxidase staining; neutrophils were almost absent.

Microcytotoxicity Test

The test was performed essentially as described in detail for rabbit hepatocytes [5]; briefly: 10 µl of lymphocyte suspension was added to the culture plate already seeded with patient hepatocytes. The target/effector (T/E) ratio ranged from 1:50 to 1:400. Since the highest cytotoxicity using total lymphocytes was obtained at the T/E ratio of 1:200–1:400, the latter ratio was used to compare the results of different cases. The same T/E ratio was also utilized when using enriched (T) and depleted (non-T) lymphocyte subpopulations since we did not find any significant difference in cytotoxicity indexes (CIs) using T/E ratio 1:100 to 1:400. 15 wells were filled with 10 µl of medium alone and used as controls, 15 wells with 10 µl of total lymphocyte suspension, 15 wells with 10 µl of T-depleted and 15 wells with 10 µl of T-enriched lymphocytes. After 48 h incubation, the plates were inverted for 2 h and gently washed with RPMI 1640 in order to detach non-viable cells and lymphocytes. Plates were then fixed and stained. The remaining hepatocytes were then counted and, by comparing the values found in control wells (C_0) with those incubated with lymphocytes (C_1), the percentage of cytotoxicity was calculated following the formula:

$$(CI) = \left(1 - \frac{C_1}{C_0}\right) \times 100.$$

Serological Studies

Serum HBV markers were detected by solid-phase radioimmunoassay (AUSRIA II, AUSAB, CORAB, HBe; Abbott, Chicago, USA).

Investigation for non-organ-specific autoantibodies was performed on cryostat sections of human stomach and thyroid and rat liver and kidney by indirect immunofluorescence.

Statistics

The non-parametric Wilcoxon rank sum test was used to compare the cytotoxicity values in the different groups studied. The χ^2 test with Yates' correction for small numbers was employed to study the relationship between membrane IgG and cytotoxic activity.

Study B (Membrane IgG and Liver-specific Autoantibodies)

Patients. Hepatocytes isolated from 101 biopsies were examined for membrane-fixed immunoglobulins. Serum samples were taken from all patients at the time of the biopsy and stored at -20°C until analysis.

CAH was diagnosed in 53 patients (28 with cirrhosis): 21 HBsAg-positive cases, 17 HBsAg−/anti-HBc$^+$ cases, 5 autoimmune CAH and 10 cryptogenic cases. Primary biliary cirrhosis (PBC) was found in 8 patients, 8 patients presented with chronic persistent hepatitis (CPH) and 14 had a prolonged course of acute viral hepatitis (PVH). Alcoholic cirrhosis was diagnosed in 11 patients. The biopsy showed normal histology or minor non-specific alterations in 7 patients who were classified as normal controls.

Immunofluorescence Test for Membrane Fixed IgG:
As in study A.

Radioimmunoprecipitation for Anti-Liver-Specific Protein Antibodies
Liver-specific protein (LSP) was isolated from normal human liver according to the technique described by *McFarlane* et al. [10] and was radiolabelled with ^{125}I with Bolton-Hunter reagent according to the method of *Krosgaard* et al. [16]. The radioimmunoprecipitation test has been described elsewhere [11].

Liver Cell Membrane Autoantibody Test
The liver cell membrane autoantibody (LMA) test was detected by indirect immunofluorescence on rabbit hepatocytes isolated by non-enzymatic method according to *Hopf* et al. [1]. This technique is described in detail elsewhere [12].

Serological Studies:
As in study A.

Statistical Analysis
The prevalence of membrane-bound IgG-positive cases and the prevalence of anti-LSP and LMA in the different groups were compared using the χ^2 test with Yates' correction for small numbers. Percentages of positive cells in different groups were compared using the non-parametric Wilcoxon rank sum test.

Results

Study A

The results of lymphocyte cytotoxicity against autologous liver cells are summarized in figure 1. The upper limit of the normal range for CIs was calculated adding 2 standard deviations to the mean values of the 9 control subjects (11.8 ± 22.2 = 34%).CIs in HBV-related and 'autoimmune' CAH cases (mean ± SD = 45.6 ± 30.9; 60.5 ± 27.1) were statistically higher than those obtained in normal controls (mean ± SD = 11.8 ± 11.1, $p < 0.005$) and in patients with inactive alcoholic cirrhosis (mean ± SD = 20.2 ± 6.3, $p < 0.05$).

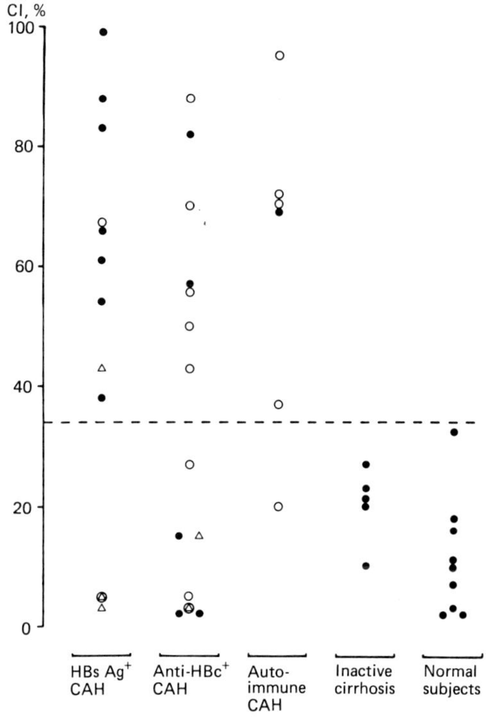

Fig. 1. Cytotoxic activity of total lymphocytes against autologous hepatocytes. ● = Untreated patients; △ = treated patients; ○ = presence of cirrhosis. Dotted line shows the upper limit of the normal range.

21 out of 31 CAH cases were positive (68%) and this percentage increases to 80% if we take into account only the untreated cases. Patients on immunosuppressive treatment had a lower mean CI than the untreated ones (13.8 ± 17.0, 54.7 ± 29.7, $p < 0.005$). No difference in cytotoxicity was found between $HBsAg^{+}$ and $HBsAg-$/anti-HBc^{+} patients. Among the latter group no significant difference was observed in cytotoxicity between the patients with or without circulating anti-HBs.

In 'autoimmune' CAH the prevalence of positive cases was 83%, while it was 68% in HBV-related cases (untreated HBV-related cases 89%). Mean CI in CAH patients with cirrhosis was not significantly different from those without cirrhosis (50.2 ±29.0; 46.6 ± 34.3). Hepat-

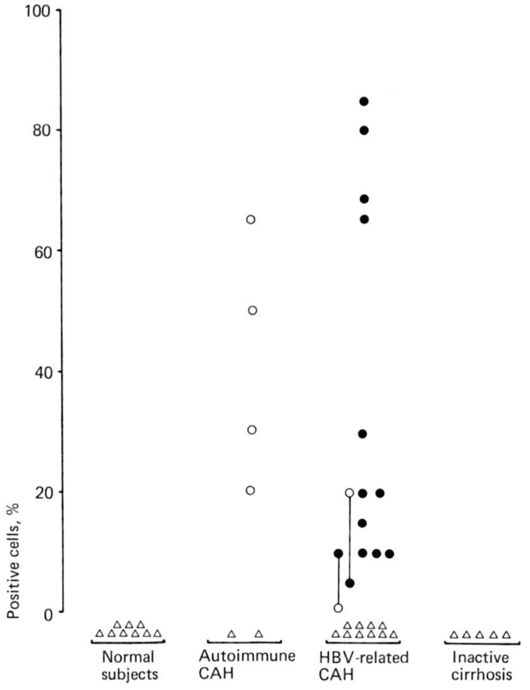

Fig. 2. Membrane-bound immunoglobulins on hepatocyte membrane. ● = Granular pattern; ○ = linear pattern; △ = absence of immunoglobulins.

ocyte membrane IgG was found only in patients with CAH (fig. 2). 4 out of 6 'autoimmune' cases, and 13/25 HBV-related cases were positive.

Granular fluorescence pattern was present only in HBV-related CAH, while 'autoimmune' cases presented linear pattern of fluorescence. The latter pattern was also present on some hepatocytes from 2 HBV-related cases together with granular fluorescence on other hepatocytes of the same patients. No significant difference was found in the prevalence of membrane IgG-positive cases in the two CAH groups.

The study of the cytotoxic activity of lymphocyte subpopulation (T-depleted and T-enriched) revealed a different behaviour of the effector cells in the two forms of CAH (fig. 3, 4).

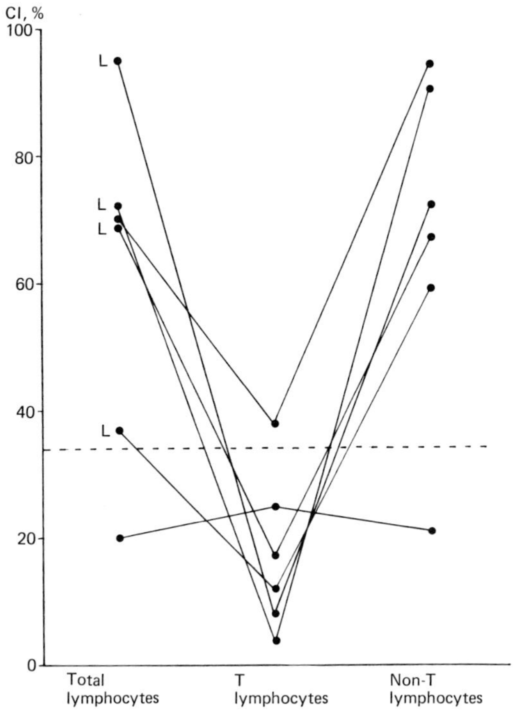

Fig. 3. Cytotoxic activity of total, T and non-T lymphocytes against autologous hepatocytes in patients with autoimmune CAH. L = Presence of membrane-bound immunoglobulins with linear pattern on target hepatocytes.

In 'autoimmune' CAH, cytotoxic activity was confined to the E rosette-forming cell-depleted subpopulation (non-T lymphocytes CI = 67.2 ± 26.3, T lymphocytes CI = 17.3 ± 12.5, $p < 0.01$) whereas in HBV-related CAH, both subpopulations exhibited cytotoxic activity (38.0 ± 27.8; 40.3 ± 31.2; n.s.).Furthermore, in the latter group of CAH there was no relationship between cytotoxicity and the presence of membrane IgG (fig. 4). On the other hand, 4 out of 5 autoimmune patients with positive cytotoxicity were positive for membrane-bound IgG (with linear pattern). It is noteworthy that the patient with normal CI was also negative for cytophilic IgG. We could not find any relationship between T lymphocytotoxicity and active viral replication measured by the presence of HBeAg in serum.

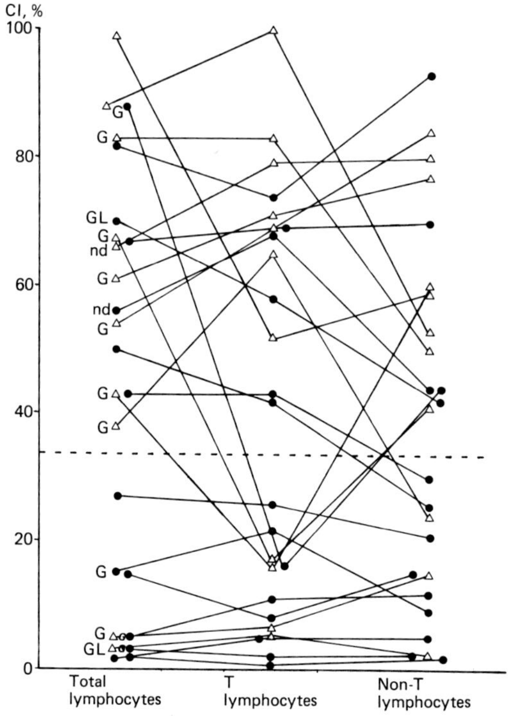

Fig. 4. Cytotoxicity of total, T and non-T lymphocytes against autologous hepatocytes in HBV-related CAH. Presence of membrane-bound immunoglobulins in granular (G) or linear (L) pattern on target hepatocytes. nd = Not determined; △ = $HBsAg^+$ patients; ● = HBsAg−, $anti\text{-}HBc^+$ patients.

Study B

The results are summarized in table I and figure 5. 41 patients presented membrane-bound IgG with granular pattern (range of positive cells 3–90%). 12 patients were positive for membrane-bound IgG with linear pattern (range of positive cells 5–65%). In 4 cases, hepatocytes positive for linear and granular membrane-bound IgG were found in the same patient (2 $HBsAg^+$ CAH, 1 HBsAg−/$anti\text{-}HBc^+$ CAH, 1 cryptogenic CAH).

Membrane IgG with granular pattern was found mainly in $HBsAg^+$ chronic active liver disease (CALD), PBC, CPH, and PVH.

Table I. Cytophilic and circulating hepatocyte membrane antibodies

Diagnosis	Number	Membrane IgG granular n	Membrane IgG granular %	Membrane IgG linear n	Membrane IgG linear %	anti-LSP n	anti-LSP %	LMA n	LMA %
HBV-related CALD									
HBsAg$^+$	21	15	71	3	14	11	52	4	19
HBsAg$^-$/anti-HBc$^+$	17	7	41	1	6	4	23	2	12
Autoimmune CALD									
CAH	5	0		4	80	3	60	4	80
PBC	8	6	75	0		7	87	2	25
Cryptogenic CALD	10	3	30	1	10	2	20	0	
CPH	8	3[a]	37	0		1	12	3	37
PVH	14	6[a]	43	1	7	3	21	4	28
ALD	11	1[b]	9	2	18	4	36	3	27
Controls	7	0		0		0		0	

[a] 1 HBsAg$^+$ case; [b] HBsAg$^+$ case.

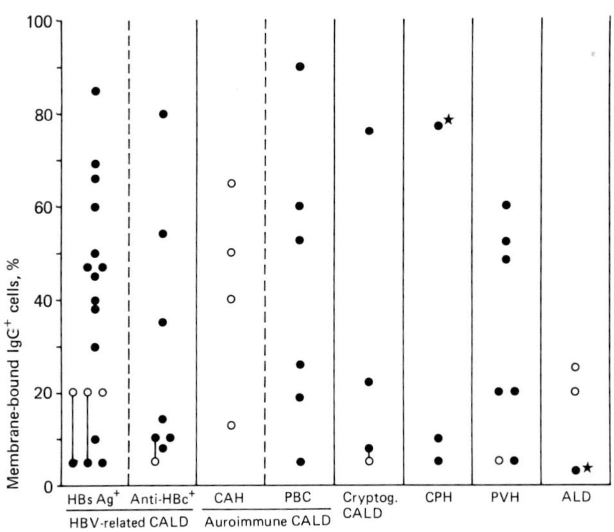

Fig. 5. Percentages of membrane-bound IgG$^+$ cells in chronic liver diseases. ○ = Linear pattern of membrane IgG; ● = granular pattern of membrane IgG; ★ = HBsAg$^+$ cases.

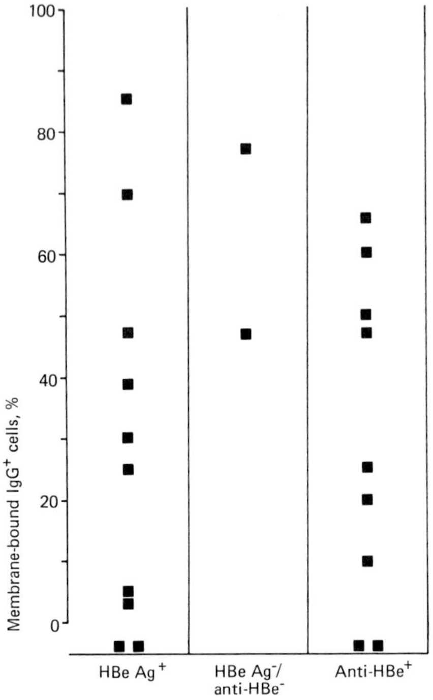

Fig. 6. Percentages of membrane-bound IgG^+ cells in relation to the HBeAg/anti-HBe system.

The fluorescence pattern was similar in the viral and non-viral forms. A significant correlation was found between the presence of serum HBsAg and the positivity for granular membrane-bound IgG (18 of 24 $HBsAg^+$ patients and 23 of 77 HBsAg− patients: 75–30% $p < 0.0005$). In the $HBsAg^+$ group no relationship was found between the percentage of granular IgG^+ cells and the presence of serum HBeAg or anti-HBe (fig. 6).

Linear membrane-fixed IgG were present mainly in autoimmune CAH, $HBsAg^+$ CAH without cirrhosis and ALD patients. 11 of the 12 patients presented a histological diagnosis consistent with CAH and moderate to severe piecemeal necrosis. The highest percentages of positive cells were found in 3 female patients with ANA (titre > 1:40) or SMA (titre $\geq$ 1:80) without HBV markers and positive for liver-specific autoantibodies (autoimmune CAH) (percent positive cells: mean ± standard deviation; autoimmune cases 42.0 ± 21.9, non-autoimmune cases 15.0 ± 8.4, $p < 0.05$).

Anti-LSP were present mainly in PBC (87%) and in autoimmune CAH (60%) while the highest prevalence of LMA was found in autoimmune CAH (80%) (table I). 9 patients were positive for both anti-LSP and LMA (2 PBC, 3 autoimmune CAH, 2 HBsAg$^+$ CAH, 1 anti-HBc$^+$ CAH, 1 alcoholic cirrhosis). In this group a strong correlation between the positivity for both liver-specific antibodies and autoimmune liver disease was found ($p < 0.0005$). There was no relationship between membrane-bound IgG-positive cases and circulating liver-specific autoantibodies. Also taking into account LMA titres and the percentage values of LSP binding, no relationship was found between the circulating and cytophilic antibodies in patients positive only for linear membrane IgG.

Discussion

IgG fixed to the liver cell surface have been found in patients with various chronic liver diseases but not in controls. As described by others [1], we also found two patterns of immunoglobulin distribution on the membrane: granular and linear (homogeneous) depositions.

The linear pattern was mainly, but not exclusively, present in autoimmune CAH patients, since 'linear' IgG was also found in HBsAg$^+$ CAH and alcoholic cirrhosis. The latter finding confirms the results obtained by *Trevisan* et al. [13] in patients with alcoholic liver disease.

The granular pattern was closely associated with chronic HBV infection, but again it was found in HBsAg$-$/anti-HBc$^+$ and cryptogenic CAH, in PVH and PBC. While in HBsAg$^+$ CAH, granular membrane IgG is probably directed against viral determinants (HBcAg) as recently demonstrated by *Trevisan* et al. [14], in the other forms of chronic liver disease this granular fluorescence could be an expression of immune complexes bound to the hepatocyte surface via the Fc receptors, particularly in PBC cases [15, 16].

In chronic HBV infection the presence of membrane IgG and the percentages of positive cells were found to correlate with active viral replication by *Trevisan* et al. [17]. HBeAg can be considered a marker of viral replication since it closely correlates with Dane particle-associated DNA polymerase activity when tested by radioimmunoassay [18, 19]. Therefore, we used the determination of HBeAg (RIA) in order to verify the assumption of the linkage between membrane IgG and viral replication. Indeed, this association was not found, given the positivity for membrane IgG in anti-HBe^+ or even HBsAg− patients. This discrepancy could be due to the use of HBeAg instead of DNA polymerase as a marker of viral replication (different sensitivity?) or to the presence of immune complexes fixed to the liver cell membrane and not related to viral replication. Interestingly, no association was found between viral replication (HBeAg positivity) and T cell cytotoxicity.

Further data on the possible involvement of membrane immunoglobulins in the immunopathogenesis of liver cell injury are provided by a parallel study on cytophilic IgG and autologous cytotoxicity. First of all, our results have shown a different behaviour in the cytotoxic mechanism of lymphocyte subpopulations (T and non-T cells) in the different forms of CAH. While cytotoxic activity, sustained by non-T lymphocytes, is present in both types of CAH, cytotoxic activity sustained by T cell-enriched populations was present only in viral CAH. This finding is particularly interesting when compared to the distribution of the IgG fixed in vivo to the hepatocyte membrane. In autoimmune CAH, a linear homogenous deposition was present and it was associated with a cytotoxic activity exclusively sustained by non-T lymphocytes. In this form of chronic hepatitis the cytotoxicity mechanism seems to be antibody-dependent with mediating antibody fixed in vivo to the surface of target cells.

Mieli Vergani et al. [20] found that hepatocytes from HBsAg− CAH were lysed by lymphocytes from normal donors and the addition of $F(ab')_2$ anti-human IgG antiserum blocked the reaction [21]. Indeed the simultaneous presence of immunoglobulins in linear pattern on the surface of target hepatocytes together with cytotoxicity suggests that these antibodies may be the mediators of the antibody dependent cell-mediated cytotoxicity (ADCC) found in these forms of CAH.

However, it is not clear to which membrane antigen these antibodies are directed. LSP or liver membrane antigen can be considered putative antigens since anti-LSP antibodies and liver membrane antibodies are found in the majority of autoimmune CAH.

In HBV-related CAH, positive cytotoxicity was found in both T-enriched and T-depleted fractions. Similar data have been recently presented by *Eddleston* et al. [21].

In our study the immunofluorescence pattern did not correlate with the cytotoxic activity presented by either T-enriched or T-depleted subpopulations, and patients were found to have membrane-bound IgG on their hepatocytes with negative cytotoxicity. A granular, but not linear, pattern was also found in CPH [12] which generally presents lower cytotoxicity indexes then CAH [5, 21, 22]. These observations suggest that membrane IgG in granular patterns are not the mediators of cytotoxic activity and this may be due to the steric configuration of these immunoglobulins as immunocomplexes on the hepatocyte membrane.

Finally, the lack of relationship between membrane IgG and anti-LSP or LMA is consistent with the different nature of the cytophilic and circulating antibodies in HBV-related cases and in PBC [14]. On the other hand, the simultaneous presence of anti-LSP, liver membrane antibodies and linear membrane IgG (with the highest percentage of positive cells) in autoimmune CAH alone suggests that liver-specific autoantibodies may contribute to the deposition of immunoglobulins on the hepatocyte surface.

Therefore, in conclusion, we suggest that in autoimmune CAH, 'linear' membrane-bound IgG may represent liver-specific autoantibodies in vivo fixed to the target antigens and able to mediate an antibody-dependent cell-mediated cytotoxic reaction against the hepatocyte membrane leading to liver cell injury. On the contrary, 'granular' membrane IgG found in viral (HBV-induced) CAH does not seem to be constituted by liver-specific autoantibodies and is probably not (directly) involved in mediating the lymphocyte aggression to liver cells also demonstrated in this form of CAH using autologous and heterologous test systems.

Summary

Two studies were performed utilizing isolated hepatocytes from patients' biopsies in order to evaluate the prevalence and pattern of membrane-bound IgG. In the first study, membrane IgG was compared with 'in vitro' lymphocytotoxicity to autologous hepatocytes. In autoimmune chronic active hepatitis (CAH), linear membrane IgG was associated with cytotoxic activity exclusively sustained by non-T cells. In hepatitis B virus (HBV)-related CAH, T and non-T lymphocytotoxicity was often present simultaneously in the same patients without any relationship to membrane IgG present with granular pattern. In the second study a larger group of patients with liver disease was studied for membrane IgG and the results compared with the prevalence of circulating anti-liver-specific protein (LSP) antibodies and liver membrane antibodies. Autoimmune CAH presented high percentages of hepatocytes with linear membrane IgG and higher prevalences of circulating anti-LSP and liver membrane autoantibodies. This association was not found in HBV-related CAH or in primary biliary cirrhosis.

References

1 Hopf, U.; Arnold, W.; Meyer zum Büschenfelde, K.H.; Forster, E.; Bolte, I.P.: Studies on the pathogenesis of chronic inflammatory liver diseases. I. Membrane-fixed IgG on isolated hepatocyte from patients. Clin. exp. Immunol. *22:*1–8 (1975).

2 Alberti, A.; Realdi, G.; Tremolada, F.; Spina, G.P.: Liver cell surface localisation of hepatitis B antigen and of immunoglobulins in acute and chronic hepatitis and in liver cirrhosis. Clin. exp. Immunol. *25:*396–402 (1976).

3 Trevisan, A.; Realdi, G.; Alberti, A.; Noventa, F.: Relationship between membrane-bound immunoglobulins and viral antigens in liver cells from patients with hepatitis B virus infection. Gastroenterology *77:*209–214 (1979).

4 Leevy, C.M.; Popper, H.; Sherlock, S.: Diseases of the liver and biliary tract: standardization of nomenclature, diagnosic criteria and diagnostic methodology, p. 9 (Year Book, Chicago 1976).

5 Facchini, A.; Stefanini, G.F.; Bernardi, M.; Miglio, F.; Gasbarrini, G.; Labo', G.: Lymphocytotoxicity test against rabbit hepatocytes in chronic liver disease. Gut *19:* 189–193 (1978).

6 Miglio, F.; Stefanini, G.F.; Bernardi, M.; Facchini, A.; Mariani, E.; Gasbarrini, G.: Lymphocytotoxicity in acute and chronic alcoholic liver disease: in vitro study on effector cells. Gastroentérol. clin. biol. *3:*531–536 (1979).

7 Moretta, L.; Webb, S.R.; Grossi, C.E.; Lydyard, P.M.; Cooper, M.D.: Functional analysis of two human T-cell subpopulations : help and suppression of B cell responses by T cells bearing receptors for IgM (T_M) or IgG (T_G). J. exp. Med. *146:* 184–192 (1977).

8 Ferrarini, M.; Moretta, L.; Arbile, R.; Durante, M.L.: Receptors for IgG molecules on human lymphocytes forming spontaneous rosettes with sheep red cells. Eur. J. Immunol. *5:*70–75 (1975).

9 Ford, W.L.: The preparation and labelling of lymphocytes; in Weir Handbook of experimental immunology, vol. 23, pp. 1–22 Blackwell, Oxford (1978).

10 McFarlane, I.G.; Wojcicka, B.; Zucker, G.M.; Eddleston, A.L.W.F.; Williams, R.: Purification and characterization of human liver-specific membrane lipoprotein (LSP). Clin. exp. Immunol. *27:*381–390 (1977).

11 Meliconi, R.; Baraldini, M.; Stefanini, G.F.; Facchini, A.; Miglio, F.; Bortolotti, F.; Alberti, A.; Realdi, G.; Amoroso, P.: Antibodies against human liver-specific protein (LSP) in acute and chronic viral hepatitis types A, B and non-A, non-B. Clin. exp. Immunol. *46:*382–390 (1981).

12 Meliconi, R.; Miglio, F.; Stancari, M.V.; Baraldini, M.; Stefanini, G.F.; Gasbarrini, G.: Hepatocyte membrane-bound IgG and circulating liver-specific autoantibodies in chronic liver disease: relation to HBV serum markers and liver histology. Hepatology (in press).

13 Trevisan, A.; Realdi, G.; Noventa, F.; Zotti, S.; Meliconi, R.: Antigenic alterations of liver cell membrane in alcoholics. Lancet *ii:* 1285 (1981).
14 Trevisan, A.; Realdi, G.; Alberti, A.; Ongaro, G.; Pornaro, E.; Meliconi, R.: Core antigen-specific immunoglobulin G bound to the liver cell membrane in chronic hepatitis B. Gastroenterology *82:* 218–222 (1982).
15 Hopf, U.; Meyer zum Büschenfelde, K.H.; Dietrich, M.P.: Demonstration of binding sites for IgG Fc and the third complement component (C3) on isolated hepatocytes. J. Immun. *117:* 633–645 (1976).
16 Krosgaard, K.; Tage-Jensen, J.; Wantzin, P.; Aldershvile, J.; Hardt, F.: Localisation of immunoglobulin on the liver cell surface in primary biliary cirrhosis. J. clin. Path. *34:* 1076–1079 (1981).
17 Trevisan, A.; Realdi, G.; Alberti, A.: Membrane-bound IgG in hepatitis B; in Bianchi, Gerok, Sickinger, Stalder, Virus and the liver, pp. 97–103 (MTP Press, New York 1980).
18 Alter, J.J.; Seef, L.R.; Kaplan, P.M; McAuliffe, V.J.; Wright, E.C.; Gerin, J.L.; Purcell, R.H.; Holland, P.V.; Zimmerman, H.J.: Type B hepatitis : the infectivity of blood positive for 'e' antigen and DNA polymerase after accidental needlestick exposure. New Engl. J. Med. *295:* 909–913 (1976).
19 Baraldini, M.; Facchini, A.; Miglio, F.; Ramadori, G.; Alberti, A.; Realdi, G.: Radioimmunoassay for hepatitis B 'e' antigen (HBeAg) and antibody (anti-HBe): correlations with viral replication and prognostic value. Vox Sang. *41:* 139–146 (1981).
20 Mieli Vergani, G.; Vergani, D.; Jenkins, P.J.; Portmann, B.; Mowat, A.P.; Eddleston, A.L.W.F.; Williams, R.: Lymphocyte cytotoxicity to autologous hepatocytes in HBsAg-negative chronic active hepatitis. Clin. exp. Immunol. *38:* 16–21 (1979).
21 Eddleston, A.L.W.F.; Mondelli, M.; Mieli Vergani, G.; Williams, R.: Lymphocyte cytotoxicity to autologous hepatocytes in chronic hepatitis B virus infection. Hepatology *2:* 1225–1275 (1982).
22 Thomson, A.D.; Cochrane, A.M.G.; McFarlane, I.G.; Eddleston, A.L.W.F.; Williams, R.: Lymphocyte cytotoxicity to isolated hepatocytes in chronic active hepatitis. Nature, Lond. *252:* 721–722 (1974).

Prof. G. Gasbarrini, MD, III Patologia Medica, Università di Bologna,
Policlinica S. Orsola, Via Massarenti 9, I-40138 Bologna (Italy)

Front. gastrointest. Res., vol. 8, pp. 106–116 (Karger, Basel 1984)

Hepatic Fibroplasia: An Evaluation of Tissue and Serological Parameters[1]

E. Pisi, F.B. Bianchi, G. Ballardini, A. Faccani, M. Fallani, G. Biagini, G. Cenacchi

Istituti di Patologia Medica 1 e di Microscopia Elettronica Clinica, Università di Bologna, Bologna, Italy

Introduction

In recent years a number of studies have been focussed on hepatic fibrosis which marks the point of no return of chronic liver disease, the assumption being that fibrosis leads to alterations of liver cell metabolism and acinar blood flow [1].

Liver fibroplasia has been studied, evaluating both tissue and serological parameters. Tissue parameters, although specific for the liver, require invasive techniques (biopsy) and can be biased by sampling errors. On the other hand, serological (and urinary) parameters, which can be studied with non-invasive procedures, are theoretically representative of the whole body.

Although a variety of cells normally present in the liver have been viewed as possible sources of both collagens and glycoproteins of the extracellular matrix, attention has been concentrated on Ito cells (Ic) or lipocytes, hepatocytes and Kupffer cells [2]. Several studies concerning qualitative and quantitative alterations of the genetic types of collagen (type I and III) have been reported both in experimental and human liver fibrosis [3]. On the contrary, few data are available on the distribution of basement membrane proteins (namely fibronectin, laminin and type IV collagen) [4, 5].

An indirect method for studying fibrogenesis is to measure products of collagen metabolism in serum and urine. Since type III collagen has been shown to appear early in the collagen deposition within

[1] Supported by the Ministero della Pubblica Istruzione, Roma, Italy, 1982.

the liver, a sensitive radio immunoassay (RIA) has been developed in order to measure serum type III aminoterminal procollagen peptide (SP III) in acute and chronic liver disease [6]. Other metabolic products, such as urinary and serum hydroxyproline and urinary glycosminoglycans, have been also investigated [7, 8].

Here are reported our data, collected over the past 2 years, in an attempt to define: (1) the cells involved in collagen synthesis, using an experimental model of liver fibrosis [9]; (2) the distribution of fibronectin, laminin and type IV collagen in experimental and human fibrosis; (3) the behaviour of SP III in primary biliary cirrhosis (PBC) and the influence of steroid treatment on SP III in chronic active liver disease (CALD); (4) the relevance of urinary hydroxyproline and glycosaminoglycans in the clinical evaluation of chronic liver disease.

Materials and Methods

Cells Involved in Collagen Synthesis

Liver fibrosis was induced in rats by intraperitoneal injections of heterologous serum [9]. The methods employed have already been reported [10, 11]. Briefly, the following parameters were evaluated by stereological techniques: volume density (VD) of fibrous septa and of lobular, septal and total Ic and Ic index. Samples of each liver were also prepared for electron microscopic (EM) studies, the determination of hydroxyproline and the evaluation of fibronectin, laminin and type IV collagen by immunofluorescent techniques (see below).

Basement Membrane Proteins

Liver biopsies were obtained from patients with chronic persistent hepatitis (5 cases), chronic active hepatitis (5 cases, 3 positive for HBsAg) and active cirrhosis (11 cases, 4 positive for HBsAg). 2 surgical samples from patients operated for gall-stones and 3 needle biopsies from HBsAg healthy carriers were used as controls.

Biopsies were snap frozen in isopentane, pre-cooled with liquid nitrogen; 5 μm cryostat were exposed to affinity chromatography-purified rabbit anti fibronectin, laminin and type IV collagen antibodies. The second antibody was a TRITC- or FITC-conjugated swine anti-rabbit Ig (Dakopatts SA, Denmark) at a dilution of 1:20. Pre-immune rabbit IgG and phosphate-buffered saline were used as controls and showed no reaction in any of the specimens. Rat liver samples were processed in an identical manner. Fluorescence microscopy was done with a Leitz Orthoplan microscope with vertical illumination. Liver samples were also prepared for EM studies.

Evaluation of SP III

SP III was determined using a recently developed RIA test (Procollageno III, Hoechst), in sera stored at –20 °C, from 2 groups of patients: (1) 32 patients with PBC at various stages of the disease; (2) 35 patients with CALD: 30 were receiving steroid treatment and 5, suffering from peptic ulcer (2 cases) and reduced glucose tolerance (3 cases), were untreated and therefore were used as controls. In the 5 untreated and 13 out of the 30 treated patients, 9–20 serum samples (median 13) were tested, corresponding to a follow-up of 25–56 months (median 43). 36 blood donors were used as controls, normal values ranging from 7.7 to 16.1 ng/ml ($\bar{x} \pm 2SD$).

Urinary Hydroxyproline and Glycosaminoglycans

They were measured in 24-hour urine samples, collected during a collagen-free diet. Total hydroxyproline and high molecular peptide-bound hydroxyproline were determined according to *Woessner* [12]. High molecular peptides were precipitated according to *Blumenkrantz and Absoe-Hansen* [13]. Glycosaminoglycans were measured according to *Bitter and Muir* [14]. 20 patients with chronic liver diseases of various aetiology – alcoholic, hepatitis B virus (HBV)-related, cryptogenic and PBC – and 14 patients hospitalized for non-organic disorders were studied.

Results

Cells Involved in Collagen Synthesis

Fibrosis was first observed after 10 injections. In this phase fibrotic changes consisted of large strands of connective tissue which outlined the periphery of zone 3 of the Rappaport acinus. The main cellular component was represented by fuse-shaped elements with various fat droplets which were positive for vitamin A autofluorescence. Apart from an elongated shape, these cells shared most of the morphological features of lobular Ic. After 20 and 40 injections fibrous septa were narrower and mainly composed of scanty cells with rare fat droplets and no vitamin A autofluorescence.

Morphometric evaluation showed that early septa were mainly composed of Ic (VD of 38.75%) while they only accounted for a 1% VD in late septa. The VD of total Ic showed a 3-fold increase after 10 injections (5.39%) with respect to saline-injected rats (1.75%). After 20 and 40 injections the VD of total Ic was lower than that of controls (1.32–1.33%). Tissue hydroxyproline, normal or slightly elevated after 5 and 10 injections (51–69 μg/100 mg dry weight) showed a 5-fold increase after 20 and 40 injections (233–290 μg/100 mg dry weight).

At the EM level cells with most of the distinctive features of Ic were found within early septa. The subplasmalemmal portion of their cytoplasm showed bundles of thin filaments parallel to the main cellular axis, with focal densities (fig. 1a, b). On the basis of their morphology these cells were regarded as myofibroblasts. In late septa the fibrous components became prominent, while the cellular component decreased and was represented by scanty typical fibroblasts.

Distribution of Basement Membrane Proteins in Experimental and Human Fibrosis

In both rat and human normal liver, fibronectin was detectable along the sinusoids and in the portal areas, where it showed a network pattern in the stroma and surrounded blood and biliary vessels.

In experimental fibrosis a positive reaction was present at the level of early fibrous septa. This reaction was stronger than that of other positive structures (sinusoids, residual portal tracts and blood vessels), which reacted as in the controls. The late fibrous septa still showed a positive reaction for fibronectin; however, this was much weaker than that of early septa (fig. 1c, d).

In human fibrosis a strong positive reaction for fibronectin was detectable within the altered portal tracts, the septa and the periportal areas with piecemeal necrosis, where the enlarged sinusoids were filled with material which was strongly positive for fibronectin (fig. 1e, f).

Laminin was strongly positive around the blood and biliary vessels, while no reaction in the sinusoids of normal liver was detectable. Structures positive for laminin were increased in chronic active hepatitis and cirrhosis. A linear positivity was often seen in sinusoids adjacent to fibrous septa, residual portal tracts and at the periphery of cirrhotic nodules. Laminin-positive structures projected with multiple stretches into the residual parenchima (fig. 2a, b).

The distribution of type IV collagen was much the same as that of laminin both in normal and fibrotic (rat and human) liver, the difference being that normal sinusoids were positive for type IV collagen. The positivity increased in the laminin-positive sinusoids (fig. 2c, d).

At the EM level a basement membrane was often seen in the wall of peripheral sinusoids of cirrhotic nodules where laminin and type IV collagen-positive material was detectable using immunohistochemical techniques (fig. 2e, f).

Biopsies from patients with chronic persistent hepatitis showed a normal distribution pattern for the basement membrane proteins studied, with the exception of a weak increase in the fibronectin positivity of the stromal matrix in the enlarged portal tracts.

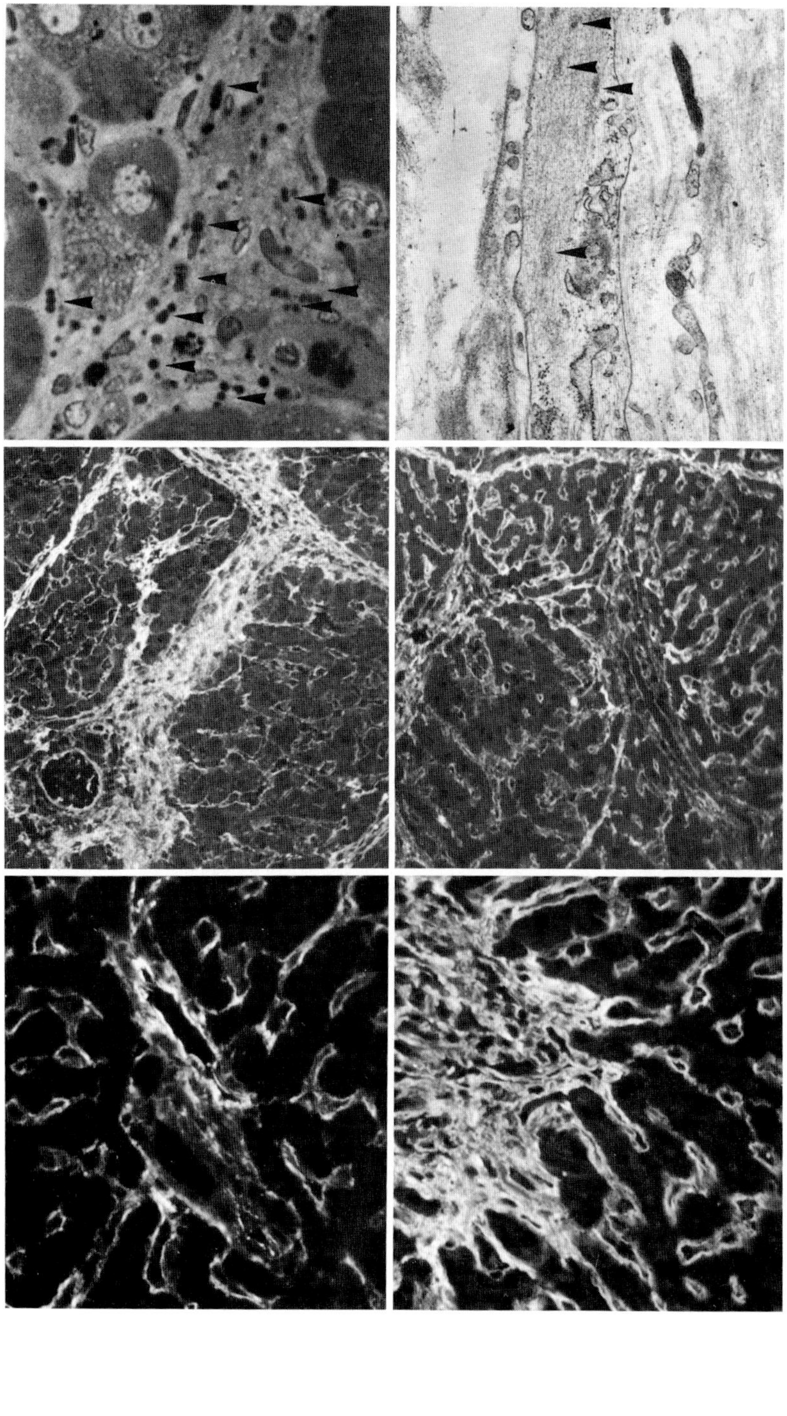

a *b* *c* *d* *e* *f*

Evaluation of SP III

PBC Cases. The sera of 32 patients with definite or probable PBC, classified using defined criteria [15], were tested. PBC cases were divided into 2 groups on the basis of liver histology: group 1 (stages 1,2) and group 2 (stages 3,4). SP III values were significantly higher in the PBC sera (median 20, range 10.5–47.3 ng/ml) with respect to the controls (median 11.5, range 8–16 ng/ml), while no difference was found between groups 1 (median 18.9, range 14–47.3 ng/ml) and 2 (median 23.5, range 10.5–39.9 ng/ml).

14 out of the 16 group 1 patients (87%) and 11 out of the 16 group 2 patients (69%) had SP III values over the upper 95% confidence limits for the controls. No clear-cut correlation was present between SP III values, possibly implying fibrogenesis, and histological stage and fibrosis.

Effect of Steroid Treatment. Treated patients had a SP III median value of 21.9 ng/ml (range 11.2–56.8) before treatment, 16 (9–32.1) 6 months later, 12 (10–40) a year later and of 12 (9–22) in the last control during treatment, 39 (19–56) months later. Untreated patients had corresponding values of 22 (16–25), 22 (22–25), 19.5 (16–28) and 18 (14–25).

Before treatment SP III values were above the normal range in 25 of the 30 cases (83%); 6 months later it was normal in 14 and decreased in 9.

Treatment was suspended in 7 out of the 13 patients with follow-up: 6 showed a SP III median increase (from 12 to 22 ng/ml); 4 of the 7 patients were treated again with steroids and this was followed by a further reduction of SP III (from 21.5 to 14 ng/ml).

Fig. 1. a, c, d Rat liver; *b, e, f* human liver. *a* Arrows point to Ic in an early septum of experimental liver fibrosis. 1-μm section. × 900. *b* Portion of cytoplasm of a myofibroblast in chronic active hepatitis. Arrows point to focal densities within a bundle of filaments. ×20,000. *c, d* Fibronectin in early *(c)* and late *(d)* septa of experimental fibrosis. ×90. *e, f* Fibronectin in a normal liver *(e)* and an area of piecemeal necrosis *(f)*. ×225.

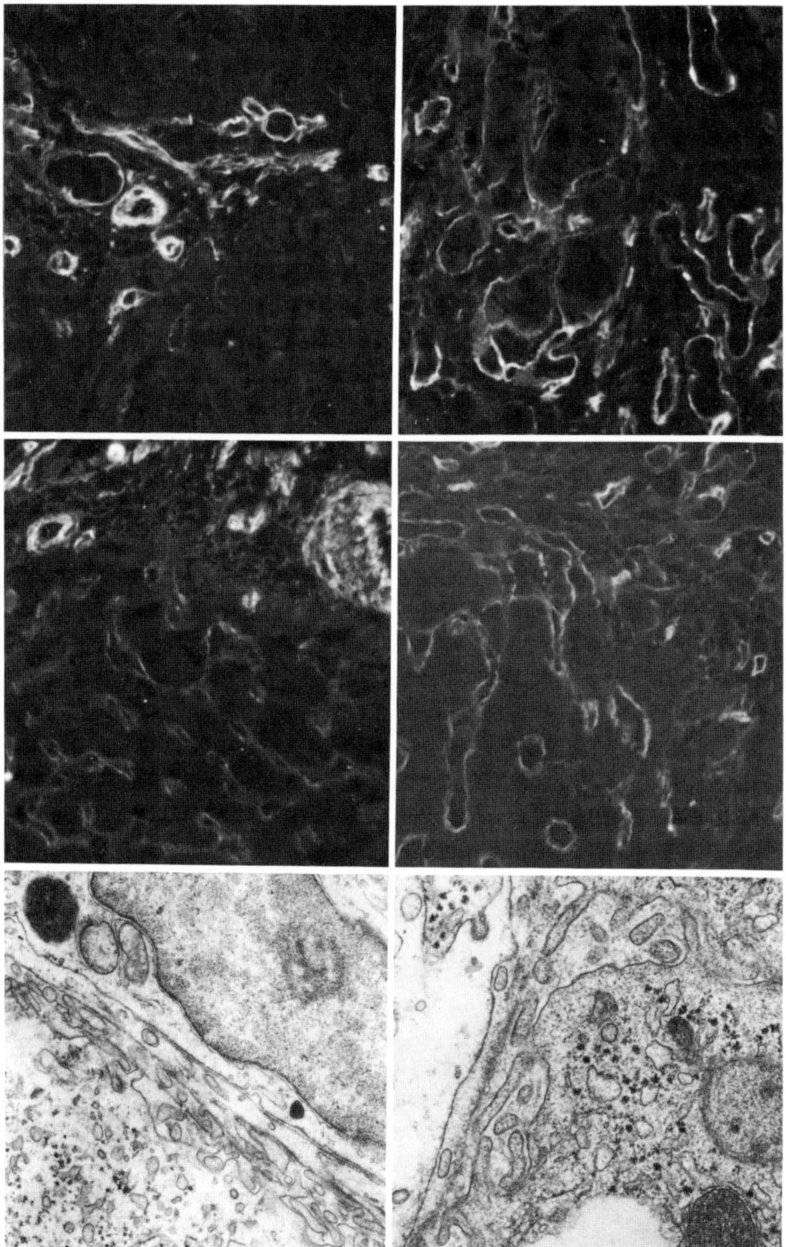

a b

c d

e f

Urinary Parameters

Total hydroxyproline was increased in chronic liver patients (median 23.2, range 11.1–44.8 mg/m/day) compared to controls (17.3; 10.9–37 mg/m/day), with no statistical difference.

High molecular peptide hydroxyproline was significantly higher in chronic liver disease patients (7; 2.7–16.4) compared to controls (4.6; 2.6–8.4) and was above the upper value of normal in 8 out of the 20 patients. Glycosaminoglycans were significantly increased in liver patients (3.4; 1.4–5.6 mg/m/day) with respect to the controls (2; 1.2–4).

Discussion

The role of Ic in hepatic fibrogenesis has already been suggested [16]. The above results are in agreement with the hypothesis that Ic are resting fibroblasts which, if properly triggered, can modulate to active fibroblasts. This is also supported by the sequential behaviour of tissue hydroxyproline in the model studied: its increase is delayed with respect to the peak of septal Ic, actually corresponding to the sharp fall of septal Ic. EM observations also suggest that myofibroblasts may represent a morpho-functional modulation of Ic.

It is well-known that the fibroblast shape may change in relation to the fibronectin contents of the plasma membrane and that fibronectin exerts a chemiotactic effect on these cells [17, 18]. Our data point to fibronectin, whatever the cell involved in its synthesis, as the trigger able to recruit and transform Ic. In human fibrosis, fibronectin is largely present at the level of active septa and piecemeal necrosis. On the basis of both experimental and human studies, tissue fibronectin must be regarded as a marker of the early events of liver fibrogenesis.

The sinusoidal wall of rat and human normal liver is negative for laminin. Sinusoids positive for laminin have been observed at the periphery of nodules in alcoholic cirrhosis [4]. Our studies extend these

---

Fig. 2. a, b Pattern of laminin distribution in normal human liver *(a)* and chronic active hepatitis (*b*, area of piecemeal necrosis). ×225. *c, d* Type IV collagen in areas as in *a* and *b*. ×225. *e, f* 'Capillarized' sinusoids of the periphery of cirrhotic nodules. *e* ×13,000; *f* ×16,000.

observations to non-alcoholic CALD where sinusoids positive for laminin can be found near the fibrous secta and in areas of peripheral piecemeal necrosis. The close correlation between laminin positivity and appearance of a defined basement membrane at the EM level (sinusoidal 'capillarization') must be emphasized and suggests that laminin might play an important role in the morphological organization of the basement membrane. Sinusoidal capillarization must be regarded as a negative event in terms of blood flow and metabolic impact at the level of cirrhotic nodules. Type IV collagen is codistributed with laminin, the difference being that it is present in the sinusoidal lining of normal rat and human liver.

Useful clinical information about active liver fibrogenesis is provided by serological and urinary parameters. The specificity of a sensitive serological test, such as the determination of type III procollagen aminoterminal peptide, is under evaluation. SP III values above the normal range have been found both in early and late stages of PBC and this suggests that fibrogenesis could be triggered by immunological agents, possibly the same (T lymphocytes?) as those involved in the biliary lesion. These data also show that fibrogenesis is not closely related to the presence of histological fibrosis.

Elevated SP III values are clearly effected by steroid treatment in patients with CALD. It is, however, difficult to establish whether this is due to a direct steroid effect on prolyl-hydroxylase or if it is secondary to a decrease of inflammation and/or liver cell necrosis. The short response rates observed seem to point to the second hypothesis.

Other parameters, such as urinary high molecular peptide-bound hydroxyproline, could theoretically provide further information. However, the low sensitivity of the urinary tests in liver disease impairs their clinical use.

Summary

Data on the behaviour of tissue and serological parameters of fibroplasia, collected from studies in an experimental model of liver fibrosis, and liver biopsies, serum and urine from patients with chronic liver disease are reported.

The relevance of Ito cells as fibroblast precursors has been investigated in experimental fibrosis. Fibronectin is largely present in sites of early fibroplasia both in human and rat liver. Laminin, absent in normal sinusoidal lining, appears in the sinusoids of cirrhotic nodules. Its presence is closely associated to a well-defined basement membrane at the electron microscopic level. Therefore, laminin seems to be a reliable marker of sinusoidal capillarization.

Serum type III procollagen aminoterminal peptide, as determined by a radio-immunoassay test, is elevated in primary biliary cirrhosis independently of the histological stage of the disease. In chronic active liver disease, steroid treatment clearly lowers its levels. Urinary tests seem to add little information because of their low sensitivity.

References

1 Rojkind, M.; Ponce-Noyola, P.: The extracellular matrix of the liver. Collagen. Rel. Res. *32:*151–175 (1982).

2 Popper, H.: Summary and outlook of the International Symposium on Fibrosis and the liver. Ital. J. Gastroenterol. *12:*116–120 (1980).

3 Rojkind, M.; Dunn, M.A.: Hepatic fibrosis. Gastroenterology *76:*849–863 (1979).

4 Hahn, E.; Wick, G.; Pencey, D.; Timpl, R.: Distribution of basement membrane proteins in normal and fibrotic human liver: collagen type IV, laminin and fibronectin. Gut *21:*63–71 (1980).

5 Voss, B.; Rauterberg, J.; Allam, S.; Pott, G.: Distribution of collagen type I and type III and of two collagenous components of basement membranes in the human liver. Path. Res. Pract. *170:*50–60 (1980).

6 Rohde, H.; Hähn, E.; Timpl, R.: Radioimmunoassay for aminoterminal procollagen peptides in liver disease. Fresenius Z. Analyt. Chem. *290:*151–152 (1978).

7 Resnick, R.H.; Cerda, J.C.; Boitnott, J.; Aron, J.; Iber, F.: Urinary hydroxyproline excretion in hepatic disorders. Am. J. Gastroent., N.Y. *60:*576–583 (1973).

8 Ballardini, G.; Biolchini, L.; Faccani, A.; Degli Esposti, S.; Bianchi, F.B.; Pisi, E.: idrossiprolina non dializzabile urinaria e glicosoaminoglicani ad alto peso molecalare come parametri di fibroplasia nella cirrosi epatica. Ital. J. Gastroenterol. *13:* suppl. F74 (1981).

9 Paronetto, F.; Popper, H.: Chronic liver injury induced by immunologic reactions. Cirrhosis following immunization with heterologous sera. Am. J. Path. *49:* 1087–1101 (1966).

10 Ballardini, G.; Degli Esposti, S.; Bianchi, F.B.; Badiali De Giorgi, L.; Faccani, A.; Biolchini, L.; Busachi, C.A.; Pisi, E.: Correlation between Ito cells and fibrogenesis in an experimental model of hepatic fibrosis. A sequential stereological study. Liver (in press).

11 Cenacchi, G.; Ballardini, G.; Badiali De Giorgi, L.; Busachi, C.A.; Del Rosso, M.; Bianchi, F.B.; Biagini, G.; Laschi, R.: Relationship between connective tissue cells and fibronectin in a sequential model of experimental hepatic fibrosis. Virchows Arch. Abt. B. Zellpath. (in press).

12 Woessner, J.F., Jr.: The determination of hydroxyproline in tissue and protein samples containing small proportions of this imino acid. Archs. Biochem. Biophys. *93:* 440–447 (1961).

13 Blumenkrantz, N.; Absoe-Hansen, G.: High molecular collagen peptide fraction in urine indicates disease activity in generalized scleroderma. Acta derm. vener., Stockh. *56:*415–421 (1976).

14 Bitter, T.; Muir, H.R.: A modified uronic acid carbazole reaction. Analyt. Biochem. *4:*330–334 (1962).

15 Taal, B.G.; Schalm, S.W.; Tenkate, F.W.J.; Feltkamp, T.E.W.; Geertzen, H.G.M.; Hermens, J.: A sensitive and specific system for the diagnosis of primary biliary cirrhosis, based on objective criteria. Abstr. 16 h Meet. EASL, Lisbon 1981, p. 85.

16 McGee, J.O.; Patrick, R.: The role of perisinusoidal cells in hepatic fibrogenesis. An electron microscopic study of acute carbon tetrachloride injury. Lab. Invest. *26:* 429–440 (1972).

17 Hynes, R.O.; Destree, A.T.; Perkins, M.E.; Wagner, D.D.: Cell surface fibronectin and oncogenic transformation. J. supramol. Struct. *11:*95–104 (1979).

18 Rennard S.I.; Hunninghake, G.W.; Bitterman, P.B.; Crystal, R.G.: Production of fibronectin by the human alveolar macrophage: mechanism for the recruitment of fibroblasts to sites of tissue injury in interstitial lung diseases. Proc. natn. Acad. Sci. USA *78:*7147–7151 (1981).

Prof. E. Pisi, Istituto di Patologia Medica 1, Policlinico S. Orsola,
V. Massarenti 9, I-40138 Bologna (Italy)

Front. gastrointest. Res., vol. 8, pp. 117–127 (Karger, Basel 1984)

Mononuclear Cell-Mediated Activation of Fibroblast Collagen Synthesis in Chronic Active Hepatitis[1]

C. Surrenti, A. Casini, S. Nieri, G. Salvadori, A. Calabrò, S. Ambu, E. Banchetti, P. Ceccatelli

Cattedra di Gastroenterologia, Università di Firenze, Florence, Italy

Introduction

Fibrosis is found in many chronic inflammatory liver diseases. Collagen accumulation in the liver may be the result of one of three processes: (a) increased collagen synthesis and deposition with normal collagen degradation; (b) decreased collagen breakdown with maintenance of normal collagen synthesis and deposition at physiologic levels, and (c) a combination of both factors.

Many investigators [1, 2] maintain that an increase in collagen synthesis is caused by specifically sensitized T lymphocytes which induce hepatocellular necrosis with inflammation and fibrogenesis. According to this hypothesis, fibroplasia is due to liver cell necrosis and the consequent inflammatory response. Recent data, however, have demonstrated that human lymphocytes produce lymphokines which are chemotactic for fibroblasts (LD CF-F) [3, 4] and which can enhance fibroblast collagen synthesis [5–9]. These data also show the possibility of interaction between cellular immune reactions and connective tissue cells.

Specifically, collagen accumulation might be the direct consequence of a cell-mediated immune reaction to hepatocyte surface membrane antigen. Among these antigens, the role of the liver-specific protein (LSP), a macrolipoprotein localized on the liver cell membrane [10–12] has been emphasized in the pathogenesis of chronic hepatitis [13–16]. In an attempt to evaluate this possibility, lymphocytes from

[1] This study was supported in part by a grant from M.P.I.

patients with chronic active hepatitis (CAH) were stimulated with LSP to produce active supernatants and the effect of these supernatants on human fibroblast collagen production was evaluated in vitro. Our results show that lymphocytes from patients with CAH triggered with LSP produce soluble factor(s) which enhance human fibroblast collagen synthesis.

Materials and Methods

Patients. 12 patients (5 males and 7 females) with HBsAg negative CAH were studied. The mean age was 53.5 years, ranging from 40 to 60 years. Diagnosis was established on the basis of clinical, laboratory and histological findings. All the patients had a histological picture of active septa. None of the patients was taking corticosteroids or other immunosuppressive drugs at the time of study. 12 healthy subjects, age- and sex-matched, were studied as controls.

Fibroblast Cultures. Human fibroblasts, obtained from normal donor skin explants, were placed in 60-mm Petri dishes and cultured in minimal Eagle's medium (MEM) supplemented with 10% inactivated fetal calf serum (FCS), 2 m*M* glutamine, 100 μg/ml ascorbic acid, 1 μg/ml $FeSO_4 \cdot 7H_2O$, 100 U/ml penicillin and 100 μg/ml streptomycin.

These primary fibroblast cultures, grown to confluency at 37 °C in a humid atmosphere of 5% CO_2 and 95% air, were then subcultured by trypsinization in 75 cm^2 tissue culture flasks (Falcon Div. Becton Dickinson and Co., Oxnard, Calif., USA). Only fibroblast cultures between 5 and 8 passages were used for the in vitro collagen production test in order to obtain a regular cell growth and to avoid variations due to senescence of the cultures.

LSP Preparation. Human LSP, isolated from normal donor liver homogenates and purified according to the method of *McFarlane* et al. [12], was kindly provided by Dr. *Eddleston* and Prof. *Gasbarrini.* The LSP-containing solution was dialysed overnight against TC 199 medium just before use in order to remove ethylenediaminetetraacetic acid (EDTA).

Preparation of Lymphocyte Supernatants. Lymphocytes were isolated from heparinized peripheral blood of patients and normal donors by centrifugation on a Ficoll-Hypaque gradient according to standard techniques [17]. Cells, containing no more than 10% of monocytes and no more than 0.5% granulocytes, were then resuspended in TC 199 medium supplemented with 10% heat-inactivated normal donor pooled AB serum, 100 U/ml penicillin and 100 μg/ml streptomycin at a concentration of 2×10^6 cells/ml. 1 ml of the cell suspension was placed in culture tubes with 1 ml of medium containing 20 μg of LSP or 1 ml of medium only as a control, to obtain a final concentration of 1×10^6 cells/ml. The tubes were incubated for 120 h at 37 °C in a humid atmosphere with 5% CO_2 in air. 16–20 h before the end of the culture period, the lymphocyte supernatants were removed by centrifugation, passed through millipore filters of 0.22-μm pore size to discard cellular debris, and stored at −70 °C for less than 1 week. The culture pellets were resuspended in 2 ml of medium and pulse-labelled with 5 μCi of (methyl-^{3}H)-thymidine (The Radiochemical Centre, Amersham, UK). The tubes were then washed three times with 10% trichloroacetic acid (TCA) and radioactivity was determined in the resulting precipitates with a Packard liquid scintillation counter (β counter).

Collagen Assay. Cell-free lymphocyte supernatants were dialysed overnight against MEM and then added at varying concentrations to non-confluent fibroblast cultures in 75 cm^2 tissue culture flasks (Falcon) in MEM enriched as previously described. The following concentrations of lymphocyte supernatant were used: 1:1, 1:2, 1:4, 1:8, 1:16.

Fibroblast cultures were also tested with supernatants from unstimulated patient lymphocytes, supernatants from LSP non-responder normal donor lymphocytes and with LSP only. Fibroblasts grown in basal conditions (medium only) were used as controls.

The treated cultures were incubated for 20 h at 37 °C in a humid atmosphere with 5% CO_2 in air and then 5 μCi of *L*-(3,4^{3}H)-proline (The Radiochemical Centre, Amersham, UK) was added to each flask for 6 h. At the end of the culture period, fibroblast supernatants were removed and collected. Cell monolayers were solubilized with 0.5 *N* NaOH, washed 3 times with warm Hank's balanced salt solution and then collected.

The supernatants and solubilized monolayers were processed by 5% TCA −0.25% tannic acid precipitation of proteins and then hydrolysed with a 6 *N* HCl solution overnight at 130 °C. ^{3}H-Hydroxyproline in the sample was assayed according to the method of *Juva and Prockop* [18] and considered as an index of fibroblast collagen synthesis.

A new method for hydroxyproline (Hyp) assay in fibroblast cultures has been subsequently performed, using a high pressure liquid chromatography (HPLC) procedure [19].

Apparatus. The HPLC system (Perkin-Elmer) consisted of a model Series 3 solvent-delivery system, a Rheodyne 7120 injection valve with a 10-μl injection loop, a model LC-100 column oven and a Model LC-55B variable wavelength UV detector operating at 486 nm. A reversed-phase RP-8 column (Brownlee; 25 cm × 4.6 mm ID, particle size 10 μm) was used. All chromatograms were recorded on a Perkin-Elmer Model 56 recorder. The mobile phases were: (A) acetonitrile, 1.66 m*M* acetic acid and 6.99 m*M* orthophosphoric acid, pH 3.5; (B) 10 m*M* sodium acetate, 1.66 m*M* acetic acid, pH 3.00, adjusted with orthophosphoric acid [20]. Other conditions were: flow-rate, 1.5 ml/min; pressure, 3.5–4 MPa; column temperature, 40 °C. Solvent programme: T_1 = 10 min (30–34% solvent A), gradient curve 1; T_2 = 10 min (34–60% solvent A), gradient curve 0.3.

Reagents. *L*-4-^{3}H(G)-Hydroxyproline (^{3}H-Hyp), specific activity 5.4 Ci/mmol, was purchased from New England Nuclear (Boston, Mass., USA). Dabsyl chloride, 3.3 mg/ml in acetone, was obtained from Pierce (Rockford, Ill., USA). *N*-Methyl-taurine (99% purity according to a liquid chromatographic assay), kindly supplied by Prof. *Adembri*, was used as the internal standard. Acetonitrile LiChrosolv, highly purified Hyp and other standard amino acids were obtained from E. Merck (Darmstadt, FRG). Solvents and other chemicals were of AR grade.

Procedure

Fibroblast Cultures. Human fibroblasts, obtained from normal donor skin explants, were cultured as described above in medium with and without ascorbic acid. The supernatants were collected. Cell monolayers were solubilized with a 0.5-*N* NaOH solution, washed 3 times with warm Hank's balanced salt solution and collected. Supernatants and solubilized cell monolayers were then hydrolysed (3 *N* HCl, overnight at 130 °C). The hydrolysates were neutralized at pH 8.9 with 4 *M* K_2CO_3.

Dabsylation. The dabsylation reaction was performed as previously described [21]. The reaction mixture consisted of 20 μl of 0.1 *M* Na_2CO_3-$NaHCO_3$ buffer solution, pH 8.9, containing 6.2 μmol/ml of internal standard and 20 μl of dabsyl chloride solution. The reaction was allowed to proceed at 70 °C for 6 min with constant shaking. A 10-μl volume of the reaction mixture was injected into the chromatograph.

Calibration Curve. Standard solutions of Hyp, ranging from 0.19 μmol/ml, were processed as described above. The calibration curve was obtained by calculating the ratio between the peak heights of Hyp and the internal standard and plotting these ratios against the concentrations. This curve was used to convert peak height ratios of unknown samples into Hyp concentration.

Results

LSP significantly enhanced ^{3}H-thymidine uptake by patient lymphocytes (6969 ± 711.7 cpm) as compared to controls (2570 ± 293.7 cpm), while this effect was not present in the normal donor lymphocyte cultures (578 ± 66.8 cpm; controls 728.5 ± 45.3).

Supernatants from LSP-activated patient lymphocytes produced a significant increase ($p < 0.0005$) in fibroblast collagen production as compared to controls. The same significant increase in collagen synthesis was observed when LSP-activated supernatants were compared with the supernatants from unstimulated patient lymphocytes, from LSP non-responder normal donor lymphocytes and with the supernatants containing LSP only (fig. 1).

Maximum stimulation was observed when fibroblasts were incubated with supernatants at a 1:4 dilution. Further dilutions resulted in a lower stimulatory effect; undiluted supernatants and the 1:2 dilution produced a slightly lower stimulatory effect, but the difference was not statistically significant (fig. 2). The enhancement of skin fibroblast collagen accumulation was particularly evident in the supernatant of the cell cultures.

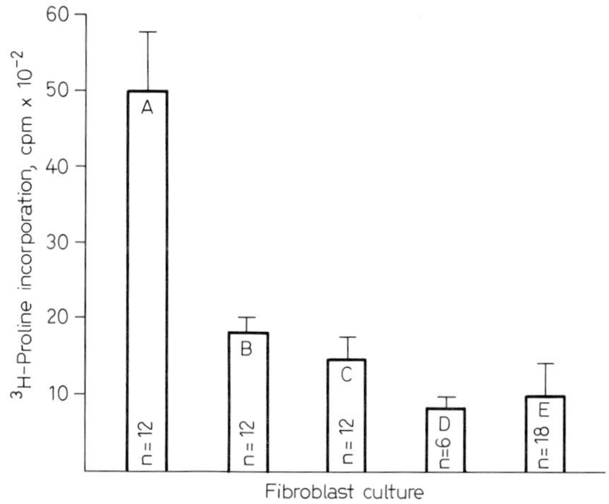

Fig. 1. Effect of lymphocyte supernatants upon collagen production by human skin fibroblast cultures. Data represent single values and the mean ± SD of total ^{3}H-proline incorporation into collagen by fibroblast cultures stimulated with a 1:4 dilution of the lymphocyte supernatants. n = number of cases. A = LSP + patients' sensitized mononuclear cells; B = unstimulated patients' mononuclear cells; C = LSP + normal donor 'non-responder' mononuclear cells; D = LSP only; E = medium only.

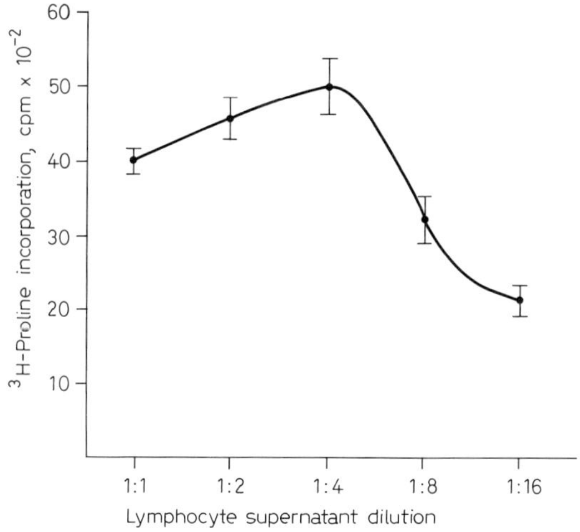

Fig. 2. Effect of varying concentrations of LSP-activated mononuclear cell supernatants upon fibroblast ^{3}H-proline incorporation. Data represent the mean value of triplicate experiments.

Finally, the supernatants obtained from unstimulated patient lymphocytes, from LSP non-responder normal donor lymphocytes and those containing LSP only did not significantly increase fibroblast ^{3}H-proline incorporation when compared to controls.

HPLC Determination of Hyp. Figure 3 shows a typical chromatogram of biological samples (supernatants + cell monolayers from fibroblast cultures) processed by the method described. It can be seen that a good separation of dabsyl chloride, *N* methyltaurine and Hyp was obtained. The retention times of the internal standard and Hyp were 8 and 15.2 min, respectively. Dabsylated standard Hyp produced a single peak.

The Hyp peak was identified on the basis of its absolute and relative retention time and by adding known amounts of standard Hyp to the samples. It was possible to identify the peaks immediately before and after Hyp by the same procedure.

In order to better identify the Hyp peak, ^{3}H-Hyp was also added to the samples at various concentrations and processed by the method described. Throughout this procedure a linear correlation ($r = 0.997$) was obtained between the single amounts of ^{3}H-Hyp added to the samples and the radioactivity found in the fractions corresponding to the Hyp peak collected from the column.

Furthermore, no radioactivity was recovered in the fractions collected before and after the one corresponding to the Hyp peak. This confirmed that Hyp gave a single derivative under the analytic conditions used. Some experiments were performed using a Perkin-Elmer LC 85-Autocontrol variable wavelength detector (equipped with a 2.4 µl flow-cell) which could record and analyse spectra of the single peak eluted from the column. The UV spectrum of the fraction corresponding to the Hyp peak overlapped (absorbance maximum at 194.5 nm) that of standard dabsylated Hyp injected into the column under the same experimental conditions. Further evidence that no other dabsyl-positive compounds might interfere with the Hyp was obtained by using a thin-layer chromatographic (TLC) procedure and a subsequent gas chromatography-mass spectrometry which confirmed the purity of this fraction [19].

Hyp was assayed in fibroblasts cultured with and without ascorbic acid in the medium in order to verify whether the HPLC method was able to measure variations in the imino acid concentration in the cul-

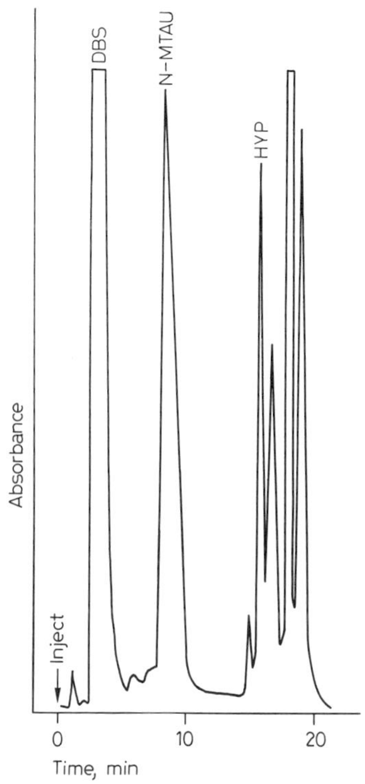

Fig. 3. Representative chromatogram of dabsyl derivative of the biological sample. DBS = Dabsyl chloride; N-MTAU = *N*-methyltaurine: Hyp = hydroxyproline.

tures [22–24]. The experiments were performed in duplicate during a period of 3 weeks. A statistically significant decrease in total Hyp content (from 0.804 ± 0.007 to 0.434 ± 0.008 µmol/ml; $p < 0.001$; n = 5) was observed when cells were cultured in ascorbic acid-free medium.

Discussion

Recent data suggest that hepatocyte damage in CAH may be correlated to a cell-mediated mechanism and particularly to LSP, which is present on the hepatocyte surface [25–29]. In our experiments, mononuclear cells from patients with CAH showed hypersensitivity to LSP and were able to produce soluble agents which can enhance in vitro ^{3}H-proline incorporation by human fibroblast cultures. These factor(s) are stable at –70 °C and are not dialysable.

Since these factor(s) are produced by a non-purified population of peripheral blood mononuclear cells, they may originate from lymphocytes or monocytes. Moreover, ^{3}H-proline incorporation was determined 20 h after lymphokine stimulation, thus the enhancement of collagen production cannot be due only to the increased number of fibroblasts.

Both extracellular and cell-associated collagen were significantly increased compared to controls, but this enhancement was particularly evident in the supernatant. This effect may be due to an active extracellular secretion of procollagen by cultured fibroblasts. Varying concentrations of lymphocyte supernatants were used to stimulate fibroblasts in vitro. The behaviour of the dose-response curve indicates that the lower stimulatory effect of the undiluted lymphocyte supernatants may be the consequence of a cytopathic effect induced by LSP or by lymphokines.

These observations suggest that cellular immune reactions can directly stimulate fibroblast collagen synthesis and may be involved in the development of active fibrosis which accompanies the progression of chronic active hepatitis to cirrhosis. In particular, soluble factors released by mononuclear cells challenged with LSP may activate fibroblast collagen synthesis without recruitment of inflammatory cells. Since hepatic fibrosis occurs in patients with agammaglobulinaemia, B cells may not be essential to the process of collagen deposition [30, 31]. On the other hand, the role of T lymphocytes in the basic process of fibrosis should be emphasized. It has been observed that wound healing is impaired when lymphocytes are absent in the peripheral blood of animals treated with cortisone or submitted to whole-body irradiation [32, 33]. In addition, the 'Hagfish', a primitive arctic species among the surviving vertebrates without a thymus gland and circulating T lymphocytes, lacks the ability to heal its wounds [34, 35].

Further information, however, is required to determine whether an immune response against collagenous components can further amplify the process of collagen deposition.

In conclusion, our findings suggest the possibility that active fibrosis associated with CAH may be directly influenced by factors released during cell-mediated immune reactions to liver-specific antigens.

Furthermore, the aim of the present study was also to produce a rapid and practical method for determining the concentration of Hyp in fibroblast cultures and other biological samples by HPLC with dabsyl derivatization. The specificity of the method was based on: (1) absolute and relative retention times of the imino acid; (2) collection of standard ^{3}H-Hyp in the fractions corresponding to the Hyp peak; (3) enrichment of biological samples by standard preparation of Hyp. The present method has the advantage of not requiring an extraction step. In fact, after hydrolysis of samples, only 6 min are needed for dabsyl derivatization and a further 20 min for chromatographic separation. Furthermore, it has a higher sensitivity than the previously described procedures for Hyp assay. Only a small volume (20 µl) is required for the analysis and, what is more, use of radioactive materials is avoided.

Summary

The present study was performed in order to verify the hypothesis that liver fibrosis is the direct consequence of cellular immunological reactions. The study was carried out using supernatant of lymphocyte culture from 12 patients with HB_sAg negative chronic active hepatitis stimulated by liver-specific protein (LSP). This supernatant was added to human skin fibroblast cultures. Collagen accumulation of the culture was measured by ^{3}H-proline incorporation into collagen. Supernatants from peripheral blood mononuclear cells of the above-mentioned patients enhanced collagen synthesis by human fibroblasts compared with control groups. This stimulatory factor was stable at -70 °C and not dialysable. This study indicates that lymphokines released during interactions between liver specific antigens and lymphocytes may be involved in the pathogenesis of liver fibrosis.

References

1 Popper, H.: Pathologic aspect of cirrhosis. Am. J. Path. *87:*228–258 (1977).
2 Thomas, H.C.: Immunological aspects of hepatic fibrosis. Ital. J. Gastroenterol. *12:* 33–36 (1980).
3 Postlethwaite, A.E.; Snyderman, R.; Kang, A.H.: The chemotactic attraction of human fibroblast to a lymphocyte-derived factor. J. exp. Med. *144:*1188–1203 (1976).
4 Postlethwaite, A.E.; Kang, A.H.: Characterization of guinea pig lymphocyte-derived chemotactic factor for fibroblast. J. Immun. *124:*1462–1466 (1980).
5 Johnson, R.L.; Ziff, M.: Lymphokine stimulation of collagen accumulation. J. clin. Invest. *58:*240–252 (1976).
6 Wahl, S.M.; Wahl, L.M.; McCarthy, J.B.: Lymphocyte-mediated activation of fibroblast proliferation and collagen production. J. Immun. *121:*942–946 (1978).

7 Surrenti, C.; Casini, A.; Ramarli, D.; Fani, A.; Nieri, S.: Correlation between hepatic fibrosis and cell-mediated immunity. Ital. J. Gastroenterol. *14:*1–4 (1982).
8 Casini, A.; Surrenti, C.; Ramarli, D.; Fani, A.; Nieri, S.: Lymphokine stimulation of hepatic fibrogenesis; in Connective tissue of the normal and fibrotic human liver (Thieme, Stuttgart 1982).
9 Surrenti, C.; Casini, A.; Nieri, S.: Immunopathogenesis of liver fibrosis; in Advances in clinical immunology of liver and kidney diseases and therapy (Piccin, Padova 1983).
10 Meyer Zum Buschenfelde, K.H.; Miescher, P.A.: Liver-specific antigens. Purification and characterization. Clin. exp. Immunol. *10:*89–102 (1972).
11 Hopf, U.; Meyer Zum Buschenfelde, K.H.; Freudenberg, J.: Liver-specific antigens of different species. Localization of a membrane antigen at cell surface of isolated hepatocytes. Clin. exp. Immunol. *16:*117–124 (1974).
12 McFarlane, G.; Wojcicka, B.M.; Zucker, G.M.; Eddleston, A.L.W.F.; Williams, R.: Purification and characterization of human liver specific membrane lipoprotein (LSP). Clin. exp. Immunol. *27:*381–390 (1977).
13 Meyer Zum Buschenfelde, K.H.; Kossling, F.K.; Miescher, P.A.: Experimental chronic active hepatitis in rabbits following immunization with human liver proteins. Clin. exp. Immunol. *10:*99–108 (1972).
14 Miller, J.; Smith, M.G.M.; Mitchell, C.G.; Reed, W.D.; Eddleston, A.L.W.F.; Williams, R.: Cell-mediated immunity to a human liver-specific antigen in patients with active chronic hepatitis and primary biliary cirrhosis. Lancet *ii:*296–297 (1972).
15 Hopf, U.; Meyer Zum Buschenfelde, K.H.: Studies on the pathogenesis of the experimental chronic active hepatitis in rabbits. Demonstration of immunoglobulin on isolated hepatocytes. Br. J. exp. Path. *55:*509–513 (1974).
16 Jensen, D.M.; McFarlane, J.G.; Portmann, B.S.; Eddleston, A.L.W.F.; Williams, R.: Detection of antibodies directed against a liver-specific membrane lipoprotein in patients with acute and chronic active hepatitis. New Engl. J. Med. *299:* 1–7 (1978).
17 Boyum, A.: Isolation of leucocytes from human blood. Scand. J. clin. Lab. Invest. *21:*suppl. 97, pp 77–89 (1968).
18 Juva, K.; Prockop, D.J.: Modified procedure for the assay of ^{3}H- or ^{14}C-labelled hydroxyproline. Analyt. Biochem. *15:*77–83 (1974).
19 Casini, A.; Martini, F.; Nieri, S.; Ramarli, D.; Franconi, F.; Surrenti, C.: High-performance liquid chromatographic determination of hydroxyproline after derivatization with 4-dimethylaminoazobenzene-4′-sulphonyl chloride. J. Chromatogr. *249:*187–192 (1982).
20 Schmidt, G.J.; Olson, D.C.; Atwood, J.G.: The determination of nanogram amounts of taurine using liquid chromatography and fluorescence detection. Chromatogr. Neswlett. *8:*13–14 (1980).
21 Lin, J.K.; Chang, J.Y.: Chromophoric labelling of amino acids with 4-dimethylaminoazobenzene-4′-sulphonyl chloride. Analyt. Chem. *47:*1634–1639 (1975).
22 Cardinale, G.J.; Udenfriend, S.: Prolyl hydroxylase. Adv. Enzymol. *45:* 245–300 (1974).
23 Blanck, T.J.; Peterkofsky, B.: The stimulation of collagen secretion by ascorbate as a result of increased proline hydroxylation in chick embryo fibroblasts. Archs Biochem. Biophys. *171:*257–259 (1975).

24 Kao, W.W.Y.; Berg, R.A.; Prockop, D.J.: Ascorbate increases the synthesis of procollagen hydroxyproline by cultured fibroblasts from chick embryo tendons without activation of prolyl hydroxylase. Biochim. biophys. Acta *411:*202–215 (1975).
25 Wands, J.R.; Isselbacher, K.J.: Lymphocyte cytotoxicity to autologous liver cells in chronic active hepatitis. Proc. natn. Acad. Sci. USA *72:*1301–1303 (1975).
26 Thomson, A.D.; Cochrane, M.A.G.; McFarlane, I.G.; Eddleston, A.L.W.F.; Williams, R.: Lymphocyte cytotoxicity to isolated hepatocytes in chronic active hepatitis. Nature, Lond. *252:*721–722 (1974).
27 Kawanishi, H.; McDermott, R.P.: K-cell-mediated antibody-dependent cellular cytotoxicity in chronic active liver disease. Gastroenterology *76:*151–158 (1979).
28 Geubel, A.P.; Keller, R.H.; Summerskill, W.H.J.; Dickson, E.R.; Tomasi, T.B.; Shorter, R.G.: Lymphocyte cytotoxicity and inhibition studied with autologous liver cells: observations in chronic active liver disease and the primary biliary cirrhosis syndrome. Gastroenterology *71:*450–456 (1976).
29 Vogten, A.J.M.; Hadzic, N.; Shorter, R.G.; Summerskill, W.H.J.; Taylor, W.F.: Cell-mediated cytotoxicity in chronic active liver disease: a new test system. Gastroenterology *74:*883–889 (1978).
30 Good, R.A.; Page, A.R.: Fatal complications of virus hepatitis in two patients with agammaglobulinemia. Am. J. Med. *29:*804–810 (1960).
31 Surrenti, C.; Casini, M.C.; Romagnani, S.; Chiarantini, E.; Caramelli, L.; Gentilini, P.: Variable immune deficiency syndrome and active chronic hepatitis. Report of two cases. Ital. J. Gastroenterol. *10:*707–711 (1978).
32 Ragan, C.; Howes, E.L.; Plotz, C.M.; Meyer, K.; Blunt, J.W.: Effect of cortisone on production of granulation tissue in the rabbit. Proc. Soc. exp. Biol. Med. *72:* 718–721 (1949).
33 Allen, J.G.; Emerson, D.M.; Landy, J.J.; Head, L.R.; Basinger, C.R.: The causes of death from total body irradiation. Ann. Surg. *146:*322–341 (1957).
34 Papermaster, B.W.; Condie, R.M.; Good, R.A.: Immune response in the California Hagfish. Nature, Lond. *196:*355–357 (1962).
35 Papermaster, B.W.; Condie, R.M.; Finstad, J.K.; Good, R.A.; Gabrielsen, A.E.: Phylogenetic development of adaptive immunity. Fed. Proc. *22:*1152–1155 (1963).

Prof. C. Surrenti, Cattedra di Gastroenterologia, Università di Firenze,
Viale Morgagni 85, I-50134 Firenze (Italy)

Front. gastrointest. Res., vol. 8, pp. 128–134 (Karger, Basel 1984)

Serum Inhibitory Factors to Cell-Mediated Immunity in Primary Biliary Cirrhosis[1]

C. Surrenti, A. Casini, D. Ramarli, S. Nieri, G. Salvadori, O. Ricci, L. Ventura, I.T. Pasquinelli

Cattedra di Gastroenterologia, Università di Firenze, Florence, Italy

Introduction

Primary biliary cirrhosis (PBC) is a chronic progressive liver disease frequently associated with humoral and cellular immunological abnormalities. These include hypergammaglobulinaemia, with high levels of IgM [1] and circulating monomeric (7S) IgM [2], the presence of autoantibodies directed to circulating T lymphocytes [3] and to mitochondrial structures [4], high levels of circulating immune complexes [5, 6], and the increase of activated complement components [6]. In addition in vitro studies of lymphocytes from PBC patients have shown a decreased phytohaemagglutinin (PHA)-induced proliferative response in liquid culture [1, 7] and an impaired suppressor cell activity [2, 8].

Recent data demonstrate that co-culture experiments with different combinations of adherent cells (AC) and peripheral blood lymphocytes (PBL) from PBC patients and normal donors provided evidence that PBC-AC function abnormally. However, this impaired function is not the sole factor involved, since the response of co-cultured PBC lymphocytes and normal AC was twice the normal value. This increase may be correlated with the state of activation of T cells demonstrated by the high number of DR-positive T cells found in the PBC patients studied [9].

It has been shown that patient serum factors can inhibit in vitro normal lymphocyte PHA responsiveness [7, 10–12] and the capacity to form E rosettes with sheep erythrocytes [11]. The nature of these factors

[1] This study was supported in part by a grant from M.P.I.

is not clear. The aim of this study is to try and define the characteristics of these factors.

Materials and Methods

The effect of sera from 13 patients with PBC on normal lymphocyte responsiveness to PHA and on the capacity to form E rosettes with sheep red blood cells (SRBC) was studied. The control group was made up of sera from normal donors. The diagnosis was established on the basis of histological, clinical, biochemical and serological findings.

Mononuclear Cells. Mononuclear cells (MC) were isolated from heparinized (Vister liophylized Na-heparin 1 mg/5 ml) peripheral blood of normal donors on a Ficoll-Hypaque gradient according to the method of *Boyum* [13]. Cells containing no more than 10% of monocytes and 0.5% granulocytes were then resuspended in TC 199 medium with 100 U/ml of penicillin and 100 μg/ml of streptomycin at a concentration of 2×10^6 cells/ml.

E Rosette-Forming Cells. Serum specimens from patients with PBC were assayed for the ability to inhibit E rosette-forming cells (ERFC) function in normal donors [14]. After a 24-hour incubation with heat-inactivated test serum, fetal calf serum (FCS) and human pooled AB serum, MC were washed with TC 199 medium and assayed for ERFC function. Experiments were performed in duplicate.

MC were suspended in TC 199 medium (2×10^6 cells/ml). SRBC, after many washings, were resuspended (1%) with TC 199. 200 μl of MC suspension, 200 μl of SRBC (1%) and 100 μl of FCS, previously heat-inactivated and adsorbed with SRBC, were mixed in a V-bottom tube. The cell suspension was incubated at 37 °C for 5 min, centrifuged at 200 *g* for 5 min and kept at 4 °C for 18 h.

A rosette was defined as a lymphocyte with 3 or more erythrocytes attached; at least 200 were counted by light microscopy.

PHA Lymphocyte Transformation Test. The effect of heat-inactivated sera from patients with PBC on PHA stimulated normal lymphocytes was evaluated. Human in vitro lymphocyte cultures were carried out using the microplate culture method [15]. Briefly, 2×10^5 lymphocytes in 100 μl of TC 199 medium containing 20% of pooled heat-inactivated normal AB serum were added to the wells of V-bottomed microtitre plates (Falcon Plastic). 100 μl of medium only (controls) or containing PHA-M (Difco) at a concentration of 1, 5 and 25 μl were added to every well. After 48 h incubation in a humid atmosphere containing 5% CO_2 in air, 0.5 μCi of ^{3}H-thymidine was added. Cultures were then harvested 18 h later by a multiple automated sample harvester (Skatron). The results were expressed as the mean value of cpm obtained in triplicate samples. Sera with 30% of inhibitory activity on normal lymphocyte PHA responsiveness were taken as positive for inhibition. The data were analysed using Student's t test.

Fractionation of Serum Factors by Sephadex G-200 Gel Filtration. Fractionation of 4 pooled sera positive for inhibitory activity and circulating immune complexes was performed by Sephadex G-200 gel filtration in order to determine the approximate molecular size of the inhibitory factors. Before use, each fraction was dialysed against PBS for 48 h and then against TC 199 medium for 24 h. After 24 h preincubation with each fraction, FCS and AB serum, normal MC were washed with TC 199 medium and assayed for ERFC function as previously described. Experiments were performed in duplicate. Each fraction was also tested for its influence on the in vitro normal lymphocyte response to PHA. Cells were incubated in a medium containing 5% of AB serum and 5% of each fraction. The ^{3}H-thymidine uptake was compared to the controls in which the lymphocytes were incubated in medium containing 10 or 5% of AB serum.

Detection of Immune Complexes. Cl_q was isolated from fresh human serum according to the modified method of *Volanakis and Stround* [16], and radiolabelled with iodine according to standard techniques [17]. The Cl_q (specific activity 1 μCi/mg) was divided into equal parts, each containing 10 mg/100 μl veronal buffer saline (VBS) and stored at -80 °C. The test was repeated in triplicate according to the method of *Zubler* et al. [18]. ^{125}I-Cl_q was allowed to clot and was diluted with 3-8 ml of VBS containing 1% of heat-inactivated human serum (30 min at 56 °C).

Aggregates of Cl_q were removed by centrifugation at 18,000 *g* for 40 min at 4 °C before use. The supernatant was used for the test. 50 μl of test serum were incubated at 37 °C in 100 μl of EDTA 0.2 *M* (pH 7.5) for 30 min. Then 50 μl of ^{125}I-labelled Cl_q and 1 ml of polyethyleneglycol (PEG) at a concentration of 3% borate buffer 0.2 *M* (pH 8.3) was added in an ice bath. After an incubation period of 60 min in the ice bath, the test samples were precipitated by centrifugation at 1,500 *g* for 20 min at 4 °C.

The radioactivity bound in the precipitate was measured. Negative controls were healthy donor sera, to which various concentrated immunoglobulin aggregates were added as a positive control.

Results

Effect of Sephadex G-200 Fractions on E Rosette Formation and PHA Stimulation. The percentage of ERFC was decreased in the presence of 4S fraction from PBC pooled sera when compared to controls. 7S and 19S fractions did not have any inhibitory effect on ERFC (fig. 1). An impaired response to PHA in AB serum 5% in comparison to controls was detected when normal lymphocytes were incubated with the 19S fraction and a lower inhibitory effect with the 7S fraction. It was not possible to demonstrate any inhibitory effect on PHA lymphocyte responsiveness by the 4S fraction (fig. 2).

Circulating Immune Complexes. Circulating immune complexes were detectable in 10 of 13 patients (76.9%) with PBC. The mean value was 10.2 (SD ± 9.7 with a range of 4.8–33.6). A correlation was found

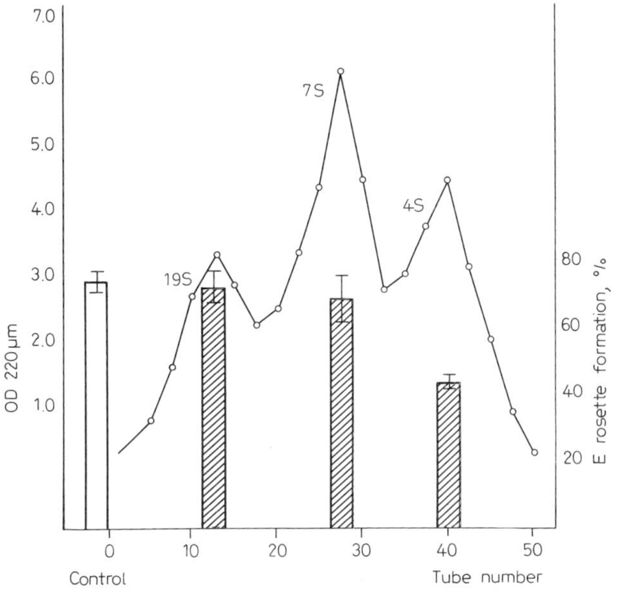

Fig. 1. Separation of pooled sera from patients with PBC on Sephadex G-200 column. The effect of three fractions (19S, 7S, 4S) on normal lymphocyte E rosette formation as compared to controls (□). The inhibitory effect is mainly present in the third peak (4S). The results represent the mean value of three experiments performed in duplicate.

between the presence of serum immune complexes and the inhibitory activity of PHA stimulation. In fact, inhibitory activity was absent in sera negative for immune complexes.

Discussion

Cell-mediated immunity was found to be impaired in PBC. This may be due to the presence of circulating inhibitory factor(s). A previous study [12] demonstrated this factor(s) was not dialysable, heat-stable at 56 °C and that this inhibitory effect was not correlated to unconjugated bilirubin. Furthermore, in the present study, separation of pooled sera on Sephadex G-200 showed that the inhibitory factors(s) of ERFC was present in the third peak which represents low molecular

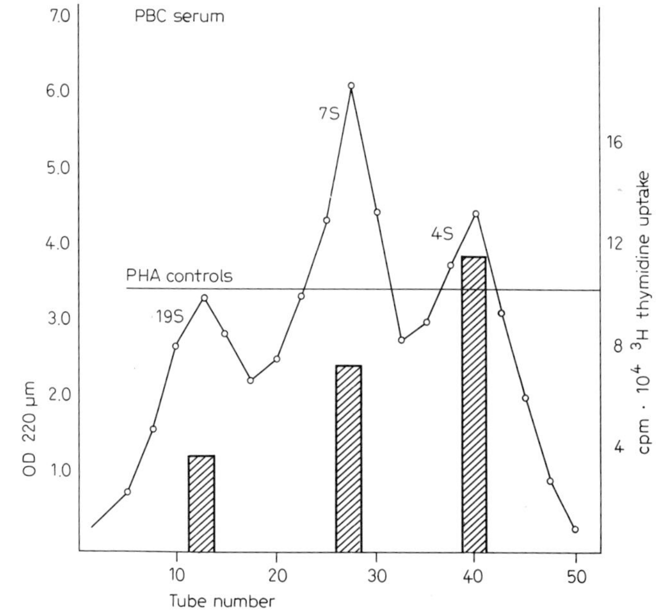

Fig. 2. Separation of pooled sera from patients with PBC on Sephadex G-200 column. The effect of three fractions (19S, 7S, 4S) on normal lymphocyte response to PHA as compared to controls (——). The inhibitory effect is mainly present in the first peak (19S). The results are expressed as mean values of CPM obtained in three experiments performed in triplicate.

weight substances. The inhibitory factor(s) of lymphocyte transformation by PHA was detectable prevalently in the first peak (substances with high molecular weight).

These results indicate that the decrease in E rosette formation may be correlated to the presence of bile acids as previously reported [19].

The finding of inhibitory factor(s) on PHA responsiveness in the peak with high molecular size suggests that there may be a correlation with large molecules (19S) such as immune complexes. Indeed, *Gupta* et al. [20] demonstrated the presence of circulating immune complexes sedimented at 19S in PBC. However, in the present study, a lower inhibitory effect on PHA responsiveness was present in the 7S peak. This may be correlated to the presence of circulating monomeric (7S) IgM [21].

Finally, it can be concluded that the study of cellular immunity carried out in these patients, using the E rosette and PHA transformation tests, has allowed the authors to demonstrate the presence of inhibitory factors of varying types, the most important of which seem to be the immune complexes.

Summary

The inhibitory effect on normal lymphocyte response to phytohaemagglutinin (PHA) and E rosette formation was evaluated using pooled sera from patients with primary biliary cirrhosis (PBC). The fractionation of the pooled sera, which were positive for inhibitory activity, was performed with a Sephadex G-200 column.

An inhibitory effect on PHA stimulation was found mainly in the 19S fraction in comparison to controls. On the contrary, inhibitory activity on E rosette formation was present in the 4S fraction. A correlation between the presence of serum immune complexes and the inhibitory activity on PHA stimulation was detected.

These results suggest that the decrease in E rosette function and the impairment of normal lymphocyte responsiveness to PHA, usually found in patients with PBC, were due to factors of varying types, the most important of which seem to be the immune complexes.

References

1 MacSween, R.N.M.; Horne, H.W.; Moffat, A.J.; Hughes, H.M.: Serum protein levels in primary biliary cirrhosis. J. clin. Path. *25:*789–792 (1972).

2 James, S.P.; Elson, C.O.; Strober, W.: Abnormal regulation of immunoglobulin synthesis in vitro in primary biliary cirrhosis. Gastroenterology *79:*242–254 (1980).

3 Ling, N.P.; Biship, S.; Jefferis R.: Use of antibody-coated red cells for the sensitive detection of antigen and in rosette tests for cells bearing surface immunoglobulins. J. immunol. Methods *15:*279–289 (1977).

4 MacSween, R.M.N.; Berg, P.A.: Autoimmune liver disease (MTP Press, Lancaster 1976).

5 Thomas, H.C.; Potter, B.J.; Sherlock, S.: Is primary biliary cirrhosis an immune complex disease? Lancet *ii:*1261–1263 (1977).

6 Wands, J.R.; Dienstag, J.L.; Bhan, A.K.; Feller, E.R.; Isselbacher, K.J.: Circulating immune-complexes and complement activation in primary biliary cirrhosis. New Engl. J. Med. *298:*233–237 (1978).

7 Fox, R.A.; James, D.J.; Samuels, M.; Milligan, J.; Sherlock, S.: Lymphocyte transformation in response to phytohaemagglutinin in primary biliary cirrhosis: the search for a plasma inhibitor factor. Gut *14:*89–93 (1973).

8 Woltjen, J.A.; Zettermann, R.K.: Suppressor cell activity in primary biliary cirrhosis (Abstract). Gastroenterology *1301:*May (1979).

9 Ramarli, D.; Canonica, G.W.; Melioli, G.; Biagioni, P.; Nieri, S.; Casini, A.; Surrenti, C.: Abnormalities of immunocompetent cells in primary biliary cirrhosis. Scand. J. Immunol. *16:*201–207 (1982).
10 MacSween, R.N.M.; Galbraith, I.; Thomas, M.A.; Watkinson, G.; Ludlam, G.B.: Phytohaemagglutinin (PHA)-induced lymphocyte transformation and *Toxoplasma gondii* antibody studied in primary biliary cirrhosis: evidence for impaired cell-mediated immunity. Clin. exp. Immunol. *15:*35–39 (1973).
11 Vierucci, A.; De Martino, M.; Novembre, E.; Surrenti, C.; Gabrielli, R.; Biadaioli, R.: Influenza del siero di pazienti con colestasi sulla formazione di rosette linfocitarie 'E'. Annali Sclavo *19:*1119–1129 (1977).
12 Surrenti, C.; Casini, A.; Ramarli, D.: Serum inhibitory activity on cell-mediated immunity in patients with cholestasis; in Problems in intrahepatic cholestasis (Karger, Basel 1979).
13 Boyum, A.: Isolation of mononuclear cells and granulocytes from human blood. Scand. J. clin. Lab. Invest. *97:*77–89 (1968).
14 Chisari, F.V.; Routemberg, J.A.; Milan, F.; Edgington, T.S.: Extrinsic modulation of human T-lymphocyte E rosette function associated with prolonged hepato-cellular injury after viral hepatitis. J. clin. Invest. *59:*134–142 (1977).
15 Romagnani, S.; Amadori, A.; Biti, G.; Bellesi, G.; Ricci, M.: In vitro lymphocyte response to phytomitogens in untreated and treated patients with Hodgkin's disease. Int. Archs. Allergy appl. Immun. *51:*378–389 (1976).
16 Volanakis, J.E.; Stround, R.M.: Rabbit Cl_q purification and structural studies. J. immunol. Methods *2:*25–34 (1972).
17 Heusser, C.; Bobsman, M.; Nordin, J.; Isliker, H.: Effect of chemical and enzymatic radioiodination on in vitro human Cl_q activities. J. Immun. *110:*820–828 (1973).
18 Zubler, R.H.; Lange, G.; Lambert, P.H.; Miescher, P.A.: Detection of immune complexes by a modified ^{125}I-labelled Cl_q binding test: effect of heating on the binding of Cl_q by immune complexes and application of the test to systemic lupus erythematosus. J. Immun. *116:*232–235 (1976).
19 Surrenti, C.; Casini, A.; Ramarli, D.: Effect of bile acids on E rosette formation and PHA stimulation. Boll. Ist. Sieroter. Milanese. *59:*133–137 (1980).
20 Gupta, R.C.; Dickson, E.R.; McDuffie, F.C.; Baggenstoss, A.H.: Circulating IgG complexes in primary biliary cirrhosis. A serial study in forty patients followed for two years. Clin. exp. Immunol. *34:*19–27 (1978).
21 Sherlock, S.: The immunology of cholestasis; in Problems in intrahepatic cholestasis (Karger, Basel 1979).

Prof. C. Surrenti, Cattedra di Gastroenterologia, Università di Firenze,
Viale Morgagni 85, I-50134 Firenze (Italy)

Front. gastrointest. Res., vol. 8, pp. 135–145 (Karger, Basel 1984)

The Occurrence of Plasma Fibronectin in Mixed Cryoglobulins from Patients with Liver Cirrhosis[1]

F. Dammacco[a], *F. Carandente*[b,c], *A. Vacca*[a], *N. Campobasso*[a], *L. Bonomo*[d]

[a] Institute of Medical Pathology, University of Bari Medical School, Bari; [b] Center of Chronomorphology and Rhythmometric Techniques, University of Milan, Medical School, Milan; [c] Chair of Chronobiology, University of L'Aquila Medical School, L'Aquila; [d] 3rd Medical Pathology, University of Rome Medical School, Rome, Italy

Fibronectin (FN) is a high molecular weight glycoprotein which may be found normally in the blood and other extracellular fluids, as well as on the surfaces of many cells. Cell surface FN is present in an insoluble form and appears to be structurally larger by approximately 10,000 daltons compared with plasma FN which, along with a shorter size, is also soluble [25]. Following the observation that neoplastic cells often lack cell surface FN and that this fact may possibly account for the invasive and metastatic properties of such cells, FN has been the object of a number of studies which have greatly clarified its functions. Indeed, FN appears to be involved in a number of seemingly different processes [21], including cell-cell and cell-substratum adhesion, malignant transformation, reticuloendothelial function and removal of colloids by liver cells and macrophages [25, 28].

Recently, *Anderson* et al. [1] reported that FN was present in all cryoprecipitates isolated from a group of patients mainly affected with connective tissue and autoimmune diseases, indicating that this glycoprotein is essential for cryoprecipitation. Although frequently found in a number of disease conditions, such as autoimmune disorders, immunoproliferative diseases, acute and chronic infections [3, 6, 7, 10], and the purpura-weakness-arthralgia syndrome [22] which is commonly de-

[1] These studies were supported in part by Grant 82/7247 from the Italian Ministry of Education and the Hoechst Foundation.

fined as essential mixed cryoglobulinemia (EMC), cryoglobulins have been repeatedly detected in patients with chronic liver disease [8, 13, 14]. On the other hand, subclinical or overt evidence of hepatic dysfunction and/or serologic findings of prior exposure to hepatitis B virus (HBV) may be revealed in over half the patients with EMC [6, 18, 26]. Finally, reduced plasma FN concentrations have been found in patients with decompensated cirrhosis [20] as well as in fulminant hepatic failure [9].

It was, therefore, thought of interest to study a group of cryoglobulins isolated from patients with liver cirrhosis (LC) and, by comparison, with EMC for the occurrence of FN and for the possible role of such a glycoprotein in cryoprecipitation.

Materials and Methods

Patients. 21 cryoglobulins were selected for this study due to accuracy in diagnosis and follow-up possibilities. Group I consisted of 10 patients with LC, diagnosis being based on accepted clinical, biochemical and histological data. A history of alcohol abuse as well as pathological lesions attributable to alcoholism, such as fatty changes, centrilobular hepatitis, Mallory's hyalin and micronodular cirrhosis were consistently absent in such cases. Group II included 11 patients who were diagnosed as having EMC. Malignant proliferative diseases, chronic infections, and autoimmune disorders were excluded in all patients by full clinical investigation and follow-up.

Isolation and Purification of Cryoglobulins. The techniques of blood sample collection, isolation and purification of cryoglobulins as well as their immunochemical characterization have been described elsewhere [7]. Separation of cryoglobulin components was achieved by gel filtration chromatography of the purified cryoprecipitates on Sephadex G-200 columns, using a Na-acetate buffer 0.2 *M*, pH 4.0, as eluent.

Tests for Hepatitis B Virus Infection. Besides routine laboratory investigations, whole sera and isolated cryoglobulins (3 mg/ml) were tested by a sensitive radioimmunoassay for the detection of both HBsAg and anti-HBs, using commercially available kits.

Fibronectin Detection and Measurement. FN was identified in both plasma samples and isolated cryoproteins by Ouchterlony double diffusion analysis and electroimmunoassay according to *Laurell's* [15] method. A monospecific rabbit antibody to human FN was purchased from Behring Institute, Scoppito, Italy. For reference purposes as well as for the preparation of standard solutions, human plasma FN was isolated by a Gelatin-Sepharose 4B (Pharmacia Fine Chemicals, Uppsala, Sweden) one-step procedure [30]. The results of this separation method were electrophoretically controlled in Sodium dodecyl sulfate (SDS)-polyacrylamide slab gels.

Quantitative determination of FN was performed by a turbidimetric assay, using a commercially available kit (Boehringer Mannheim GmbH, Milan, Italy). The test principle is based on the kinetic turbidimetric measurement of the antigen-antibody reaction according to the 'fixed time' method, which allows a measuring range of approximately 0–1,000 μg FN/ml plasma.

Effect of FN on Cryoprecipitation. In order to evaluate the role of FN in both the formation and amount of a cryoprecipitate, FN was selectively removed from aliquots of each serum sample using the same affinity chromatography approach on Gelatin-Sepharose 4B columns already described for the isolation of FN. Controls included the same samples which had been applied to simple Sepharose columns in the absence of gelatin. Cryoprecipitation was then comparatively verified in each case before and after removal of the FN. Conversely, the amount of cryoprecipitates was again calculated following the addition of increasing quantities of FN.

Anti-FN Activity of Cryoprecipitates. The possibility that the occurrence of FN in the cryoglobulins might be related to the anti-FN activity of cryoproteins and/or their isolated constituents was investigated by an indirect solid-phase radioimmunoassay essentially as described by *Beaulieu* et al. [2]. Briefly, plate wells were coated with FN (1 h at 37 °C), washed with buffered saline, filled with 1% bovine serum albumin (BSA), and then washed again with buffered saline. Afterwards, solubilized whole cryoglobulins and their fractions were added and incubated overnight. Following a further washing, a ^{125}I-labelled goat antiserum to human IgG was added to the system and incubated for 18 h at room temperature. A positive control included a rabbit antibody to human FN, followed by a ^{125}I-labelled goat antiserum to rabbit IgG. By this assay, about 40 ng/ml of the affinity-purified rabbit antibody to human FN can be detected.

Other Procedures. Protein measurements were determined by the Folin-Lowry method using human albumin as a standard, and expressed as cryoprecipitable protein/ 1 ml of starting serum. Fibrinogen concentrations were measured by single radial immunodiffusion using plates commercially purchased (Behring Institute), whereas fibrinogen degradation products were detected by a latex agglutination procedure.

Results

As reported in table I, immunochemical typing of the isolated cryoglobulins showed that all cases included in this study were of the mixed type, cryoglobulins with a monoclonal component being more numerous in the EMC group and mixed polyclonal-type cryoglobulins in the LC group. In addition, the amount of the cryoprecipitable protein exceeded 5 mg/ml in over half of the mixed cryoglobulins with a monoclonal component and none of the mixed polyclonal cryoproteins. It should also be mentioned that the monotypic component was an

Table I. Summary of clinical and laboratory features of patients studied

Patient, No.	Age, years	Sex	Diagnosis	Serum cryoglobulin			
				concen-tration (mg/ml)	immuno-chemical analysis	FN	fibrinogen and FDP
1	52	F	LC	3.7	IgMk-IgG	+	–
2	56	M	LC	5.0	IgM -IgG	–	–
3	47	F	LC	4.2	IgM -IgG	+	–
4	45	F	LC	3.3	IgMk-IgG	+	+
5	63	M	LC	3.0	IgMk-IgG	+	–
6	42	F	LC	2.8	IgM -IgG	+	–
7	57	M	LC	3.5	IgM -IgG	–	–
8	60	F	LC	3.5	IgM -IgG	+	–
9	64	F	LC	4.0	IgM -IgG	+	–
10	49	M	LC	4.2	IgM -IgG	+	–
11	61	F	EMC	8.0	IgM -IgG	+	–
12	40	M	EMC	5.4	IgMk-IgG	+	–
13	38	F	EMC	9.2	IgMk-IgG	+	–
14	45	M	EMC	5.0	IgMk-IgG	+	–
15	31	F	EMC	5.5	IgM -IgG	+	–
16	40	F	EMC	4.6	IgMk-IgG	+	–
17	38	F	EMC	3.8	IgM -IgG	+	+
18	35	F	EMC	10.0	IgM -IgG	+	–
19	52	M	EMC	4.2	IgMk-IgG	+	–
20	37	M	EMC	4.0	IgMk-IgG	+	–
21	41	F	EMC	7.5	IgM -IgG	+	–

IgM-κ in all cases and that the IgM fractions, regardless of their light chain typing, consistently exhibited antihuman γ-globulin (rheumatoid factor) activity. Finally, the occurrence of serum HBsAg and/or anti-HBs was detected in 45% of the EMC sera and in 60% of the LC group. While characteristic clinical, laboratory and histological features of advanced liver impairment were obviously striking in group I LC patients, it should be emphasized that biochemical signs of hepatic dysfunction were also detected in 45% of the patients with EMC although reaching clinical relevance in only 2 out of 11 such patients (18%). Liver biopsy specimens were available for 2 patients showing laboratory features of liver impairment, and in both chronic persistent

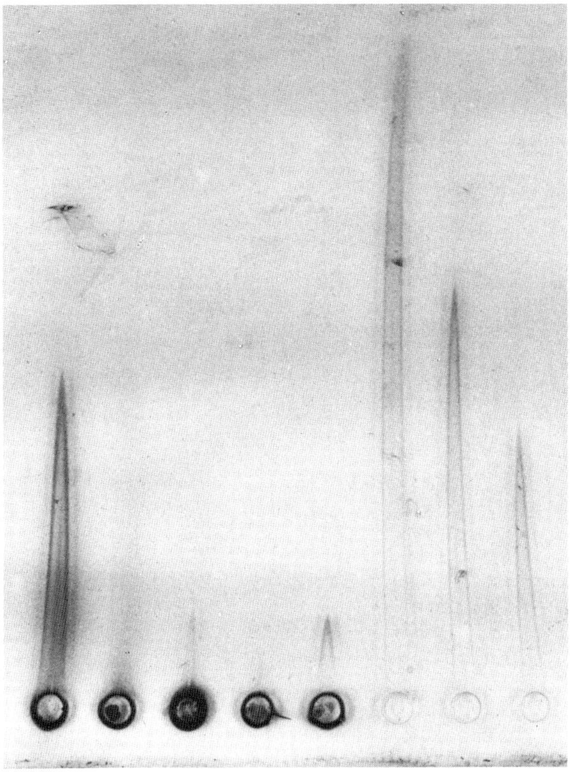

Fig. 1. Laurell's electroimmunoassay in 1% agarose gel containing a rabbit antiserum to human FN. Five different purified cryoglobulins (3 mg/ml) were placed in wells 1–5, whereas wells 6–8 were filled with standard FN solutions (1,000, 750, and 500 μg/ml, respectively). Despite poor solubility of cryoproteins, a 'rocket' is clearly visible with samples 1, 3 and 5.

hepatitis was diagnosed by the pathologist despite the absence in the past medical history of any previous conditions which could be related to the present liver status.

When tested by both Ouchterlony double diffusion analysis and Laurell's electroimmunoassay, FN was detected in all of 11 EMC-cryoprecipitates and in 8 of 10 LC cryoproteins. Figure 1 shows examples of FN 'rockets' revealed by electroimmunoassay. In addition, an identity reaction was always obtained between the FN lines of human plasma and of the cryoprecipitates by double diffusion examination.

Table II. Total protein (TP) and FN concentration in sera and cryoprecipitates

Patient No.	Diagnosis	TP/FN, mg/100 ml		Enrichment factor (ratio 1:2)
		serum (1)	cryoprecipitates (2)	
1	LC	6,900/28	370/24	16
4	LC	6,600/17	330/21	25
5	LC	7,000/22	300/44	47
8	LC	5,800/12	265/21	38
10	LC	5,700/23	61/15	62
13	EMC	7,400/19	920/32	13
15	EMC	7,000/12	550/36	38
16	EMC	8,200/32	460/19	10
20	EMC	7,800/29	400/28	19

Immunoreactive serum and cryoglobulin FN were quantitated by an in vitro turbidimetric assay in 4 serum-cryoprotein samples of the EMC group and in 5 paired samples of the LC group (table II). By calculating the ratios of total protein (TP) over FN for each of the specimens, significantly lower ratios were found in cryoprecipitates from both patient groups, suggesting a relatively higher FN content in the cryoglobulins.

Since mixed cryoglobulins are usually believed to represent antigen-antibody complexes [3, 10, 12], which might conceal antibody-combining sites for FN, 4 cryoprecipitates (2 for each patient group) were fractionated into their IgM and IgG fractions by gel filtration chromatography, and whole cryoproteins as well as their constituent fractions were tested for antibody activity to FN by an indirect solid-phase radioimmunoassay. No antibodies to FN could be clearly detected in either unfractionated or fractionated cryoprecipitates.

In order to assess whether and to what extent FN is essential for the formation of cryoglobulins, a number of FN removal and reconstitution experiments were undertaken. Preliminary results indicate that, when serum FN is selectively removed by affinity chromatography on Gelatin-Sepharose columns, a significant (though largely variable from case to case) reduction in the cryoprecipitate formation was usually observed. Conversely, stepwise addition of human purified FN to FN-depleted EMC and LC sera appeared to result in a gradual increase in cryoprecipitation. Further details of these experiments will be reported elsewhere [5]. It should also be mentioned that both fibrinogen and fibrinogen degradation products could not, with rare exceptions, be detected in the cryoprecipitates (table I), suggesting that interactions between FN and fibrinogen products are not likely to account for the serum cryoprecipitate formation.

Discussion

A variety of clinical and immunochemical studies may be traced in the literature in the last 15–20 years, dealing with relatively large series of patients with cryoglobulinemia [3, 6, 10, 12]. Cold precipitable proteins in serum and possibly other biological fluids have been detected in a number of disease conditions, including LC [13, 14, 27] and EMC [10, 22]. As regards their pathogenetic significance, there is presently no doubt as to the immune complex nature of most cryoglobulins. Systemic vasculitis and renal damage, which may be frequently observed in patients with cryoglobulinemia, are indeed typical features of an immune complex disease. In addition, an expanding spectrum of clinical features and pathogenetic clues has been recently emphasized in EMC, and justifies our choice of a group of EMC patients to be compared with the LC patients. Actually, liver involvement, though reaching clinical relevance in a minority of instances, appears strikingly common in EMC and should therefore be considered as a consistent feature of its peculiar clinical picture [6, 17]. To account for the relatively high prevalence of HBsAg and anti-HBs in both EMC sera and isolated cryoglobulins, it has been suggested that the purpura-weakness-arthralgia syndrome typical of EMC might be related to an immune complex vasculitis secondary to HBV exposure [18, 24]. However, apparently indistinguishable liver involvement may also be detected in EMC patients with no evidence of HBV infection [6, 23] and, on the other hand, 40% of the selected LC patients included in this study had serum mixed cryoglobulins unassociated with specific serological markers of HBV infection.

Plasma FN has been recently reported to be a regular component of serum cryoglobulins from patients with connective tissue diseases [1, 2] and acute viral hepatitis [16], as well as of synovial fluid cryoprecipitates from patients with rheumatic disease [4, 19]. Similar findings were also detected in isolated instances of EMC [2] and chronic active hepatitis [16]. The results reported in this paper confirm and extend such findings, in that plasma FN has been demonstrated to be a constituent of serum cryoproteins from all patients with EMC and from the large majority of those with LC. Whether failure of FN detection in 2 LC cryoglobulins might be related to the relatively low cryocrit of both patients, or to the actual lack of this glycoprotein, remains to be ascertained. In addition, FN in cryoglobulins always gave a reaction of complete identity with a sample of purified human FN by double diffusion analysis, thus suggesting a single immunoreactive species of FN in all cases.

Comparative FN measurements in both sera and cryoprecipitates, as well as ratios of total protein over FN clearly showed a quantitative enrichment of FN in the cryoproteins, indicating that it was not passively entrapped into such cryoglobulins. The possibility that FN could be actively engaged in the formation of antigen-antibody complexes seemed unlikely, in that no anti-FN activity could apparently be demonstrated in either whole cryoglobulins or their chromatographically isolated constituents. Although a possible interaction of FN with the IgG and/or IgM fractions as well as with other serum components cannot be ruled out with certainty, failure to demonstrate fibrin-fibrinogen products in the cryoprecipitates would also seem to exclude FN interaction with this system, in spite of the fact that such a mechanism has been strongly suspected to account for the cryoprecipitate formation in synovial fluids [4].

For a better elucidation of the actual role which FN plays in the formation of cryoprecipitates, a number of FN removal and reconstitution experiments were undertaken, which will form the basis of a future report. However, similarly to the findings reported by *Beaulieu* et al. [2], preliminary results indicate that, though to a variable extent from case to case, removal of plasma FN by affinity chromatography may remarkably affect cryoprecipitation, this property being progressively recovered upon addition of increasing amounts of purified FN.

According to recent in vitro experiments [11], cryoprecipitability seems a property of a group of large normal serum proteins, rather than of immune complexes themselves. In the course of inflammatory diseases some of these proteins, acting as acute-phase reactants, may significantly increase making intermolecular attractive forces prominent, especially in the cold. Larger and less soluble molecular aggregates might then interact with FN and/or immune complexes, resulting in the formation of cryoprecipitates. On the other hand, since FN has been found to enhance clearance of particles (fibrin and collagen debris, bacterial products, etc.) by the reticuloendothelial and macrophage systems [21, 29], it seems reasonable to hypothesize that in mixed cryoglobulinemia FN might on the one hand favor precipitation of cryoproteins, and on the other hand enhance the removal of cryoprecipitable immune complexes by the clearance system [16].

Summary

A total of 21 IgM-IgG mixed cryoglobulins from patients with LC and EMC were tested for the presence of plasma FN. This glycoprotein was identified in all EMC and in 8 out of 10 LC cryoprecipitates; in addition, quantitation studies showed a relative enrichment of FN in the cryoproteins as compared with corresponding serum samples. No anti-FN activity could be demonstrated in the isolated IgM or IgG fractions from 4 different cryoglobulins and, on the other hand, an interaction of FN with fibrin-fibrinogen products could also be excluded. Preliminary experiments of FN depletion and reconstitution showed a remarkable, though variable effect on cryoprecipitation. It is postulated that FN may enhance both precipitation and removal of cryoprecipitable immune complexes by the reticuloendothelial system.

References

1 Anderson, B.; Rucker, M.; Entwistle, R.; Schmid, F.R.; Wood, G.W.: Plasma fibronectin is a component of cryoglobulins from patients with connective tissue and other diseases. Ann. rheum. Dis. *40:*50–54 (1981).

2 Beaulieu, A.D.; Valet, J.P.; Strevey, J.: The influence of fibronectin on cryoprecipitate formation in rheumatoid arthritis and systemic lupus erythematosus. Arthritis Rheum. *24:*1383–1388 (1981).

3 Brouet, J.-C.; Clauvel, J.-P.; Danon, F.; Klein, M.; Seligmann, M.: Biologic and clinical significance of cryoglobulins. A report of 86 cases. Am. J. Med. *57:*775–788 (1974).

4 Carsons, S.; Mosesson, M.W.; Diamond, H.S.: Detection and quantitation of fibronectin in synovial fluid from patients with rheumatic disease. Arthritis Rheum. *24:* 1261–1267 (1981).

5 Dammacco, F.; Carandente, F.; Pitari, G.; Vacca, A.: The influence of fibronectin depletion and reconstitution on serum cryoprecipitation (in preparation).

6 Dammacco, F.; Gnisci, G.; Silvestris, F.; Bonomo, L.: New clinical and immunological trends in cryoglobulinemia. Ricerca Clin. Lab. *10:*51–57 (1980).

7 Dammacco, F.; Lucivero, G.; Miglietta, A.; Antonaci, S.; Bonomo, L.: Cryoglobulinemia: a model of immune complex disease. Immunochemical and structural studies. Ricerca Clin. Lab. *4:*569–590 (1974).

8 Florin-Christensen, A.; Roux, M.E.B.; Arana, R.M.: Cryoglobulins in acute and chronic liver diseases. Clin. exp. Immunol. *16:*599–605 (1974).

9 Gonzales-Calvin, J.; Scully, M.F.; Sanger, Y.; Fok, J.; Kakkar, V.V.; Hughes, R.D.; Gimson, A.E.S.; Williams, R.: Fibronectin in fulminant hepatic failure. Brit. Med. J. *285:*1231–1232 (1982).

10 Gorevic, P.D.; Kassab, H.J.; Levo, Y.; Kohn, R.; Meltzer, M.; Prose, P.; Franklin, E.C.: Mixed cryoglobulinemia: clinical aspects and long-term follow-up of 40 patients. Am. J. Med. *69:*287–308 (1980).

11 Hardin, J.A.: Cryoprecipitogogue from normal serum: mechanism for cryoprecipitation of immune complexes. Proc. natn. Acad. Sci. USA *78:*4562–4565 (1981).

12 Invernizzi, F.; Pioltelli, P.; Cattaneo, R.; Gavazzeni, V.; Borzini, P.; Monti, G.; Zanussi, C.: A long-term follow-up study in essential cryoglobulinemia. Acta haemat. *61:*93–99 (1979).

13 Jori, G.P.; Buonanno, G.: Chronic hepatitis and cirrhosis of the liver in cryoglobulinaemia. Gut *13:*610–613 (1972).

14 Jori, G.P.; Buonanno, G.; D'Onofrio, F.; Tirelli, A.; Gonnella, F.; Gentile, S.: Incidence and immunochemical features of serum cryoglobulin in chronic liver disease. Gut *18:*245–249 (1977).

15 Laurell, C.B.: Quantitative estimation of proteins by electrophoresis in agarose gel containing antibodies. Analyt. Biochem. *15:*45–53 (1966).

16 Levo, Y.: Presence of fibronectin in cold precipitates of patients with cryoimmunoglobulinemia. Int. Archs Allergy appl. Immun. *68:*179–181 (1982).

17 Levo, Y.; Gorevic, P.D.; Kassab, H.J.; Tobias, H.; Franklin, E.C.: Liver involvement in the syndrome of mixed cryoglobulinemia. Ann. intern. Med. *87:*287–292 (1977).

18 Levo, Y.; Gorevic, P.D.; Kassab, H.J.; Zucker-Franklin, D.; Franklin, E.C.: Association between hepatitis B virus and essential mixed cryoglobulinemia. New Engl. J. Med. *296:*1501–1504 (1977).

19 Lu-Steffes, M.; Iammartino, A.J.; Schmid, F.R.; Castor, C.W.; Davis, L.; Entwistle, R.; Anderson, B.: Fibronectin in rheumatoid and non-rheumatoid arthritic synovial fluids and in synovial fluid cryoproteins. Ann. clin. Lab. Sci. *12:* 178–185 (1982).

20 Matsuda, M.; Yamanaka, T.; Matsuda, A.: Distribution of fibronectin in plasma and liver in liver diseases. Clinica chim. Acta *118:*191–199 (1982).

21 McDonagh, J.: Fibronectin, a molecular glue. Archs Path. lab. Med. *105:*393–396 (1981).

22 Meltzer, M.; Franklin, E.C.: Cryoglobulinemia. A clinical and laboratory study. II. Cryoglobulins with rheumatoid factor activity. Am. J. Med. *40:*837–856 (1966).

23 Popp, J.W.; Dienstag, J.L.; Wands, J.R.; Bloch, K.J.: Essential mixed cryoglobulinemia without evidence for hepatitis-B virus infection. Ann. intern. Med. *92:* 379–383 (1980).

24 Realdi, G.; Alberti, A.; Rigoli, A.; Tremolada, F.: Immune complexes and Australia antigen in cryoglobulinemic sera. Z. Immunforsch. *147:*114–126 (1974).

25 Ruoslahti, E.; Engvall, E.; Hayman, E.G.: Fibronectin: current concepts of its structure and functions. Collagen Res. *1:*95–128 (1981).

26 Shalit, M.; Wollner, S.; Levo, Y.: Cryoglobulinemia in acute type-A hepatitis. Clin. exp. Immunol. *47:*613–616 (1982).

27 Tiku, M.L.; Beutner, K.R.; Ogra, P.L.: Immunological aspects of cryoprecipitates from the sera of chronic HBsAg carriers. Clin. exp. Immunol. *36:*54–62 (1979).
28 Tsukamoto, Y.; Helsel, W.E.; Wahl, S.M.: Macrophage production of fibronectin, a chemoattractant for fibroblasts. J. Immun. *127:*673–678 (1981).
29 Zardi, L.; Carnemolla, B.; Siri, A.; Santi, L.; Accolla, R.S.: Somatic cell hybrids producing antibodies specific to human fibronectin. Int. J. Cancer *25:* 325–329 (1980).
30 Zardi, L.; Siri, A.; Carnemolla, B.; Cosulich, E.; Viale, G.; Santi, L.: A simplified procedure for the preparation of antibodies to serum fibronectin. J. immunol. Methods *34:*155–165 (1980).

F. Dammacco, MD, Istituto di Patologia Medica, Policlinico,
I-70124 Bari (Italy)

Front. gastrointest. Res., vol. 8, pp. 146–157 (Karger, Basel 1984)

Liver Cirrhosis during Idiopathic Hemochromatosis

The Importance of an Early Diagnosis[1]

Claudio Tiribelli[a], *Mauro G. Marin*[b], *Roberto Vallini*[a], *Maria Grazia Zaramella*[a], *Tullio Zacchi*[c]

[a] Instituto di Patologia Medica, Università di Trieste, Trieste; [b] II Divisione Medicine, Ospedale Pordenone; [c] Centro Immunotrasfusionale, Ospedali Riuniti di Trieste, Trieste, Italy

Idiopathic hemochromatosis (IHC) is an inherited disease characterized by a progressive iron overload in various tissues, secondary to an increased intestinal absorption, the mechanism of which is still unknown [1, 2]. Within the cell, the excess iron is mostly stored in lisosomes as hemosiderin [3]. The subsequent damage is thought to be due either to lipid peroxidation induced by free radicals [4–6] or to enzyme release after disruption of the lisosomal membrane [7]. In addition, iron overload stimulates the collagen synthesis [8, 9] causing fibrosis, the grade of which is roughly proportional to the amount of metal in the tissue [10]. Interestingly, it has been reported that fibrosis can be stopped [9, 11, 12], and even reversed [13, 14], when the tissue iron concentration is reduced by specific treatment.

IHC is a chronic, long-lasting, slow-progressing disease. The clinical onset could be hepatic cirrhosis [15, 16], diabetes [17], skin hyperpigmentation [18], cardiac failure [19, 20], hypogonadism [21, 22] and/or artropathy [23]. A constant feature is hepatic cirrhosis that is complicated, in more than 35% of the cases, by hepatic carcinoma [15]. It must be stressed, however, that such a complication has never been reported in those cases in which the iron overload has been removed by venosections before cirrhosis developed fully [1, 2]. This finding

[1] This investigation has been partially supported by a grant of Ministero Pubblica Istruzione. The authors are deeply indebted to Miss *Luisella Vici* for her excellent help in preparing and typing the manuscript.

strongly indicates the importance of early diagnosis of IHC, especially during the asymptomatic stage of the disease.

The diagnosis of IHC is based mainly on the serum determination of iron concentration, percentage of transferrin saturation and serum ferritin as well as on the HLA typing [23-26]. Liver biopsy should be performed in cases of evident iron overload [15] so that the siderosis can be graded histologically [27] and the iron concentration in the tissue measured [28].

IHC is linked to an autosomal recessive inheritance [29, 30] with a gene located on chromosome 6, near locus A [31, 32]. Whereas the involvement of HLA-A3 antigen in the disease is well-known [33, 34], the B locus is still a matter for discussion since the association of A3 with both B7 [35] and B14 [36] has been reported. In a recent study, a close association with BW35 was demonstrated, suggesting that, at last, in North-East Italy, A3,BW35 could be the main haplotype making up the hemochromatosis allele [37]. The close relationship between IHC and HLA antigens allows the pattern of transmission of the disease to be followed in each family in which a proband is discovered. Relatives with the same HLA phenotype of the proband are considered homozygous for IHC allele, with a high disease risk. On the contrary, relatives with other HLA antigens associated with those of the proband are classified as heterozygotes [26]. The present investigation is aimed to verify the biochemical expression of iron overload, and possible tissue damage, in both homozygous and heterozygous subjects for IHC genotype.

Patients and Methods

50 subjects of 4 unrelated families with IHC were studied. All the families come from the same geographical area in North-East Italy. Diagnosis of IHC was based on criteria described above [37]. At the time of the diagnosis only 2 probands showed liver cirrhosis, diabetes and hypogonadism, whereas the remaining 3 were asymptomatic, the diagnosis being suspected by an occasional finding of a high serum iron concentration. Serum iron concentration [38], percentage of transferrin saturation [39] and serum ferritin [40] were determined in all the subjects. In addition, serum aminotransferase activity was determined and special attention was paid to rule out excessive ethanol intake. Possible treatment with iron medication and cause of bleeding, either chronic or acute, were investigated and eventually recorded. HLA antigens were determined with the microcytotoxicity test on purified lymphocytes [41] with sera provided by the 'Gruppo Collaborativo Italiano per la Tipizzazione Tessutale', determining 12 antigens for locus A and 18 for locus B. Liver biopsy was carried out in subjects with clear signs of iron overload and

hepatic siderosis graded according to *Walker* et al. [27], after Perl's stain. Chronic alcohol intake and/or hepatic damage of other origin was excluded in all the subjects on the basis of history and adequate blood tests.

From the family tree of each family, HLA genotype was derived from HLA phenotype. In each family, the proband was considered homozygote and relatives were classified as homozygous and heterozygous on the basis of the presence of one or both HLA haplotypes in common with the proband. Subjects with both HLA haplotypes different with the proband were classified as normal [26].

The results are expressed as mean ± SD and Student's t test was employed for statistical analysis. Differences of less than 5% were considered to be statistically significant.

Results

Table I reports the clinical and laboratory data of homozygous patients. In addition to the 4 probands, 6 subjects were classified in this group on the basis of HLA typing. In all these patients, the 3 markers of iron deposits are increased with the exception of subject A-II-6, a 76-year-old female. In this subject, liver biopsy was not performed due

Table I. Clinical, laboratory and histological data of the homozygous subjects (XX). H = Hepatomegaly; D = diabetes; P = cutaneous pigmentation; T = testicular atrophy; A = arthritis; C = cirrhosis; F = fibrosis

Subject	Sex	Age years	Clinical signs	Serum iron μg/dl	Transferrin sat. %	Serum ferritin ng/ml	HLA phenotype	Liver biopsy	
								siderosis	histology
Probands									
A-III-10	M	42	H	228	100	1,000	A3,BW35,W6	4	F
B-II-3	M	42	HPDTA	221	100	860	A3,BW35,W6	4	C
C-III-2	M	26	H	220	100	466	A3,BW35,W6	4	F
D-I-1	M	67	HPDTA	192	87	540	A3,BW35,W6	4	C
Relatives									
A-IV-2	M	24	–	220	100	700	A3,BW35,W6	4	F
A-II-10	F	68	H	171	67	225	A3,BW35,W6	4	F
A-II-6	F	76	–	64	32	78	A3,BW35,W6	–	–
C-III-3	F	23	–	230	81	236	A3,BW35,W6	3	F
C-III-4	M	19	–	212	68	415	A3,BW35,W6	3–4	F
C-II-1	F	51	–	184	76	323	A3,BW35,W6	3	F

Table II. Clinical and laboratory data of the heterozygous subjects (XO)

Subject	Sex	Age years	Clinical signs	Serum iron μg/dl	Transferrin sat. %	Serum ferritin ng/ml	HLA phenotype
A-II-2	F	79	-	81	29	93	A1,A9,B5,B8,W6
A-II-8	F	72	-	48	15	67	A3,A9,B8,BW35,W6
A-II-11	M	75	-	110	27	84	A3,A9,B18,BW35, W6
A-III-1	F	57	-	105	26	76	A3,A9,B8,BW35, W4,W6
A-III-7	F	48	-	65	23	52	A1,A3,B8,BW35,W6
A-IV-1	F	28	-	63	15	49	A3,A9,B5,BW35,W6
A-IV-3	F	23	-	160	48	60	A1,A3,B8,BW35,W6
A-IV-4	F	16	-	72	13	53	A1,A3,B8,BW35,W6
A-IV-5	M	10	-	128	23	38	A1,A3,B8,BW35,W6
B-II-1	F	54	-	140	48	65	A3,BW35,W6
B-I-2	F	71	-	92	27	25	A3,B14,BW35,W6
C-II-7	F	57	-	158	45	173	A3,A26,BW35,W6
C-II-8	F	53	-	104	29	40	A3,A10,BW35,W6
C-II-12	F	49	-	77	24	24	A3,A10,BW35,W6
C-III-1	F	28	-	55	26	39	A3,A10,BW35,W6
C-III-13	F	23	-	120	32	29	A3,A26,BW35, BW51,W4,W6
D-II-3	M	32	-	141	62	315	A3,B27,BW35,W4, W6
D-II-10	F	55	-	135	43	152	A3,A26,BW35,W6

to severe heart disease which contraindicated such a diagnostic procedure. From table I it can also be seen that the clinical signs are mostly present in probands, 2 of which (B-II-3 and D-I-1) show a fully developed disease. Interestingly, in these 2 patients, liver cirrhosis is present at liver biposy. No symptoms and/or signs were recorded in the relatives with the exception of subject A-II-10, in whom hepatomegaly was present at physical examination.

Table II shows the data of heterozygous subjects. 17 do not show any evidence of iron overload, whereas in subject E-II-3, a slight increase both in the percentage of transferrin saturation and serum ferri-

Table III. Iron status of normal subjects (00)

Subject	Sex	Age years	Serum iron µg/dl	Transferrin sat. %	Serum ferritin ng/ml
A-II-1	M	90	81	29	93
A-III-6	F	47	76	21	47
A-III-11	F	40	104	31	10
B-II-2	F	50	88	28	52
B-II-4	F	40	68	14	37
B-II-5	M	32	144	29	89
C-II-3	F	49	92	27	14
C-II-4	M	59	157	41	80
C-II-5	F	54	48	26	32
C-II-6	M	56	130	33	40
C-II-13	M	60	149	28	191
C-III-5	F	28	108	31	20
C-III-7	F	28	127	27	20
C-III-8	M	26	103	29	168
C-III-9	F	24	72	22	20
C-III-10	M	20	100	31	60
C-III-14	M	20	99	27	47
C-III-15	F	18	71	24	21
D-I-1	F	64	55	24	20
D-II-1	F	33	48	15	22
D-II-2	M	34	86	37	80
D-II-4	F	30	70	26	24

tin is present. Liver biopsy showed a grade 3 hepatic siderosis associated with modest fibrosis. Of particular interest is subject B-II-1, a 54-year-old woman with HLA phenotype identical to that of the proband B-II-3. Because it was impossible to test her deceased father, the two blanks in the loci A and B could not be determined and, consequently, she was classified as heterozygote.

Table III reports the data of family members with different HLA antigens on chromosome 6 and, therefore, classified as normals.

Figure 1 shows the serum iron concentration in the three groups. Except for the case discussed above, homozygotes show increased serum iron levels, whereas all the heterozygotes fall within the normal range.

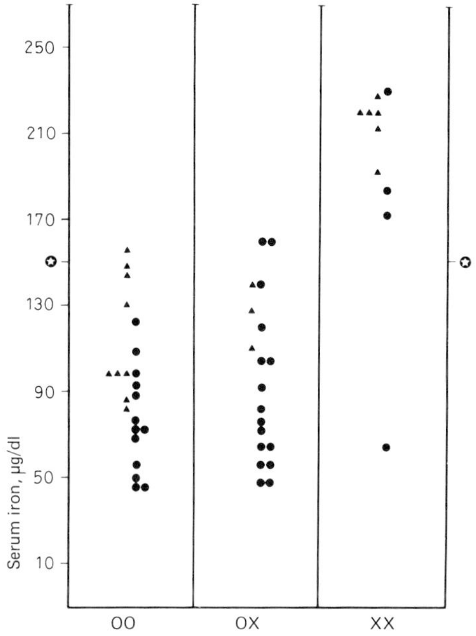

Fig. 1. Serum iron concentration (µg/dl) in normals (00), heterozygous (0X) and homozygous (XX) subjects. Triangles refer to males and circles to females. Normal value is indicated by a star within circle.

Figure 2 shows the distribution among the groups of the percentage of transferrin saturation. Again, all the homozygotes, but one, have an increased percentage of saturation, and all the heterozygotes, on the contrary, are in the normal range. Only subject E-II-3, with histologicaly proved hepatic siderosis, shows an increased value of transferrin saturation.

Figure 3 gives the values of serum ferritin. Although a quite large scattering and some overlap of the data exists, especially for homozygous females, higher values are present in homozygotes when compared to heterozygotes and normals.

In table IV, mean ± SD for the three considered parameters is reported. A highly statistically significant difference ($p < 0.001$) is pres-

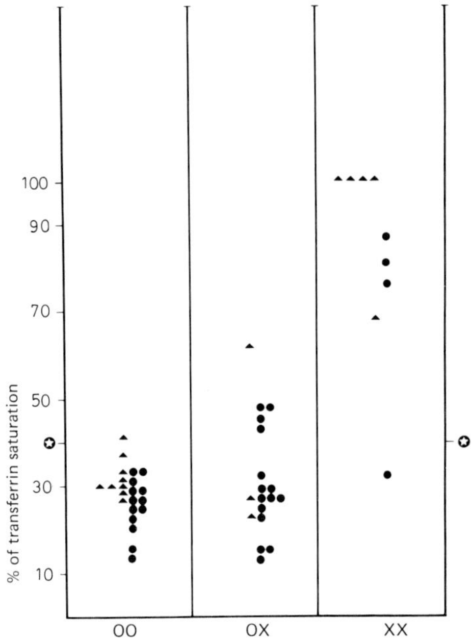

Fig. 2. Percentage of transferrin saturation in normals (00), heterozygous (0X) and homozygous (XX) subjects. For explanation of symbols, see figure 1.

ent between homozygotes and the two other groups whereas no difference could be observed between heterozygotes and normals. It is interesting to note that the mean age of the groups is superimposable whereas the male/female ratio is higher in homozygotes than in the remaining two groups.

Discussion

Homozygous subjects with HLA genotype identical to that of patients with IHC clearly show the presence of iron overload. On the con-

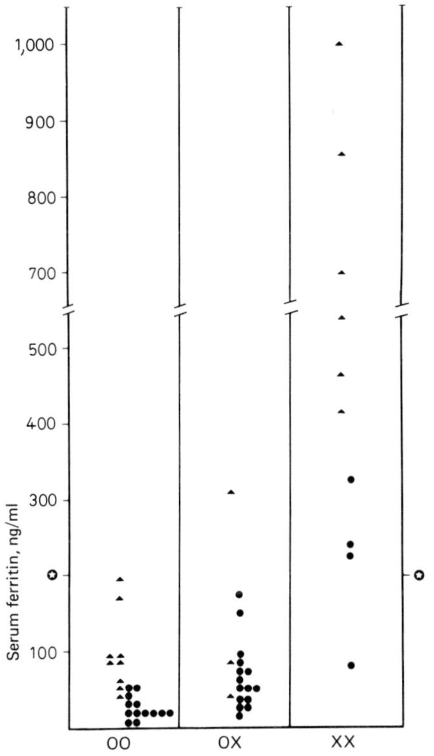

Fig. 3. Serum ferritin (ng/ml) in normals (00), heterozygous (0X) and homozygous (XX) subjects. For explanation of symbols, see figure 1.

Table IV. Iron status in homozygotes, heterozygotes and normals. The results are expressed as mean ± SD. * $p < 0.001$ vs heterozygotes and normals

	Number	M/F	Age years	Serum iron µg/dl	Transferrin sat. %	Serum ferritin ng/ml
Homozygotes	10	6/4	44 (19–76)	194.80 ± 49.88*	81.10 ± 21.78*	484.30 ± 294.74*
Heterozygotes	18	3/15	46 (10–79)	103.00 ± 35.72	30.83 ± 13.25	79.67 ± 71.26
Normals	22	9/13	41 (18–90)	94.36 ± 31.75	27.27 ± 6.12	54.02 ± 47.80

trary, subjects sharing only one HLA haplotype with the proband (heterozygotes), usually present only minor signs, if any of the disease and, in the majority of the cases, no difference could be detected between heterozygotes and control population. These data agree with previous reports [26, 42, 43]. It appears, therefore, that HLA typing of families of patients with IHC allows early diagnosis of the disease, also when the clinical and biochemical expression of IHC is not fully evident. This fact is particularly relevant both in menstruating females and/or in young males. Moreover, attention should be paid to the possibility of generating homozygous sons from genetically predisposed parents.

The most important aspect, however, seems to us to be the identification and treatment of subjects with clinically silent disease. In this study, abnormal parameters of iron status are present in all the relatives associated with a marked hepatic iron content, in spite of a silent clinical picture. As indicated in table IV, even in the presence of quite a large degree of variability, a highly statistical significance in the level of serum iron, ferritin and percentage of transferrin saturation existes between homozygotes and heterozygotes. In contrast, no difference could be detected between heterozygotes and normals. Therefore, the association of these three parameters of body iron content with HLA typing is able to detect the presence of IHC in an asymptomatic stage and, subsequently, to treat it. This fact is particularly relevant considering that early treatment has been shown to significantly modify the prognosis of IHC [11–13]. In those areas in which a high incidence of IHC has been reported (the mountain area in the North-East of Italy in our case), the execution of these three tests seems therefore particularly recommended, considering that IHC is symptomless for quite a long period [23]. Once the diagnosis is established, HLA typing of the family could be of great help in the eventual detection of other relatives sharing the proband with the same HLA phenotype and, therefore, with a high risk of developing the disease and, in particular, the complications associated with the disorder. It is important to underline that hepatic carcinoma has been reported to complicate only IHC in which a liver cirrhosis has developed [1, 44].

In conclusion, the determination of serum iron, percentage of saturation of transferrin and serum ferritin, associated with a familial study by HLA typing, seems to be a sensitive and specific method for an early diagnosis of IHC, able to greatly influence the duration and the quality of life of these subjects.

Summary

Early recognition and subsequent diagnosis of tissue iron overload is important since it allows appropriate treatment while reducing the progression to idiopathic hemochromatosis. This could be achieved with the determination of three markers of iron body content (serum iron, serum ferritin and percentage of transferrin saturation) associated with familial HLA typing, especially in those families in which a patient with the disease has been diagnosed. HLA typing allows the identification of both homozygous and heterozygous subjects.

In 50 subjects of 4 unrelated families, 9 out of 10 homozygotes with the A3, BW3, W6 genotype have clear signs of iron overload in 2 cases associated with clinical symptoms. On the contrary, no markers of increased iron body stores could be found in 17 out of 18 heterozygotes.

References

1 Halliday, J.W.; Powell, L.W.: Iron overload. Semin. Hematol. *19:*42 (1982).
2 Powell, L.A.; Bassett, M.L.; Halliday, J.: Hemochromatosis: 1980 update. Gastroenterology *78:*377–381 (1980).
3 Trump, B.F.; Valigorsky, J.M.; Arstila, A.U.: The relationship of intracellular pathways of iron metabolism to cellular iron overload and iron storage diseases. Am. J. Path. *72:*295–336 (1973).
4 Hunter, F.E.; Gebicki, J.M.; Hoffstein, P.E.: Swelling and lysis of rat liver mitochondria induced by ferrous ions. J. biol. Chem. *238:*828–835 (1963).
5 Arstila, A.U.; Smith, M.A.; Trump, B.F.: Microsomal lipid peroxidation: morphological characterization. Science *175:*530–533 (1972).
6 Graziano, J.H.: Potential usefulness of free radical scavanger in iron overload. Birth Defects *12:*135–142 (1976).
7 Peters, T.J.; Seymour, C.A.: Acid hydrolase activities and lysosomal integrity in liver biopsies from patients with iron overload. Clin. Sci. mol. Med. *50:* 75–78 (1976).
8 Iancu, T.C.; Neustein, H.B.; Landing, B.H.: The liver in thalassemia major: ultrastructural observations; in Iron metabolism. Ciba Fdn Symp., No. 51 (new series) (Elsevier North-Holland, Amsterdam 1977).
9 Barry, M.; Flynn, D.M.; Letsky, E.A.: Long-term chelation therapy in thalassemia major: effect on liver iron concentration, liver histology and clinical progress. Br. med. J. *ii:*16–20 (1974).
10 Isaacson, C.; Seftel, H.C.; Helley, K.J.: Siderosis in the bantù: the relationship between iron overload and cirrhosis. J. Lab. Clin. *58:*845–853 (1961).
11 Bomford, A.; Williams, R.: Long-term results of venesection therapy in idiopathic haemochromatosis. Q. Jl. Med. *45:*611–623 (1976).
12 Powell, L.W.: Changing concepts in haemochromatosis. Post-grad. med. J. *46:* 200–209 (1970).
13 Rojkind, M.; Dunn, M.A.: Hepatic fibrosis. Gastroenterology *76:*849–863 (1979).

14 Knauer, C.M.; Gamble, C.N.; Monroe, L.S.: The reversal of hemochromatosic cirrhosis by multiple phlebotomies. Gastroenterology *49:*667 (1965).
15 Powell, L.W.; Isselbacher, K.J.: Hemochromatosis; in Principles of internal medicine, pp. 488–491 (Harrison's Mc Grow Hill, New York 1980).
16 Grace, N.D.; Powell, L.W.: Iron storage disorders of the liver. Gastroenterology *67:* 1257–1283 (1974).
17 Dymock, I.W.; Cassar, J.; Pyke, D.A.; Oakley, W.G.; Williams, R.: Observations on the pathogenesis, complications and treatment of diabetes in 115 cases of haemochromatosis. Am. J. Med. *52:*203–210 (1971).
18 Chevrant-Breton, J.; Simon, M.; Bourel, M.: Cutaneous manifestations of idiopathic hemochromatosis. Archs. Derm. *113:*161–165 (1977).
19 Short, E.M.; Winkle, R.A.; Billingham, M.E.: Myocardial involvement in idiopathic hemochromatosis. Morphologic and clinical improvement following venesection. Am. J. Med. *70:*1275–1279 (1981).
20 Leclercq, J.F.; Passa, P.; Warnet, A.: La myocardiopathie hémochromatosique: rôle bénéfique de la déplétion ferrique. Ann. Cardiol. Angéiol. *24:*291–295 (1980).
21 Charbonnel, B.: Pituitary function in idiopathic haemochromatosis: hormonal study in 36 male patients. Acta endocr., Copenh. *98:*178–183 (1981).
22 Walsh, C.H.; Wright, A.D.; Williams, J.E.: A study of pituitary function in patients with idiopathic hemochromatosis. J. clin. Endocr. Metab. *43:*866–872 (1976).
23 Powell, L.W.; Halliday, J.W.: The detection of early hemochromatosis. Dig. Dis. *23:*377–379 (1978).
24 Valberg, L.S.; Lloyd, D.A.; Ghent, C.N.: Clinical and biochemical expression of the genetic abnormality in idiopathic haemochromatosis. Gastroenterology *79:* 884–892 (1980).
25 Halliday, J.W.; Russo, A.M.; Cowlishaw, J.L.; Powell, L.W.: Serum ferritin in diagnosis of haemochromatosis. Lancet *i:*621–624 (1977).
26 Bassett, M.L.; Halliday, J.W.; Powell, L.W.: HLA typing in idiopathic hemochromatosis; distinction between homozygotes and heterozygotes with biochemical expression. Hepatology *1:*120–126 (1981).
27 Walker, R.J.; Miller, J.P.; Dymock, I.; Shilkin, K.; Williams, R.: Relationship of hepatic iron concentration to histochemical grading to total chelatable iron in conditions associated with iron overload. Gut *12:*1011–1014 (1971).
28 Barry, M.; Sherlock, S.: Measurement of liver-iron concentration in needle-biopsy specimens. Lancet *i:*100–103 (1971).
29 Cartwright, G.E.; Edwards, C.Q.; Kravitz, K.: Hereditary hemochromatosis: phenotypic expression of the disease. New Engl. J. Med. *301:*175–179 (1979).
30 Simon, M.; Bourel, M.; Genetet, B.: Idiopathic hemochromatosis: demonstration of recessive transmission and early detection by family HLA typing. New Engl. J. Med. *297:*1007–1021 (1977).
31 Simon, M.; Hespel, H.J.P.; Beaumont, C.: Idiopathic hemochromatosis: a study of biochemical expression in 247 heterozygous members of 63 families: evidence for a single major HLA-linked gene. Gastroenterology *78:*703–708 (1980).
32 Edwards, C.Q.; Cartwright, G.E.; Skolnick, M.H.; Amos, D.B.: Genetic mapping of the hemochromatosis locus on chromosome six. Hum. Immunol. *1:*19–22 (1980).
33 Simon, M.; Pawlotsky, Y.; Bourel, M.: Hémochromatose idiopathique: maladie associée à l'antigène HLA-A3? Nouv. Presse méd. *4:*1432 (1975).

34 Simon, M.; Alexandre, J.L.; Bourel, M.: Heredity of idiopathic hemochromatosis: a study of 106 families. Clin. Genet. *11*:327–341 (1977).
35 Shewan, W.G.; Mouat, S.A.; Alban, T.M.: HLA antigens in hemochromatosis. Br. med. J. *i*:281–282 (1976).
36 Simon, M.; Bourel, M.; Fauchet, R.: Association of HLA-A3 and HLA-B14 antigens with idiopathic haemochromatosis. Gut *17*:332–334 (1976).
37 Zacchi, T.; Marin, M.G.; Galli, G.; Tiribelli, C.: HLA antigens and idiopathic hemochromatosis in North-East of Italy: a familial study. Ital. J. Gastroenterol. *14*: 205–207 (1982).
38 Ramsay, M.N.M.: The determination of iron in blood plasma or serum. Clinica chim. Acta *2*:214–220 (1957).
39 Ramsay, M.N.M.: The determination of the total iron-binding capacity of serum. Clinica chim. Acta *2*:221–226 (1957).
40 Addison, G.M.; Beamish, M.R.; Hales, C.N.: An immunoradiometric assay for ferritin in the serum of normal subjects and patients with iron deficiency and iron overload. J.clin. Path. *25*:326–329 (1972).
41 Mittal, K.K.; Mickey, M.R.; Singla, D.F.: Serotyping for homotransplantation. Transplantation *6*:913–927 (1968).
42 Cartwright, G.E.; Edwards, C.O.; Kravitz, M.: Hereditary hemochromatosis. Phenotype expression of the disease. New Engl. J. Med. *301*:175–179 (1976).
43 Basset, M.L.; Halliday, J.W.; Powell, L.W.; Doran, T.: Early detection of idiopathic haemochromatosis: relative value of serum ferritin and HLA typing. Lancet *ii*:4–7 (1979).
44 Hines, C.; Davis, W.D.; Ferrante, W.A.: Hepatoma developing in hemochromatosis in spite of adequate treatment by multiple phlebotomies. Am. J. dig. Dis. *16*:349–355 (1971).

C. Tiribelli, MD, Istituto di Patologia Medica, Università di Trieste,
I-34129 Trieste (Italy)

Front. gastrointest. Res., vol. 8, pp. 158–168 (Karger, Basel 1984)

Clinical Features and Diagnostic Markers in Non-A, Non-B Hepatitis[1]

G. Realdi, F. Tremolada, F. Bortolotti, A. Alberti, A.M. Rigoli, G. Diodati, C. Valfrè, A. Bertaggia, A. Ruol

Istituto di Medicina clinica, Patologia medica dell'Università di Padova, Padova, Italy

Introduction

The term 'non-A, non-B hepatitis' is currently used to define acute and chronic inflammatory liver disease thought to be caused by transmissible viral agents distinct from known hepatitis viruses. These agents and their serological markers have not yet been identified and the diagnosis of non-A, non-B hepatitis must be based upon exclusion criteria. This limitation has at least two relevant implications for the diagnostic approach. Firstly, while serological markers are available to reasonably exclude a role of known hepatitis viral agents (represented in our region mainly by the hepatitis A and B viruses and, less frequently, by the cytomegalovirus (CMV) and by Epstein-Barr virus), it may be more difficult in the individual case to also rule out any other possible non-viral cause of liver damage. Secondly, the use of exclusion criteria in the diagnosis of non-A, non-B hepatitis precludes the identification of distinct non-A, non-B agents that may be involved in the etiology of the illness and may be responsible for important differences in the clinical course and outcome.

Despite these limitations, a diagnosis of non-A, non-B hepatitis may be reasonably made in the presence of acute liver damage, when an accurate history of the patient is obtained and serial serum samples are collected during the acute phase for serology. When the illness

[1] This work was supported in part by a grant from the Ministero della Pubblica Istruzione, Italy. Electron microscopy studies of liver biopsies were carried out in collaboration with Prof. *C.A. Busachi* of the Istituto di Patologia medica I° dell'Università di Bologna.

develops after blood transfusion, then the diagnosis is almost certain. If such cases do not recover after the acute phase and transaminases remain persistently elevated, chronic non-A, non-B hepatitis is highly probable. On the contrary, the diagnostic approach is more difficult in cases of chronic hepatitis when an acute onset has not been recognized. In fact, although we may suspect that a large proportion of HBsAg-negative, non-alcoholic chronic liver diseases could be due to infection with non-A, non-B hepatitis agents, this may not be more than presumptive, as other unrecognized causes of liver damage, not necessarily of viral nature, cannot be excluded.

On the basis of these considerations, it seems conceivable that today the only way to explore the natural history of non-A, non-B hepatitis is to rely on prospective studies of patients who develop acute hepatitis of the non-A, non-B type with or without evidence of parenteral exposure, and are then followed longitudinally over a prolonged period of time.

We have recently conducted two distinct prospective studies of this type and will review here the results obtained in relation to the epidemiological and clinical features of acute and chronic non-A, non-B hepatitis and to the possibility of identifying specific diagnostic markers in the diseased liver.

Patients and Methods

Selection of Patients. Two separate groups of patients with acute non-A, non-B hepatitis were considered in this study. The first group included all 94 patients with acute symptomatic non-A, non-B hepatitis admitted to our Infectious Diseases Department during a survey period of 30 consecutive months [1]. They represented 15% of all acute hepatitis cases seen during the same period of time. The diagnosis of non-A, non-B hepatitis was based upon exclusion criteria. The 94 patients were observed during the acute phase and then followed longitudinally up to complete normalization of transaminase levels. Those patients in whom transaminases did not normalize after 12 months of follow-up were considered to have developed chronic disease. Some of them underwent percutaneous liver biopsy for histological evaluation of the chronic liver lesions.

The second group of patients considered were cases of post-transfusion hepatitis acquired during open-heart surgery. These patients were identified during a prospective study of post-transfusion hepatitis recently conducted in our Institute [2]. Among 297 transfused patients, 63 (21.2%) developed post-transfusion hepatitis, defined as a rise in alanineaminotransferase (ALT) levels on at least two consecutive occasions between 14 and 180 days after transfusion. Non-viral causes of ALT elevations were excluded and etiology was defined by serological tests for hepatitis A (anti-HAV IgM), for hepatitis B

(HBsAg and anti-HBc) and for CMV infection (anti-CMV IgM seroconversion). Patients with no evidence of recent infection with these agents were considered affected by non-A, non-B hepatitis. All patients were followed prospectively at regular intervals after the acute phase to assess the long-term evolution of the disease. Those patients who showed persistence of abnormal ALT levels 12 months after onset were considered to have developed chronic liver disease. Many of these patients underwent a liver biopsy to assess the severity of histological lesions.

Viral Markers in Serum. HBsAg, anti-HBs, anti-HBc, anti-HAV IgM (Abbott Radioimmunoassay Kits) and anti-CMV IgM (Cytomegalia Enzygnost Behringwerke) were tested in serum according to the manufacturer's instructions.

Non-A, Non-B Hepatitis Markers in the Liver. Liver biopsies, when available, were studied for two tissue markers, recently identified as linked to non-A, non-B hepatitis. A non-A, non-B nuclear antigen was studied on cryostatic liver sections by direct immunofluorescence according to *Alberti* et al. [3]. Intranuclear particles of the type described by *Shimizu* et al. [4] in chimpanzees with non-A, non-B hepatitis were studied by electron microscopy on liver biopsies fixed in 2.5% glutaraldeyde, as described in detail elsewhere [5].

Statistical Analysis. The χ^2 test and the Student's t test were used for statistical evaluation of the results.

Results

Acute 'Sporadic' Non-A, Non-B Hepatitis. Among the 94 patients with acute symptomatic non-A, non-B hepatitis, various risk factors could be identified as possible sources of infection. When patients with a history of blood transfusion (25 cases) or of frequent contact with blood samples (2 cases) and those admitting parenteral abuse of illicit drugs (25 cases) were excluded, 42 cases with no evidence of overt parenteral exposure remained. These 'sporadic' cases of acute non-A, non-B hepatitis (6.7% of all acute hepatitis cases observed during the period of the survey) dit not differ clinically during the acute phase from those who could have acquired the infection by the parenteral route, while a significant difference was noted in relation to the long-term outcome of the illness (table I). Indeed, by considering only those cases who have completed a 12-month follow-up, more than half of the patients with parenteral exposure (22 out of 34) showed evidence of progression to chronic hepatitis compared to only 16.2% of the sporadic cases. 12 patients of the former group and 3 of the latter underwent liver biopsy that showed a wide spectrum of liver lesions. Fea-

Table I. Acute phase clinical features and evolution of liver disease in 94 consecutive patients with acute symptomatic non-A, non-B hepatitis admitted to the Department of Infectious Disease

Exposure	Symptoms		Jaundice		ALT[1] peak mean ± SD	Bilirubin peak mean mg/100 ml ± SD	Increased ALT levels at 12 months (%)	Liver histology during chronic phase
	cases	%	cases	%				
Post-transfusion (25 cases)	15	60	15	60	1,233 ± 974	12.1 ± 8.6	7/14 (50)	CAH = 2 CPH = 2
Drug addiction (25 cases)	20	80	10	40	1,418 ± 694	7.9 ± 2.3	13/18 (72)	CAH = 1 CPH = 5 CLH = 1
Health-care staff (2 cases)	2		–	–	760 ± 40	–	2/2	CPH = 1
Unknown ('sporadic hepatitis') (42 cases)	38	90	33	78	1,475 ± 888	11.4 ± 7.0	6/37 (16)[2]	CPH = 1 CLH = 2

[1] ALT: normal values < 50 IU/l. [2] $p < 0.01$ vs post-transfusion and drug-addicts cases. CAH = Chronic active hepatitis; CPH = chronic persistent hepatitis; CLH = chronic lobular hepatitis.

tures of chronic active hepatitis (CAH) were observed only in patients thought to have acquired the illness by the parenteral route. When patients with chronic hepatitis were then followed up to 24 months, none of those with parenteral exposure normalized transaminase levels, while this event persistently occurred in 2 of the 6 'sporadic' cases.

Post-Transfusion Non-A, Non-B Hepatitis. Among the 297 consecutive open-heart surgery patients, post-transfusion hepatitis occurred in 63 (21.2%) cases, with a significantly higher hepatitis attack rate in 51 recipients of commercial clotting factor concentrates compared to 246 recipients of blood units from single volunteer donors (table II). Of the 63 patients with post-transfusion hepatitis, 2 (3%) had evidence of hepatitis B virus infection, 8 (13%) showed seroconversion to anti-CMV IgM during acute phase and 53 (84%) were considered cases of non-A, non-B hepatitis. When patients with CMV infection and those with non-A, non-B hepatitis were compared for several parame-

Table II. Prospective study of post-transfusion hepatitis: incidence of acute hepatitis in open-heart surgery patients receiving only blood or also blood products (clotting factors)

Product transfused	Number of cases	Blood units transfused mean ± SD	Recipients with post-transfusion hepatitis		
			number	%	Nr/1,000 U
Only blood	246	6.1 ± 3.1	34	13.8	22.5
Blood units and clotting factors	51	10.5 ± 8.8*	29	56.8*	54.0

* $p < 0.001$.

Table III. Prospective study of post-transfusion hepatitis: acute phase clinical features of non-A, non-B hepatitis and of CMV hepatitis in the group of open-heart surgery patients receiving only blood or also blood products (clotting factors)

	CMV (8 cases)	Non-A, Non-B (53 cases)	p
Number of blood units transfused, mean ± SD	14.7 ± 11.7	9.1 ± 4.6	n.s.
Incubation period, mean weeks ± SD	5.4 ± 1.8	7.1 ± 4.8	< 0.05
Jaundice (%): All cases	2 (25)	17 (32)	n.s.
Receiving only blood units	1	5	n.s.
Number of cases with symptoms of liver disease (%)	5 (62)	36 (68)	n.s.
Number of cases with fever or pleural effusion (%)	4 (50)	4 (7.5)	< 0.01
ALT peak, mean ± SD	823 ± 450	876 ± 679	n.s.
Number of patients with increased transaminase (ALT) levels after 12 months (%)	2[1]/8 (25)	30/40 (75)	< 0.01

ALT: normal values < 50 IU/l. [1] 1 of these 2 patients had received blood units and clotting factor concentrates during operation.

Table IV. Prospective study of non-A, non-B hepatitis: acute phase clinical and biochemical features in relation to outcome of the disease

	Patients recovered (10 cases)	Patients with chronic evolution (30 cases)	p
Age, mean years ± SD	52.9 ± 7.17	42.2 ± 13.3	< 0.001
Sex, % male	60	75	n.s.
Incubation period, mean weeks ± SD	5.7 ± 2.3	6.3 ± 2.1	n.s.
Follow-up, mean months ± SD and range	18.3 ± 7.8 (12–34)	18.5 ± 5.7 (12–32)	n.s.
Jaundice, %	30	33	n.s.
ALT peak, mean ± SD	763 ± 581	1,067 ± 735	n.s.
Gammaglobulin, mean ± SD	1.51 ± 0.33	1.64 ± 0.37	n.s.
ALT pattern, (number of cases):			
Monophasic	7	2	
Polyphasic	3	25	< 0.02
Plateau	0	3	
Receiving blood and also clotting factors	4/10	18/30	n.s.

ALT: normal values < 50 IU/l.

ters (table III), the incubation period was significantly shorter in the CMV cases, who also had a higher incidence of fever and of pleural effusion during the acute phase. The two groups also differed significantly in relation to the long-term evolution that was benign in most CMV cases in contrast to the non-A, non-B cases, 75% of whom showed persistence of abnormal transaminase levels for more than 12 months after onset.

Because of the high rate of chronicity following post-transfusion non-A, non-B hepatitis, we tried to identify acute phase clinical and biochemical features of the illness that could predict its outcome. As shown in table IV, only the ALT pattern during the acute phase was discriminative between the two groups, being mainly monophasic in patients who subsequently recovered and polyphasic in those with chronic evolution. These latter cases had a significantly younger mean age, but the significance of this finding remains unknown. No statistical difference was noted between the two groups in relation to the percentage of cases who had received only blood or blood and clotting

Table V. Liver biopsies taken during chronic phase of non-A, non-B hepatitis in 14 patients included in the prospective study and in further 19 open-heart surgery patients referred to our Institution for acute non-A, non-B hepatitis before starting the prospective study

	Histology	Years after acute hepatitis (1 2 3 4 5 6)	Total cases
Patients with chronic post-transfusion non-A, non-B hepatitis included in the prospective study	CLH	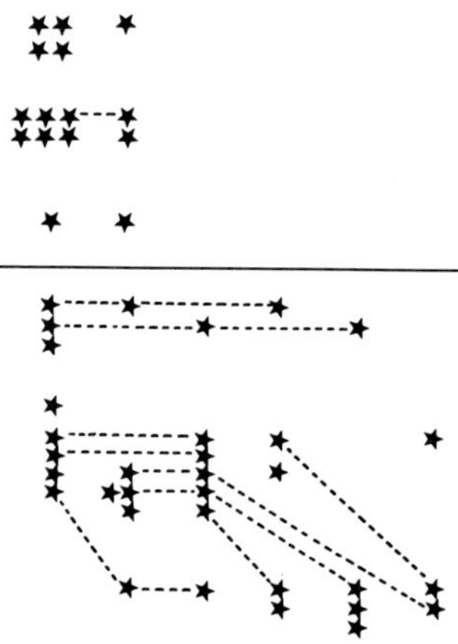	/
	CPH		5
	CAH		7
	liver cirrhosis		2
Other patients with chronic post-transfusion non-A, non-B hepatitis	CLH		3
	CPH		1
	CHA		7
	liver cirrhosis		8

CAH cases with subsequent evolution to liver cirrhosis have been considered only in this last group for the total cases number. Dotted lines indicate subsequent liver biopsies in the same patient.

factors, indicating that the type of transfusion did not influence the outcome of the illness.

Histological Findings in Chronic Post-Transfusion Non-A, Non-B hepatitis. 14 patients who showed chronic evolution of post-transfusion non-A, non-B hepatitis during the prospective study and 19 further non-consecutive patients who had developed chronic hepatitis after acute post-transfusion non-A, non-B hepatitis, underwent liver biopsy at varying time intervals after acute onset. The results of histologi-

Table VI. Non-A, non-B nuclear antigen, detected by immunofluorescence (IF), and 15–27 nm intranuclear particles, detected by electron microscopy (EM), in the liver of patients with different types of chronic hepatitis

Non-A, non-B nuclear antigen (IF)	+	+	–	–
Intranuclear particles (EM)	+	–	+	–
Chronic hepatitis following acute post-transfusion non-A, non-B hepatitis	8	2	1	1
Chronic hepatitis following acute 'sporadic' non-A, non-B hepatitis	0	0	0	3
Cryptogenic chronic hepatitis	0	0	0	7

cal assessment of these liver biopsies are shown in table V. A wide spectrum of liver lesions was observed, ranging from features of CLH to those of established liver cirrhosis. Most cases of cirrhosis were diagnosed between the second and the fifth year of follow-up. Progression from CAH to liver cirrhosis was documented in 5 patients who underwent more than one liver biopsy during the period of observation.

Candidate Non-A, Non-B Hepatitis Markers in the Liver. Liver biopsies from patients who developed chronic liver lesions during our prospective study of non-A, non-B hepatitis were used to better define the diagnostic significance of two tissue markers, recently identified as linked to non-A, non-B infection. As shown in table VI, most patients who had developed chronic hepatitis following post-transfusion non-A, non-B hepatitis were positive in the liver for the non-A, non-B nuclear antigen identified by immunofluorescence, and for 15–27 nm intranuclear particles detected by electron microscopy. 8 of the 12 patients were positive for both markers, indicating a close association of the non-A, non-B nuclear antigen with the 15–27 nm particles. In contrast, neither of the two markers was observed in the liver of patients who had progressed to chronic hepatitis after acute 'sporadic' non-A, non-B hepatitis. 7 additional patients with HBsAg-negative, non-alcoholic chronic liver disease and no evidence of acute onset and of parenteral exposure were also negative for the two non-A, non-B markers in the liver.

Discussion

Our prospective study of the long-term evolution of acute non-A, non-B hepatitis represents the first experience of this type in Italy. Although the survey is still in progress, the data obtained so far and summarized in this report allow some considerations.

We have confirmed that, as reported by *Alter* [6], non-A, non-B infection is frequently acquired by the parenteral route, with a hepatitis attack rate that is particularly high in multitransfused patients, especially if also treated with clotting factor concentrates [2]. There are, however, other forms of acute non-A, non-B hepatitis that occur in patients with no overt evidence of parenteral exposure to blood or blood derivates [7]. These forms, which are often incorrectly defined as 'sporadic' non-A, non-B hepatitis, may not differ clinically during the acute phase from post-transfusion cases. However, a clear difference is seen between the two forms in relation to the long-term outcome [1]. While most 'sporadic' cases recover after acute liver damage, more than 50% of post-transfusion cases may develop chronic liver disease that in a significant number of patients (70% in our experience) is associated with severe histological lesions, including liver cirrhosis [8].

These clinical differences between the two types of non-A, non-B hepatitis may be due to several factors, such as a different virus load, that is certainly higher in post-transfusion cases, or variations in the host immune system, which may be damaged following massive transfusion. However, the most likely explanation for the differences observed in post-transfusion and 'sporadic' cases of non-A, non-B hepatitis is an involvement of distinct non-A, non-B agents. This would also explain the results obtained by immunofluorescence and by electron microscopy.

Indeed, the non-A, non-B nuclear antigen and the 15–27 nm intranuclear particles may represent two markers of the same non-A, non-B agent, responsible for the large majority of parenterally acquired non-A, non-B hepatitis infection, but not involved in the 'sporadic' form of the illness. If this is the case, the two markers would be diagnostic for one type of non-A, non-B hepatitis that is acquired mainly from blood transfusion and frequently shows chronic evolution with development of severe liver lesions. On the other hand, these two markers may not be of great diagnostic relevance in identifying other forms of chronic liver disease, possibly the large majority of cases of HBsAg-negative, non-alcoholic liver disease seen in the clinical practice, due to other, not yet identified, etiological agents.

Summary

In this report we describe some epidemiological and clinical features of non-A, non-B hepatitis, as they were identified during two distinct prospective studies conducted in our Institute.

Two types of non-A, non-B hepatitis could be identified: one form is acquired by the parenteral route, being frequently observed in multitransfused patients and in drug addicts, while the other occurs in patients with no evidence of overt parenteral exposure. This latter form has been incorrectly termed 'sporadic' non-A, non-B hepatitis. Acute phase clinical and biochemical features are not dissimilar between these two forms of non-A, non-B hepatitis although they clearly differ in relation to the long-term outcome. Indeed, non-A, non-B hepatitis following parenteral exposure shows a high incidence of progression to chronic hepatitis, often with severe histological lesions, including cirrhosis, while 'sporadic' non-A, non-B hepatitis is more benign, with complete resolution in the majority of cases. Patients with chronic post-transfusion non-A, non-B hepatitis are frequently positive in the liver for a non-A, non-B nuclear antigen, detected by immunofluorescence, and for 15-27 nm intranuclear virus-like particles, visualized by electron microscopy. These two tissue markers are not detected in cases of 'sporadic' chronic non-A, non-B hepatitis nor in cases of cryptogenic chronic liver disease, suggesting the existence of distinct etiological agents.

References

1 Bortolotti, F.; Realdi, G.; Cadrobbi, P.; Crivellaro, C.; Pornaro, E.: Clinical features and evolution of non-A, non-B hepatitis in relation to the epidemiological background. Ital. J. Gastroenterol. *14:*86–89 (1982).
2 Tremolada, F.; Chiappetta, F.; Noventa, F.; Valfrè, C.; Ongaro, G.; Realdi, G.: Prospective study of post-transfusion hepatitis in cardia surgery patients receiving only blood or also blood products. Vox Sang. *44:*25–30 (1983).
3 Alberti, A.; Realdi, G.; Bortolotti, F.; Cadrobbi, P.; Barbieri, R.; Tremolada, F.; Ongaro, G.: Detection by immunofluorescence of an antigen-antibody system in patients with acute and chronic non-A, non-B hepatitis. Liver *1:*183–190 (1981).
4 Shimizu, J.K.; Feinstone, S.M.; Purcell, R.H.; Alter, K.J.; London, W.T.: Non-A, non-B hepatitis: ultrastructural evidence of two agents in experimentally infected chimpanzees. Science *205:*197–200 (1979).
5 Busachi, C.A.; Realdi, G.; Alberti, A.; Badiali De Giorgi, L.; Tremolada, F.: Ultrastructural changes in the liver of patients with chronic non-A, non-B hepatitis. J. Med. Virol. *7:*205–212 (1981).
6 Alter, H.J.: The dominant role of non-A, non-B in the pathogenesis of post-transfusion hepatitis: a clinical assessment. Clin. Gastroenterol. *9:*155–170 (1980).

7 Villarejos, V.M.; Visona, K.A.; Eduarte, C.A.; Provost, P.J.; Hilleman, M.R.: Evidence for viral hepatitis other than type A or type B among persons in Costa Rica. New Engl. J. Med. *293:* 1350–1352 (1975).

8 Realdi, G.; Alberti, A.; Rugge, M.; Rigoli, A.M.; Tremolada, F.; Schivazappa, L.; Ruol, A.: Long-term follow-up of acute and chronic non-A, non-B post-transfusion hepatitis: evidence of progression to liver cirrhosis. Gut *23:*270–275 (1982).

Dr. G. Realdi, MD, Istituto di Medicina clinica, Patologia medica,
Via Giustiniani 2, I-35100 Padova (Italy)

Front. gastrointest. Res., vol. 8, pp. 169–179 (Karger, Basel 1984)

Familial Clustering of HBV Infection and Chronic Liver Disease[1]

M. Coltorti, C. Del Vecchio-Blanco, N. Caporaso, R. Suozzo, F. Servillo

Istituto di Semeiotica Medica, I Facoltà di Medicina e Chirurgia, Università di Napoli, Napoli, Italy

A high prevalence of serological markers of hepatitis B virus (HBV) infection has previously been documented in the general population and even more in chronic liver disease in Italy, particularly in the south of the country, where alcoholic liver disease is infrequent [1–7].

A high degree of familial clustering of HBV infection has been reported particularly from countries with a large diffusion of HBsAg [8–12]. Both environmental and genetic factors may contribute to this phenomenon [9, 13–15].

However, the following points need to be more thoroughly investigated: (1) The influence of the presence or absence of liver disease among the index cases on the intrafamilial spread of HBV infection and chronic liver disease (CLD); (2) the connection between the degree of family relationship and the response of the household contacts to HBV infection; (3) the correlation between familial clustering of HBV infection and of CLD.

The purpose of the present study was to evaluate the risk of HBV infection and CLD in relation to the diagnosis and serological status of the index cases as well as to age, sex and family position of the relatives.

[1] This study was supported by grants from the Consiglio Nazionale delle Ricerche (No. 81.00028.04) and from the Ministero della Pubblica Istruzione, Progetto Nazionale 'Cirrosi Epatica'.

Table I. Study population

Index cases	Families	Family contacts investigated				
		parents	spouses	siblings	offspring	total
$HBsAg^+$ CALD	78	40	45	80	105	270
HBsAg chronic healthy carriers	82	21	54	46	121	242
$HBsAg^-$ CALD	38	4	26	5	102	137
Total	198	65	125	131	328	649
Age of contacts: Median	49	34	20	10	21	
Range	24-67	22-73	2-45	1-46	1-73	

Materials and Methods

Study Sample

During the years 1978–1980, 200 voluntary blood donors, found to be healthy carriers of HBsAg, and 360 consecutive patients, all coming from Naples and hospitalized for chronic nonalcoholic liver disease, were asked permission to carry out a personal interview and to test their family contacts for hepatitis B markers and liver disease.

82 blood donors and 116 patients with chronic active liver disease (CALD), 78 $HBsAg^+$ and 38 $HBsAg^-$ agreed to participate with their families in this study. 649 household relatives of the 198 index cases were studied.

Data of family contacts in relation to index cases are shown in table I. None of the studied subjects had received a blood transfusion, was a drug addict or heavy alcohol drinker. 6 index cases and 35 households had history of acute icteric hepatitis dating back 2–10 years before entering this study.

Methods

The diagnosis of a chronic healthy carrier was based on persistent normality of both clinical and biochemical data (transaminases, bilirubin, alkaline phosphatase, γ-globulins) on bi-monthly checks for at least 1 year in all cases. In 20 index cases the absence of liver disease was confirmed by liver histology. CALD with or without cirrhosis was diagnosed in all cases according to the accepted histological criteria [16–18].

The diagnosis of CLD in household contacts was based on clinical and biochemical abnormalities (transaminase levels at least twice upper maximum normal values and/or γ-globulin levels $\geqq$ 1.8 g/100 ml) persisting for at least 1 year. In 21 cases the diagnosis was confirmed by liver biopsy (2 chronic persistent hepatitis, 2 chronic lobular hepatitis, 12 chronic active hepatitis and 5 active cirrhosis).

Table II. Prevalence of serological markers of HBV infection and CLD in household relatives of index cases with $HBsAg^+$ CALD

Relation to index cases	$HBsAg^+$		$Anti\text{-}HB^+$		Negative		total	CLD
	total	CLD	total	CLD	total	CLD		
Parents	6	3	30	2	4	0	40	5
Spouses	4	2	34	0	7	0	45	2
Siblings	40	24	31	0	9	0	80	24
Offspring	23	14	24	1	58	0	105	15
Total	73	43	119	3	78	0	270	46

Table III. Prevalence of serological markers of HBV infection and CLD in household relatives of index cases HBsAg chronic healthy carriers

Relation to index cases	$HBsAg^+$		$Anti\text{-}HB^+$		Negative		total	CLD
	total	CLD	total	CLD	total	CLD		
Parents	2	0	11	0	8	1	21	1
Spouses	4	0	34	0	16	1	54	1
Siblings	21	2	12	0	13	0	46	2
Offspring	20	4	23	0	78	0	121	4
Total	47	6	80	0	115	2	242	8

HBsAg, anti-HBs and anti-HBc were tested in all cases, and HBeAg and anti-HBe in 129/160 $HBsAg^+$ index cases and in all their $HBsAg^+$ household contacts by radioimmunoassay (AUSRIA II, AUSAB, CORAB, HBe, Abbott Laboratories). Statistical analyses were performed by the χ^2 test, with Yates's correction, when necessary.

Results

A high prevalence of HBsAg positivity was found in family contacts of index cases with $HBsAg^+$ CALD (27.0%) and of chronic healthy carriers (19.4%) (table II, III). The difference in the overall prevalence between these two groups was slightly significant ($p < 0.05$), but no significant differences were observed when the relatives with the same family position were compared (table IV). The prevalence of HBsAg in

Table IV. Prevalence of HBsAg positivity in family members of $HBsAg^+$ index cases (with CALD or chronic healthy carriers)

Relation to index cases	Index cases				χ^2
	$HBsAg^+$ CALD		chronic healthy carriers		
	n	%	n	%	
Parents	6/40	15.0	2/21	9.5	n.s.
Spouses	4/45	8.9	4/54	7.4	n.s.
Siblings	40/80	50.0	21/46	45.6	n.s.
Offspring	23/105	21.9	20/121	16.5	n.s.
Total	73/270	27.0	47/242	19.4	$p < 0.05$

Table V. Prevalence of serological markers of HBV infection and CLD in household relatives of index cases with $HBsAg^-$ CALD

Relation to index cases	$HBsAg^+$		$Anti\text{-}HB^+$		Negative		total	CLD
	total	CLD	total	CLD	total	CLD		
Parents	1	0	3	0	0	0	4	0
Spouses	0	0	14	1	12	0	26	1
Siblings	10	0	32	1	60	1	102	2
Offspring	1	1	2	2	2	0	5	3
Total	12	1	51	4	74	1	137	6

the contacts of patients with $HBsAg^-$ CALD (table V) was only 8.8%, this figure slightly exceeding that (≃ 5%) found in our general population [1–3].

The greatest prevalence of HBsAg positivity was observed among siblings (48.2%), whereas the lowest was found in spouses (8.1%). In this latter group, the highest prevalence of anti-HB positivity was detected (68.7%). No significant differences of HBsAg or anti-HB were observed between males and females in all groups of households studied.

The prevalence of CLD among $HBsAg^+$ contacts of patients with

Table VI. Overall prevalence of CLD in relation to the kinds of index cases

Serological pattern of CLD cases identified among contacts		Index cases		
		HBsAg⁺ CALD	chronic healthy carriers	HBsAg⁻ CALD
$HBsAg^+$	n	43/73*	6/47*	1/12
	%	58.9	12.8	8.3
Anti-HB^+	n	3/119	0/80	4/51
	%	2.5	0.0	7.8
Negative	n	0/78	2/115	1/74
	%	0.0	1.7	1.3
Total	n	46/270	8/242	6/137
	%	17.0	3.3	4.4

* $\chi^2 = 27.1$; $p < 0.001$.

Table VII. Prevalence of serological markers of HBV infection and CLD in relation to the presence or absence of at least 1 $HBeAg^+$ member in the family

Family groups	Serological pattern of members			Total
	$HBsAg^+$	anti-HB^+	negative	
With at least 1 $HBeAg^+$ member (n = 35)				
n	88*	66	20	174
CLD, %	70.4**	1.5	0.0	36.2
With all members $HBeAg^-$ (n = 94)				
n	128*	131	161	420
CLD, %	34.3**	4.6	0.6	12.1

* $\chi^2 = 21.1$; $p < 0.001$.
** $\chi^2 = 27.1$; $p < 0.001$.

$HBsAg^+$ CALD (58.9%) was significantly higher than that observed among $HBsAg^+$ relatives of chronic healthy carriers (12.8%, $p < 0.001$). Only a few cases of CLD were observed in the absence of HBsAg (table VI).

In families of $HBsAg^+$ index cases, the prevalence of both HBsAg and of CLD was significantly higher when at least 1 member was $HBeAg^+$ (table VII).

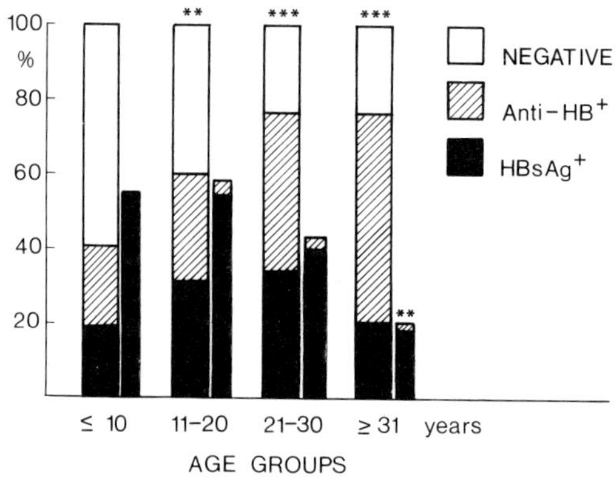

Fig. 1. Wide columns: prevalence of HBsAg and anti-HB positivity in contacts, grouped according to age, of $HBsAg^+$ index cases (healthy and ill). Narrow columns: Prevalence of CLD in $HBsAg^+$ and anti-HB^+ contacts. Significance of differences in HBV markers and CLD in comparison with ≦ 10-year-old group: ** $p < 0.01$; *** $p < 0.001$.

Figure 1 shows the exposure to HBV and the prevalence of CLD in relation to age of the contacts. The overall prevalence of HBV markers significantly increases with increasing age, whereas the frequency of CLD is significantly higher in the younger group of $HBsAg^+$ subjects than in the ≧ 31-year-old group ($p < 0.005$).

As shown in tables VIII and IX, all $HBsAg^+$ siblings and offspring of chronic healthy carriers cluster in only 27/70 families and all 63 $HBsAg^+$ siblings and offspring of index cases with $HBsAg^+$ CALD in 31/69 families. All 44 cases of CLD among siblings and offspring group in only 23/58 families with at least 1 member $HBsAg^+$, regardless of clinical status of the index case.

Discussion

The spread of HBV infection among the relatives of $HBsAg^+$ subjects has been reported from many countries [8–15]. Our data confirm those reports and indicate that HBV infection is particularly frequent in family contacts of patients with $HBsAg^+$ CALD. Such a finding can be reasonably explained on the basis of a high incidence in this group

Table VIII. Inhomogeneous family aggregation of HBsAg positivity and CLD in siblings and offspring of index cases with $HBsAg^+$ CALD

	Siblings + offspring		
	with $HBsAg^+$ CLD	HBsAg healthy carriers	$HBsAg^-$
Number of families	20	11	38
$HBsAg^+$ cases: n (% of total)	49 (77.7)	14 (58.4)	0
Cases of CLD: n (% of total)	38 (60.3)	0	1
Total number of siblings + offspring	63	24	98

Table IX. Inhomogeneous familial aggregation of HBsAg positivity and CLD in siblings and offspring of index cases HBsAg chronic healthy carriers

	Siblings + offspring		
	with $HBsAg^+$ CLD	HBsAg healthy carriers	$HBsAg^-$
Number of families	3	24	43
$HBsAg^+$ cases: n (% of total)	6 (60.0)	35 (76.1)	0
Cases of CLD: n (% of total)	6[1] (60.0)	0	0
Total siblings + offspring	10	46	111

[1] 2 cases in each family.

of HBeAg, a marker of viral replication and high infectivity [19–21]. Accordingly, a considerably higher prevalence of HBsAg positivity was detected in families with at least 1 $HBeAg^+$ member, as compared with that found in families without HBeAg positivity among their components. It is noteworthy, however, that a high prevalence of HBV infection markers was also observed in households of chronic healthy carriers, though to a lesser extent than that found in contacts of $HBsAg^+$ CALD. According to previous observations [22, 23] we have never documented HBeAg positivity in chronic healthy carriers or in their households. The spread of HBV infection among household relatives of $HBeAg^-$ cases may be related to: (a) infectivity of $HBsAg^+$ subjects even if $HBeAg^-$ [24, 25]; (b) infection of household contacts before seroconversion HBeAg → anti-HBe of the index cases. The latter hypothesis is supported by reports from other authors, who observed HBeAg → anti-HBe seroconversion during the course of CLD [26].

The high frequency of HBsAg positivity in siblings indicates a prevalent horizontal diffusion of HBV infection in our family groups. However, the high prevalence of HBV markers in children demonstrates that in our country the infection is acquired early in life, especially in the presence of a household relative with ongoing HBV infection. In spouses the risk of acquiring the infection was similar to that observed in siblings. However, the majority of spouses shows anti-HB positivity. The low HBsAg and the high anti-HB positivity rate among spouses as compared with the serological pattern of siblings, suggests that genetic factors, different between blood and nonblood relatives, can modulate the response to HBV infection.

Familial clustering of CLD and HBV infection has been reported by some authors [27, 28], but the respective role of infection and genetic factors remains unknown. Our data show a high prevalence of CLD in $HBsAg^+$ household contacts of patients with $HBsAg^+$ CALD. On the contrary a relatively low incidence of CLD was found in household contacts of HBsAg chronic healthy carriers, in spite of a high frequency of HBsAg positivity.

Such findings indicate that the discovery of a healthy carrier in a family is an important risk indicator of HBV infection, but not so much of CLD, in the households. The contemporary presence of CALD and HBV infection in a family member highly increases the risk of CLD among the relatives. In fact, the frequency of CLD among $HBsAg^+$ blood relatives of index cases with $HBsAg^+$ CALD proved to be 4.6 times higher than in $HBsAg^+$ blood relatives of chronic healthy carriers. The lower frequency of CLD in older than in younger relatives appears to be due to the prevalence in the first group of parents and spouses, more frequently healthy and $HBsAg^-$.

The following points suggest that different genetic factors may predispose to development of HBsAg carrier state with or without CALD: (1) the higher prevalence of CLD in $HBsAg^+$ relatives of index cases with $HBsAg^+$ CALD than in $HBsAg^+$ relatives of healthy carriers; (2) the presence of many cases of HBsAg positivity and/or CLD only in a few families and their absence in many others, in spite of the same serological and clinical stage of the index cases.

Alternatively, we cannot exclude that the development of CLD in a few families with many $HBsAg^+$ members might be due to other environmental factors, for example superinfection by δ agent [29].

In conclusion, we agree with the hypothesis [9] that both environmental and genetic factors, separately or more probably together, contribute to familial clustering of HBV infection and $HBsAg^+$ CLD.

Summary

649 household relatives of 198 index cases (of these 78 with $HBsAg^+$ CALD, 82 HBsAg chronic healthy carriers and 38 with $HBsAg^-$ CALD) were studied to evaluate family aggregation of HBV infection and CLD. A high and similar spread of HBV infection was observed among family contacts of $HBsAg^+$ index cases with CALD or chronic healthy carriers, whereas a significantly higher rate of CLD was detected in household relatives of index cases with $HBsAg^+$ CALD. Siblings and, to a lesser extent, offspring showed a high frequency of HBsAg positivity, while spouses were found to be predominantly anti-HB^+. There were no consistent sex-related differences in prevalence of HBsAg, anti-HB and CLD. A particularly high spread of HBsAg was found in family groups with at least 1 $HBeAg^+$ member. CLD and HBsAg positivity were found to cluster only in a few family groups, while they were absent in others, regardless of the type of index case. The data presented support the hypothesis that, in the families studied, an interplay between genetic and environmental factors may be involved in favoring persistence of HBV infection and development of CLD.

References

1 Del Vecchio-Blanco, C.; Caporaso, N.; De Carlo, R.; Murino, P.: Studi popolazionistici di HBsAg in centri trasfusionali. II. Risultati di un'inchiesta policentrica. Riv. Pat. Umana. *3:*373–391 (1976).

2 Del Vecchio-Blanco, C.; Stroffolini, T.; Caporaso, N.; Albanese, R.: Diffusione di HBsAg e HBsAb nella popolazione campana. Igiene Moderna *69:*461–464 (1976).

3 Ascione, A.; Forte, G.B.; Amitrano, L.; De Bellis, G.; Pandolfi, C.; Iaccarino, L.; Mariconda, M.; Petringa, D.; Utech, W.; Vacca, C.: Serological markers of hepatitis B virus infection in healthy volunteer blood donors in Campania (Southern Italy). Vox Sang. *41:*146–150 (1981).

4 Bianchi, P.; Bianchi Porro, G.; Coltorti, M.; Dardanoni, L.; Del Vecchio-Blanco, C.; Fagiolo, U.; Farini, R.; Menozzi, I.; Naccarato, R.; Pagliaro, L.; Spanò, C.; Verme, G.: The occurrence of the Australia antigen in chronic hepatitis in Italy. Gastroenterology *63:*482–485 (1972).

5 Del Vecchio-Blanco, C.; Caporaso, N.; Coltorti, M.: Contribution à la connaissance des différences géographiques dans l'étiologie des hépatopaties chroniques. Ann. Gastroent. Hépatol. *9:*423–433 (1973).

6 Coltorti, M.; Del Vecchio-Blanco, C.; Surrenti, C.; Caporaso, N.; Rinaldi, M.; Radaeli, E.; Meliconi, F.: Distribution of HBsAg and HBsAb in patients with chronic hepatitis in three areas of Italy; in Gentilini, Popper, Teodori, Chronic hepatitis. Int. Symp. Problems of Chronic Hepatitis, Montecatini 1975, pp. 86–90 (Karger, Basel 1976).
7 Coltorti, M.; Lucchelli, P.: e gruppo policentrico Associazione Italiana per lo Studio del Fegato (A.I.S.F.). Patologia cronica del fegato in Italia. Inchiesta retrospettiva multicentrica. Recent Prog. Med. *63:*447–469 (1977).
8 Singleton, J.W.; Kohler, P.F.; Merrill, D.A.: Infectivity risk of household contact with chronic carriers of hepatitis B antigen (HBAg). Gastroenterology *64:* 172 (1973).
9 Szmuness, W.; Prince, A.M.; Hirsch, R.L.; Brotman, B.: Familial clustering of hepatitis B infection. New Engl. J. Med. *289:*1162–1166 (1973).
10 Bruguera, M.; Bosch, J.; Rodes, J.; Pedreira, J.: Familial clustering of hepatitis B antigen: a study in relatives of patients with liver diseases and hepatitis B antigenaemia. Br. med. J. *iii:*495–497 (1974).
11 Eliakim, M.; Ligumski, M.; Sandler, S.G.; Zlotnick, A.: Familial clustering and immune response in family contacts of patients with HBsAg positive liver cirrhosis. Am. J. dig. Dis. *23:*407–412 (1978).
12 Hess, G.; Born, M.; Dormeyer, H.; Zöller, B.; Arnold, W.; Knolle, J.: Hepatitis B virus markers among family contacts of asymptomatic HBsAg carriers. Scand. J. Gastroent. *14:*373–378 (1979).
13 Lindberg, J.; Lindholm, A.; Iwarson, S.: Genetic factors in the development of chronic active hepatitis. Lancet *i:*67–68 (1977).
14 Nasrallah, S.M.; Nassar, V.H.; Skammaa, M.H.: Genetic and immunological aspects of familial chronic active hepatitis (type B). Gastroenterology *75:* 302–306 (1978).
15 Chiaramonte, M.; Giani, G.; Floreani, A.; Venturi-Pasini, C.; DeLazzari, F.; Fagiolo, U.; Naccarato, R.: Aggregazione familiare di portatori cronici di HBsAg: ruolo di fattori genetici ed ambientali; in Gasbarrini, Miglio, Bernardi, IIIrd Int. Congr. Gastroent., Bologna 1979, pp. 99–102 (Compositori, Bologna 1980).
16 De Groote, J.; Desmet, V.J.; Gedigk, P.; Korb, G.; Popper, H.; Poulsen, H.; Scheuer, P.J.; Schmid, M.; Thaler, H.; Uehlinger, E.; Whepler, W.: A classification of chronic hepatitis. Lancet *ii:*626–628 (1968).
17 Boyer, J.L.: Chronic hepatitis: a perspective on classification and determinants of prognosis. Gastroenterology *70:*1161–1171 (1976).
18 Acute and chronic hepatitis revisited. Review by an international group. Lancet *ii:* 914–919 (1977).
19 Schweitzer, I.L.; Edwards, V.M.; Brezina, M.: e antigen in HBsAg-carrier mothers. New Engl. J. Med. *293:*940–944 (1975).
20 Okada, K.; Kamiyama, I.; Inomata, M.; Imai, M.; Miyakawa, Y.; Mayumi, M.: e antigen and anti-e in the serum of asymptomatic carrier mothers as indicators of positive and negative transmission of hepatitis B virus to their infants. New Engl. J. Med. *294:*746–749 (1976).
21 Perrillo, R.P.; Gelb, L.; Campbell, C.; Wellinghoff, W.; Ellis, F.R.; Overby, L.; Aach, R.D.: Hepatitis B e antigen, DNA polymerase activity, and infection of household contacts with hepatitis B virus. Gastroenterology *76:*1319–1325 (1979).

22 Nielsen, J.O.; Dietrichson, O.; Juhl, E.: Incidence and meaning of the e determinant among hepatitis B antigen positive patients with acute and chronic liver diseases. Lancet *ii:*913–920 (1974).
23 El Sheikh, N.; Woolf, I.L.; Galbraith, R.M.; Eddleston, A.L.W.F.; Dymock, I.W.; Williams, R.: e antigen - antibody system as indicator of liver damage in patients with hepatitis B antigen. Br. med. J. *iv:*252–253 (1975).
24 Berquist, K.R.; Maynard, J.E.; Murphy, B.L.: Infectivity of serum containing HBsAg and antibody to e antigen. Lancet *i:* 1026–1027 (1976).
25 Zanetti, A.R.; Magliano, E.; Pirovano, P.; Moro, M.; Tanzi, E.; Riva, C.; Ferroni, P.: Trasmissione perinatale del virus dell'epatite B. Vth Riun. Ann. Informazione Epatologica 'Il Portatore di HBsAg', Chianciano 1980, pp. 87–88 (Compositori, Bologna 1981).
26 Realdi, G.; Alberti, A.; Rugge, M.; Bortolotti, A.M.; Rigoli, A.M.; Tremolada, F.; Ruol, A.: Seroconversion from hepatitis B e antigen to anti-HBe in chronic hepatitis B virus infection. Gastroenterology *79:*95–99 (1980).
27 Ohbayashi, A.; Okochi, K.; Mayumi, M.: Familial clustering of asymptomatic carriers of Australia antigen and patients with chronic liver disease or primary liver cancer. Gastroenterology *62:*618–626 (1972).
28 Coltorti, M.; Del Vecchio-Blanco, C.; Caporaso, N.; Ambrogio, G.; Mattera, D.: Prevalence of HBsAg, HBsAb and CALD in families of HBsAg- or Ab-positive subjects. Int. Congr. Intern. Med., Roma 1978. Excerpta Med. Int. Congr. Ser., No. 502, p. 559.
29 Smedile, A.; Farci, P.; Verme, G.; Caredda, F.; Cargnel, A.; Caporaso, N.; Dentico, P.; Trepo, C.; Opolon, P.; Gimson, A.; Vergani, D.; Williams, R.; Rizzetto, M.: Influence of delta infection on severity of hepatitis B. Lancet *ii:*945–947 (1982).

Prof. M. Coltorti, Istituto di Semeiotica Medica, I Facoltà di Medicina e Chirurgia, II Policlinico, Pad. III, Via Pansini, I-80131 Napoli (Italy)

Front. gastrointest. Res., vol. 8, pp. 180–190 (Karger, Basel 1984)

Late Stages of Chronic Liver Disease

I. Pattern of Hepatitis B Virus Infection [1]

Antonio Craxí[a], *Paolo Pasqua*[a], *Gandolfo Giannuoli*[a], *Carmen Di Franco*[b], *Rosa Giovanna Simonetti*[a], *Luigi Pagliaro*[a]

[a] Cattedra di Patologia Medica R, c/o Divisione di Medicina Interna, Ospedale V. Cervello, Palermo; [b] Servizio di Anatomia Patologica, Ospedale V. Cervello, Palermo, Italia

Introduction

Chronic infection with the hepatitis B virus (HBV) is closely associated with hepatocellular carcinoma (HCC), as demonstrated by epidemiological studies [1, 2] and by the integration of specific HBV-DNA sequences in the genome of neoplastic hepatocytes [3]. This relationship is also evident in Italy [4], where both the chronic HBsAg carrier state and HCC have an intermediate prevalence between high- and low-risk areas [5, 6]. HBsAg seronegative patients with HCC have a higher prevalence of other HBV markers in comparison to the general population [1] and to appropriate controls with chronic liver disease (CLD) [4]. This points to an oncogenic role of HBV in these cases, in accordance with recent reports of small series of patients with HCC who are seronegative for HBsAg (with or without other HBV markers) but who have integrated HBV-DNA sequences in their liver cells [7].

Patients with chronic hepatitis or cirrhosis often display HBsAg and/or HbcAg in the liver even in the absence of circulating HBsAg [8–15]. This is considered as evidence of continuing 'minimal' HBV replication and of a causal role of this virus in the etiology of the liver disease [16]. In particular, a high prevalence of intrahepatic HBcAg in anti-HBc$^+$ subjects with CLD and the presence of HBcAg in 2/2 patients with anti-HBc$^+$ HCC have been reported by *Omata* et al. [12, 17].

[1] This work was partly supported by the Ministero Pubblica Istruzione.

However, no data are available on the prevalence of tissue HBV markers in a large population of patients with HCC, in relation to serological markers. The only study investigating such a relationship in HCC patients deals exclusively with HBsAg$^+$ cases [18].

The pattern of distribution of tissue HBV markers has been reported [13, 14] to correlate with the type of liver disease. This relationship has been questioned, and the use of monoclonal antibodies against HBsAg and HBcAg in an immunofluorescence system [19] has shown that HBV marker expression in the liver correlates only with the level of viral replication.

We studied the distribution of serum and tissue markers of HBV infection in HCC patients, in comparison to an appropriate control group of subjects with inactive cirrhosis, in whom the initiating factors of liver disease should be similarly remote in time. In addition we also studied a group of patients with active cirrhosis, among whom the HBV-related cases should represent a more recent phase of chronic infection.

Our aims were to assess whether the frequency of tissue markers of HBV infection in HCC is different than in other advanced, non-neoplastic liver diseases, and if a specific pattern of distribution of these markers is present in association with neoplastic transformation. Because of contradictory reports about the ability of neoplastic hepatocytes to sustain the synthesis of viral proteins [1, 18, 20–24], we also investigated whether the expression of HBV markers differed among cancerous and non-cancerous areas in the same liver.

Materials and Methods

Patients. We studied a total of 154 patients admitted to our Department between January 1977 and June 1982. A liver biopsy was performed in all patients for diagnostic purposes (evaluation of liver disease; clinical suspicion of HCC). In all cases serum samples were collected at the time of diagnostic evaluation. Only patients whose biopsy was at least 10 mm long, well-fixed and embedded, were selected for the study. Patients with HCC were selected only if the biopsy showed non-neoplastic liver cells (usually in a sufficient amount to include or exclude cirrhosis) as well as neoplastic tissue. Necrotic tissue was not analyzed. Inflammatory activity in the biopsy was assessed according to standard criteria.

Patients wer classified as follows:

(1) HCC: 49 cases (44 with clearcut histologic evidence of associated cirrhosis).
(2) Inactive cirrhosis (IC), 52 cases.

(3) Active cirrhosis (AC), 53 cases. Almost all patients in groups 2 and 3, except those observed during the last months of the study, had a post-biopsy follow-up of 1 year or more, to exclude undiagnosed concurrent HCC. None of the patients in groups 1 and 2 and about one fourth in group 3 were on steroids and/or immunosuppressants at the time of the study.

Serological markers of HBV only were evaluated in a further group of 137 patients with HCC, in whom the biopsy was not suitable for immunohistology because of poor preservation or lack of non-neoplastic tissue, or without a morphologic confirmation of HCC but having compatible clinical features and α-fetoprotein levels consistently above 1,000 ng/ml by radioimmunoassay (RIA). Details of patients are given in table I.

Serology. All sera were tested for HBsAg, anti-HBs, and anti-HBc by commercial RIA (Abbott). All $HBsAg^+$ samples were tested by RIA (Abbott) for the HBeAg/anti-HBe system. α-Fetoprotein determinations were carried out by RIA (Abbott).

Immunohistology: Liver cores taken at biopsy were immediately fixed in 10% phosphate-buffered formalin (pH 7.2), processed by standard histological methods and embedded in paraffin. For immunofluorescence, 5-μm thick paraffin sections were cut and adhered to the slides by heating at 45 °C for 60 min. Sections were deparaffinized and dehydrated by xylene/ethanol washing and then digested by incubation in 0.025% solution of type III Bovine trypsin (Sigma Chemicals) in Tris buffer, pH 7.8, with 0,1% $CaCl_2$ for 1 h at 37 °C.

HBsAg was detected by indirect immunofluorescence, using a murine IgG_1 class monoclonal antibody , RF-HBs1, directed against a common epitope on HBsAg and recognizing all the subtypes of the antigen [25]. A dilution of 10^1 was used throughout the study. A FITC-conjugated rabbit anti-mouse Ig (Meloy Inc.) was used as second layer. Blocking experiments were performed using purified HBsAg, prepared by affinity chromatography on Sepharose 4B/RF-HBs1 [26].

HBcAg was detected by direct immunofluorescence using the FITC-conjugated IgG fraction of a high-titer human anti-HBc serum. Blocking experiments were performed by adding an undiluted, high-titer human anti-HBc prior to incubation with FITC-conjugated anti-HBc.

A Leitz Orthoplan microscope, with Ploemopak filter system, was used for reading. Slides were interpreted blindly by two observers. When there was discordance between

Table I. Clinical features of patients

Disease	Age, years	M/F Ratio	Alcohol ≥ 80 g/day, %
HCC (biopsy) (49 patients)	59.2	6.8	33.8
HCC (no biopsy) (137 patients)	63.2	6.2	34.2
IC (52 patients)	54.7	2.2	36.2
AC (53 patients)	51.3	1.8	32.4

observers a second reading was performed on newly prepared sections, in order to reach an agreement. To evaluate the loss of sensitivity due to embedding and trypsin digestion, 15 biopsies from HBsAg$^+$ patients (7 HBeAg$^+$), 8 anti-HBe$^+$ with chronic liver disease and 10 biopsies from similar patients negative for all markers of HBV infection were divided into two parts: one was processed for normal histology, and the other immediately snap-frozen in liquid nitrogen-cooled isopenthane, embedded in OCT (Ames Co.), cut in a Leitz oryostat, and fixed, after air drying for 10 min, by immersion in a 50:50 mixture of acetone and ethanol at 4 °C for 5 min. Both standard and cryostat sections were used as they were, or submitted to trypsin digestion before incubation with the anti-HBs or anti-HBc.

Results

Sensitivity of Detection of Tissue Markers before and after Treatment. None of the patients with negative HBV serology showed HBsAg or HBcAg in the liver, whatever treatment was used on the tissue. Results for HBsAg$^+$ patients are shown in figure 1. Both trypsin digestion of the fresh tissue and the embedding procedure reduced the sensitivity of HBsAg detection, mainly when the antigen was present in small amounts on the liver cell membrane. Trypsin digestion of embedded sections restored sensitivity of antigen detection by reducing background autofluorescence. The 'membranous' pattern, however was lost. HBcAg expression was substantially unaffected (fig. 1) at least on trypsin-digested sections. On the basis of these results, trypsin digestion on embedded biopsies was used throughout the rest of the study, although a slight reduction in sensitivity in comparison to frozen biopsies (because of the loss of the 'membranous' pattern) could be expected.

Serological Markers of HBV. The serological pattern of the various groups of patients is shown in figure 2. Results for the two subgroups of patients with HCC were very similar. Subjects with HCC had a higher rate of positivity for HBsAg than those with inactive cirrhosis, but were comparable to those with AC. Anti-HBs, alone or in combination with anti-HBc, was found more frequently in subjects with IC than in those with AC or HCC. Positivity for HBeAg, an indicator of high-level HBV replication, was most frequent in the AC group.

Viral Markers in the Liver. HBsAg (table II). When present, HBsAg appeared in all groups in the form of a homogeneous cytoplas-

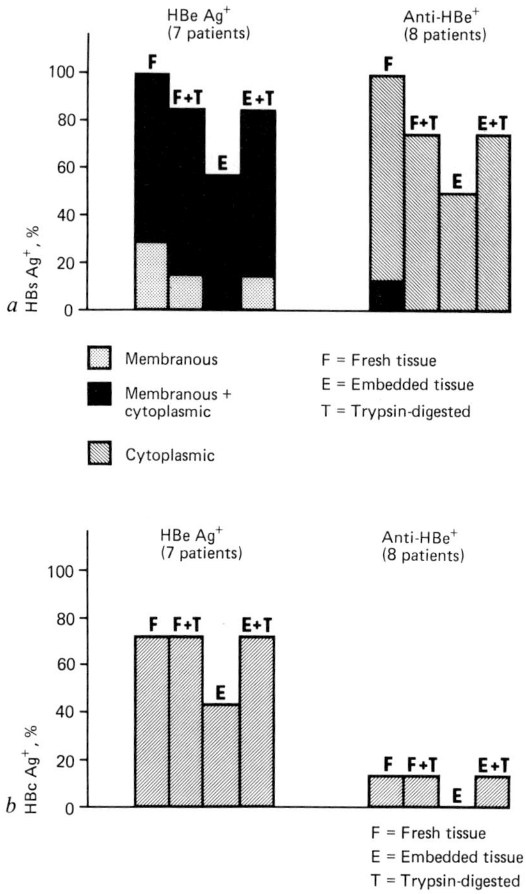

Fig. 1 a,b. Effect of embedding and trypsin digestion on the sensitivity of detection of intrahepatic HBsAg and HBcAg.

mic staining sometimes associated with a membranous staining, or as inclusion-like intracytoplasmic bodies. A pure membranous pattern was never observed. The antigen was usually distributed in small clusters of hepatocytes or in scattered cells; more rarely, it was present in the majority of cells.

Overall sensitivity of detection was good, HBsAg being found in 81.3% (26/32) of HBsAg seropositive patients. 11 out of 80 subjects seronegative for HBsAg, but who were anti-HBc+ (13%), were found to

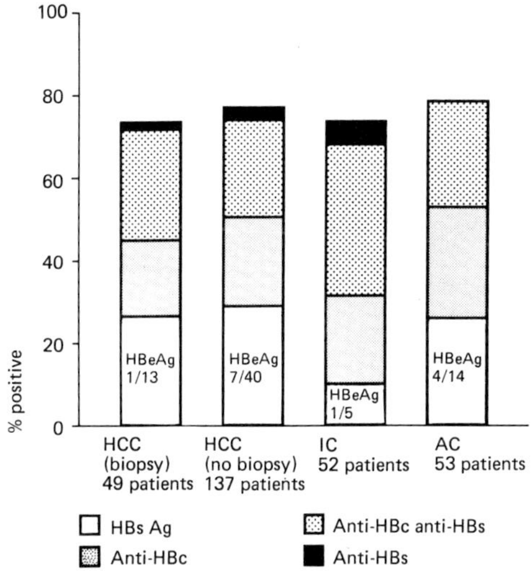

Fig. 2. Serological markers of HBV infection.

Table II. Intrahepatic HBsAg in relation to HBV serology

	Serum HBsAg+	Anti-HBc+	Anti-HBc+ Anti-HBs+
HCC[1] (49 patients)	10/13	1/9	2/13
IC (52 patients)	5/5	4/11	2/19
AC (53 patients)	11/14	4/14	1/14

[1] Non-neoplastic tissue.

have a comparable amount of HBsAg in the liver. No expression of HBsAg was found in the 69 subjects without serological features of HBV infection or in the 9 cases with anti-HBs alone. HBsAg was observed with the same frequency in the various types of disease. In the liver of patients with HCC, the antigen was present exclusively in non-neoplastic cells, except in 1 case. In this patient, who was $HBeAg^+$ in serum and who also had HBcAg in non-neoplastic and neoplastic liver cells, almost all hepatocytes in cancerous and non-cancerous tissue

Table III. Intrahepatic HBcAg in relation to HBV Serology

	Serum HBsAg⁺	Anti-HBc⁺	Anti-HBc⁺ Anti-HBs⁺
HCC[1] (49 patients)	1/13	2/9	0/13
IC (52 patients)	1/5	5/11	3/19
AC (53 patients)	3/14	7/14	0/14

[1] Non-neoplastic tissue.

Table IV. Correlation between tissue markers of HBV

	$HBsAg^+$ $HBcAg^+$	$HBsAg^+HBcAg^-$	$HBsAg^-HBcAg^+$	$HBsAg^-HBcAg^-$
HCC[1] (49 patients)	1	12	2	34
IC (52 patients)	1	9	8	34
AC (53 patients)	6	10	4	33

[1] Non-neoplastic tissue.

contained large inclusion-like cytoplasmic bodies reacting with anti-HBs. This pattern could be due to a recent superinfection with HBV in a patient with HBV-unrelated liver disease.

HBcAg (table III). Staining for HBcAg was always exclusively nuclear and usually focal. HBcAg was found more rarely than HBsAg in the liver of $HBsAg^+$ patients, being present only in 5/32. A close correlation between isolated positivity for anti-HBc in serum and the presence of HBcAg in the nuclei of hepatocytes was noted in all groups. However the correlation was stronger for AC (7/14) and IC (5/11) than for HCC (2/9). HBsAg subjects who were negative for all HBV markers or positive for anti-HBs alone did not display HBcAg in the liver. Neoplastic liver cells never expressed HBcAg in their nuclei, except in the aforementioned case. A pattern of high level viral replication ($HBsAg^+$, $HBcAg^+$) was rarely seen in HCC or IC but was frequent in AC (6 cases out of 14 HBsAg seropositive, 4 of whom HBeAg seropositive) (table IV).

Discussion

Subjects in this study are representative of the whole population of patients with CLD seen in our area [27] as regards mean age, M/F ratio, and alcohol consumption. The methods employed in the study have a high specificity for viral components. None of the patients with negative HBV serology had HBsAg or HBcAg in the liver, and blocking tests always confirmed the specificity in the positive cases. The sensitivity of detection of HBV proteins in formalin-fixed and embedded tissue was relatively reduced in comparison to frozen tissue. Trypsin digestion restored the sensitivity of the assay for HBsAg and HBcAg to that reported for immunoperoxidase by other workers [12, 19].

The overall prevalence of serum markers of HBV infection was similar among all groups. However when subjects with HCC are compared with the more appropriate controls (e.g. patients with IC) a marked reduction in the frequency of HBsAg seropositivity and an increase in anti-HBs positivity is observed in the latter group. Subjects with AC are similar in this respect to those with HCC. Assuming that in both HCC and IC, infection with the virus had occurred long before clinical presentation, the above findings could be due to a relatively shorter duration of disease for AC and to a more persisting, albeit incomplete, viral replication in HCC compared to IC.

Positivity for intrahepatic HBsAg was evenly distributed among the various groups of patients. Even if absent from serum, HBsAg was demonstrated in the liver in the presence of serum anti-HBc, alone or with anti-HBs, in 16% of cases. HBcAg was found in liver tissue in a small number of HBsAg seropositive cases, in close relation to features of high-level viral replication (HBeAg positivity). In the presence of isolated positivity for anti-HBc in serum a substantial proportion of patients had intrahepatic HBcAg expression, this being more frequent in AC and IC than in HCC. HBcAg synthesis might indicate a peculiar replicative cycle of the virus, with exclusive synthesis of the proteins of viral core and possibly of small amounts of complete HBV virions, as demonstrated by the infectivity of high-titer anti-HBc^{+} blood [28]. The fact that this pattern is more frequently observed in the more active stage of CLD is in keeping with the recent observation [29] that the cellular immune response in $HBsAg^{+}$ liver disease is mainly evoked by HBcAg expression on the liver cell membrane. Patients with HCC, in whom primary infection is likely to be more remote, have the least frequent intrahepatic display of HBcAg. No specific pattern of expression of HBsAg or HBcAg was observed in non-neoplastic liver tissue from HCC, thus confirming the previous observation [19] that HBV marker display in the hepatocytes correlates only with the level of viral replication.

Expression of viral antigens in neoplastic liver cells was always absent, except in a case with high-level viral replication both in the serum and the liver. This is in agreement with the findings by other groups and suggests that the loss of cellular differentiation is usually accompanied by inability to synthesize detectable amounts of viral proteins. HBsAg synthesis occurs in cultured HCC cell lines, but these cells produce small amounts of the protein and never support complete replication of the virus. The integrated genome of HBV in neoplastic cells is at least in part in a nonfunctional form.

We have not specifically assessed the relationship between the presence of 'dysplastic' cells and the presence of HBsAg and HBcAg in such cells because the methods we used do not allow a direct comparison with standard HE staining on the same sections. The similar frequency of viral antigens in non-neoplastic tissue from adjoining areas to foci of HCC (which almost invariably show severe dysplasia) and tissue from patients with IC or AC should rule out this correlation.

Our results indicate, in accordance with *Omata* et al. [12, 18], that even the more sensitive serological methods can miss a substantial proportion of subjects with 'minimal' HBV infection, in whom synthesis of viral proteins is below the level of sensitivity of the assay or a circulating antibody blocks the detection of the immunoreactivity. The frequency of detection of intrahepatic HBsAg, which appears to be higher than in other studies, could be due to the enzyme digestion of the biopsies and/or to the use of an anti-HBs monoclonal antibody, with superior sensitivity and specificity of staining.

Summary

49 patients with HCC (13 HBsAg$^+$), 52 patients with IC (5 HBsAg$^+$), and 53 patients with AC (14 HBsAg$^+$) were studied in order to assess the prevalence of tissue markers of HBV infection (HBsAg) in HBcAg seropositive and seronegative HCC as compared to other advanced liver diseases (inactive cirrhosis, IC, and active cirrhosis, AC). Among HBsAg seropositive patients intrahepatic HBsAg was frequently found (26/32 cases), while HBcAg was present more rarely (5/32 cases) and correlated with serological fea-

tures of high-level viral replication. HBsAg seronegative, anti-HBc$^{\pm}$ subjects had intrahepatic HBsAg in 8/34 cases and HBcAg in liver cell nuclei in 14/34 cases. HBcAg was more frequent in cirrhosis than in HCC. No other differences in the intrahepatic display of HBV markers were observed, nor was a specific pattern identified for HCC. Viral components were never found in the liver in the absence of serum HBsAg or anti-HBc. Neoplastic hepatocytes did not usually support the synthesis of HBsAg or HBcAg.

References

1 Kew, M.: Hepatoma and the HBV; in Viral hepatitis - 1978. Int. Symp. (Franklin Institute Press, Philadelphia 1979).

2 Palmer Beasley, R.; Hwang, L.Y.; Lin, C.C.; Chien, C.S.: Hepatocellular carcinoma and the hepatitis B virus. Lancet *ii:* 1129-1133 (1981).

3 Shafritz, D.A.; Shouval, D.; Sherman, H.I.: Integration of hepatitis B virus into the genome of liver cells in chronic liver disease and hepatocellular carcinoma. Studies in percutaneous liver biopsies and post-mortem tissue specimens. New Engl. J. Med. *305:* 1067-1073 (1981).

4 Pagliaro, L.; Simonetti, R.G.; Craxí, A.: Alcohol and HBV infection as risk factors for hepatocellular carcinoma in Italy: a multicentric controlled study. Hepatogastroenterology (in press).

5 Sobeslavsky, O.: HBV as a global problem; in Viral hepatitis - 1978. Int. Symp. (Franklin Institute Press, Philadelphia 1979).

6 Giusti, G.: Rapporti virus-ospite e dimensioni del problema; in Il portatore cronico di HBsAg (sano e malato) (Editrice Compositori, Bologna 1981).

7 Bréchot, C.; Nalpas, B.; Couroucé, A.M.; Duhamel, G.; Callard, P.; Carnot, F.; Tiollais, P.; Berthelot, P.: Evidence that hepatitis B virus has a role in liver cell carcinoma in alcoholic liver disease. New Engl. J.Med. *306:* 1384-1387 (1983).

8 Huang, S.N.: Immunohistochemical demonstration of HBcAg and HBsAg in paraffin sections. Lab. Invest. *33:* 88-95 (1975).

9 Yamada, G.; Nakane, P.K.: HBcAg and HBsAg in liver tissue. Lab. Invest. *36:* 649-659 (1977).

10 Ray, M.B.; Desmet, V.J.; Fevery, J.: HBsAg in the liver of patients with hepatitis: a comparison with serological detection. J. Clin. Path. *29:* 89-93 (1976).

11 Kojima, M.; Udo, K.; Takahashi, Y.: Correlation between titer of anti-HBc and presence of viral antigens in the liver. Gastroenterology *73:* 664-667 (1977).

12 Omata, M.; Afroudakis, A.; Liew, C.T; Ashcavai, M.; Peters, R.L.: Comparison of serum HBsAg and serum anti-HBc with tissue HBsAg and HBcAg. Gastroenterology *75:* 1003-1009 (1978).

13 Gudat, F.; Bianchi, L.; Sonnabend, W.: Pattern of core and surface expression in liver tissue reflects state of specific immune response in hepatitis B. Lab. Invest. *32:* 1-9 (1975).

14 Ray, M.B.; Desmet, V.J.; Bradburne, A.F.: Differential distribution of HBsAg and HBcAg in the liver of hepatitis B patients. Gastroenterology *71:* 462-467 (1976).

15 Akeyama, T.; Kamada, T.; Koyama, M.: Distribution patterns of HBsAg in the liver: immunofluorescence studies. Archs Path. *98:* 252-256 (1974).

16 Dienstag, J.L.: Hepatitis B virus infection: more than meets the eye. Gastroenterology *75:*1172–1174 (1978).

17 Omata, M.; Ashcavai, M.; Liew, C.T.; Peters, R.L.: Hepatocellular carcinoma in the USA, etiologic considerations. Localization of hepatitis B antigens. Gastroenterology *76:*279–287 (1979).

18 Omata, M.; Mori, J.; Yokosuka, O.; Iwama, S.; Ito, Y.; Okuda, K.: Hepatitis B virus antigens in liver tissue in hepatocellular carcinoma and advanced chronic liver disease. Relationship to liver cell dysplasia. Liver *2:*125–132 (1982).

19 Montano, L., Waters, J.; Goodall, A.H.; Weller, I.V.D.; Miescher, G.C.; Craxì, A.; Janossy, G.; Thomas, H.C.: Detection of HBsAg and HBcAg in human liver using monoclonal and heterologous antibodies: relationship to hepatitis B viral replication and inflammatory activity. Liver (in press).

20 Chen, D.S.; Sung, J.L.: Cellular localization of HBsAg in the non-cirrhotic liver with hepatocellular carcinoma. Ital. J. Gastroenterol. *10:*81–84 (1978).

21 Shikata, T.: Australia antigen in hepatoma and cirrhosis tissue. Tumor Res. *8:* 127–130 (1973).

22 Gerber, M.A.; Hadziyannis, S.; Vernace, S.; Vissoulis, C.: Incidence and nature of cytoplasmic HBsAg in hepatocytes. Lab. Invest. *32:*251–256 (1975).

23 Okuda, K.; Nakashima, T.; Obata, H.: Hepatitis B virus and primary liver cell carcinoma; in Virus and the liver (MTP Press, Lancaster 1980).

24 Popper, H.; Thung, S.N.; Gerber, M.A.: Chronic sequelae of viral hepatitis; in Viral hepatitis - 1981. Int. Symp. (Franklin Institute Press, Philadelphia 1982).

25 Goodall, A.H.; Miescher, G.C.; Meek, F.M.; Janossy, G.; Thomas, H.C.: Monoclonal antibodies in a solid-phase radiometric assay for HBsAg. Med. Lab. Sci. *38:* 349–354 (1981).

26 Godall, A.H.; Waters, J.A.; Meek, F.M.; Janossy, G.; Thomas, H.C.: Affinity purification of HBsAg using RF-HBs-1 monoclonal antibody; in XXXth colloquium. Protides of the biological fluids, vol. 30 (Pergamon Press, Oxford 1982).

27 Pagliaro, L.; Saracci, R.; Bardelli, D.; Filippazzo, G.: Chronic liver disease, alcohol consumption, and HBsAg in Italy: a multiregional case-control study. Ital. J. Gastroenterol. *14:*90–95 (1982).

28 Hoofnagle, J.H.; Seef, L.B.; Bales, Z.B.: Type B hepatitis after transfusion with blood containing anti-HBc. New Engl. J. Med. *298:*1379–1383 (1978).

29 Eddleston, A.L.W.F.; Mondelli, M.; Mieli-Vergani, G.; Williams, R.: Lymphocyte cytotoxicity to autologous hepatocytes in chronic hepatitis B virus infection. Hepatology *2:*122S–127S (1982).

30 Macnab, G.M.; Alexander, J.J.; Lecatsas, G.; Bey, E.M.; Urbanowicz, J.M.: HBsAg produced by a human hepatoma cell line. Br. J. Cancer *34:*509–515 (1976).

31 Mann, J.F.; Copeland, J.A.; Skelly, J.; Howard, C.R.; Zuckerman, A.J.: In vitro studies on the PLC/PRF/5 hepatoma cell line; in Viral hepatitis - 1981. Int. Symp. (Franklin Institute Press, Philadelphia 1982).

Dr. A. Craxí, Divisione di Medicina Interna, Ospedale V. Cervello,
Via Trabucco 180, I-90100 Palermo (Italy)

Front. gastrointest. Res., vol. 8, pp. 191–196 (Karger, Basel 1984)

Late Stages of Chronic Liver Disease

II. Occurrence of Delta Agent Infection[1]

Antonio Craxí[a], *Giovanni Raimondo*[b], *Paolo Pasqua*[a], *Gandolfo Giannuoli*[a], *Carmen Di Franco*[c], *Luigi Pagliaro*[a]

[a] Cattedra di Patologia Medica R, c/o Divisione di Medicina Interna, Ospedale V. Cervello, Palermo; [b] Istituto di Clinica Medica II, Policlinico, Messina; [c] Servizio di Anatomia Patologica, Ospedale V. Cervello, Palermo, Italia

Introduction

Delta agent, a defective virus infectious only in the presence of concurrent HBV infection [1, 2], is widespread in Italy, particularly in the Southern part of the country [3]. 38% of 560 chronic HBsAg carriers seen in our Department between 1977 and 1982 had anti-delta in their serum [unpublished data]. Superinfection with the delta agent is believed to enhance the severity of $HBsAg^+$ liver disease by converting an asymptomatic chronic HBsAg carrier state into a fulminant or chronic active hepatitis [4, 5].

Seropositivity for anti-delta cannot, by any means, be considered as an indicator of *present* infection with delta agent, since a number of $HBsAg^-$ individuals with a history of past $HBsAg^+$ hepatitis circulate low-titer anti-delta [3]. Evaluation of the IgM anti-delta response assesses more accurately the state of delta infection [6]. The absolute parameter for present infection however, remains the demonstration of delta Ag in liver cells. No information is available on the possible role of delta agent in neoplastic transformation of hepatocytes chronically infected with HBV.

We investigated the occurrence of markers of delta infection in the serum and the liver of HBsAg seropositive and negative patients with HCC, of patients with IC in whom the HBV infection thought to be the

[1] This work was partly supported by the Ministero Pubblica Istruzione.

initiator of liver damage should be comparably remote, and of patients with AC, whose HBV infection is probably more recent. Our aims were: (1) to analyze the rate of positivity and the correlation between serum and tissue markers of delta in the more advanced stages of chronic liver disease; (2) to evaluate the role of delta agent in the development of HCC.

Materials and Methods

Patients. The same group of subjects evaluated in the preceding chapter of this volume, consisting of 186 patients with HCC (53 HBsAg$^+$, 49 of whom had a biopsy suitable for the immunofluorescence study of tissue markers), of 52 patients with IC (5 HBsAg$^+$), and of 53 patients with AC (14 HBsAg$^+$), all with a suitable biopsy, was tested for serum and tissue markers of delta infection.

Serology. All sera were tested for anti-delta by a solid-phase micro-ELISA system [7]. The reagents used were a gift from Dr. *M. Rizzetto.*

Immunohistology. Details of the procedures used throughout the study are reported in the preceding chapter. Delta Ag was sought by direct immunofluorescence using a FITC-conjugated total γ-globulin fraction of a high-titer human anti-delta serum (also a gift from Dr. *M. Rizzetto*) at a dilution of 10^2. Specificity of staining was confirmed in all positive cases by preincubation of the slides with a high-titer ($> 10^5$) anti-delta, low-titer ($< 10^2$) anti-HBc human serum at a dilution of 10^3.

In agreement with other reports [8] sensitivity of detection of delta Ag in tissue was not reduced after embedding and trypsin digestion in 15 HBsAg$^+$ patients (7 HBeAg$^+$, 8 anti-HBe$^+$). 1 out of 7 HBeAg$^+$ and 3 out of 8 anti-HBe$^+$ subjects were positive for delta Ag and remained so after embedding and enzyme digestion.

Results

Serum Anti-Delta. Although HBV serology was similar in the AC and HCC group, marked difference in the incidence of anti-delta was found among them (fig. 1), the antibody being present in only 5 out of 53 HBsAg$^+$ cases of HCC but in 10 out of 14 cases of AC. Among the small number of HBsAg$^+$ patients with IC [5], 2 were positive for anti-delta. A proportion of HBsAg$^-$, anti-HBc$^+$ and/or anti-HBs$^+$ subjects (fig. 2) in all groups was also positive for anti-delta. When tested at a dilution above 10^2 for the antibody, these patients were uniformly negative, while the HBsAg$^+$ anti-delta$^+$ subjects had higher titers.

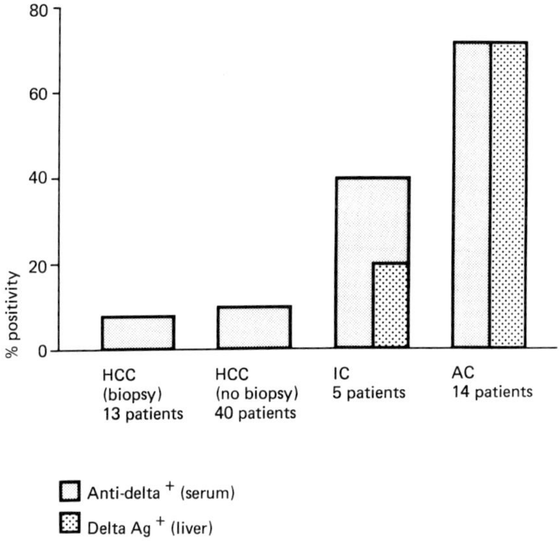

Fig. 1. Markers of delta infection in $HBsAg^+$ subjects.

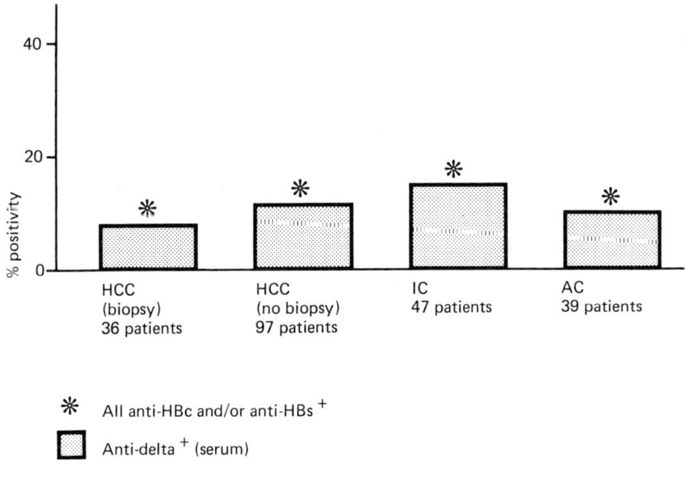

Fig. 2. Markers of delta infection in $HBsAg^-$ subjects.

Liver Delta Ag. Delta Ag was found, in a nuclear fluorescence with occasional perinuclear granules, only in $HBsAg^+$ subjects. None of the $HBsAg^-$, anti-$delta^+$ subjects had delta Ag in the liver. More than 50% of nuclei were positive in 6 cases, while a scattered and scarce distribution of delta Ag was found in 4 patients.

No patient with HCC had delta Ag in the liver (fig. 1), and only 1 out of 5 $HBsAg^+$ subjects with IC was delta Ag^+. On the contrary, 10/14 (71.4%) of the HBsAg seropositive patients with AC had delta Ag in the liver. Delta Ag positivity was usually, but not exclusively, associated to low-level HBV replication, being present in 9 $HBsAg^+$, $HBcAg^-$, and in 2 $HBsAg^+$ $HBcAg^+$ cases.

Discussion

Delta agent infection, as assessed by the antibody response, was a frequent event in our patients. Overall, 24% of HBsAg seropositive patients and 11% of the HBsAg seronegative subjects had serological evidence of an infectious contact with the delta agent. The positivity for anti-delta at a low titer in the absence of circulating HBsAg always occurred in the presence of other markers of HBV infection (anti-HBc and/or anti-HBs). This concurrence may be interpreted as evidence of a previous combined infection with HBV and delta. On disappearance of HBsAg synthesis (sometime followed by persistence of HBcAg in the liver) delta infection resolved, leaving only the low-titer antibody response, presumably of IgG class. This is confirmed by the similar distribution of anti-delta in $HBsAg^-$ subjects in all groups of disease, by the lack of intrahepatic expression of delta Ag in these patients, and by the fact that some of them have HBsAg and/or HBcAg in the liver [*Craxì* et al., this volume]. On the contrary, not only was anti-delta positivity much more frequent in $HBsAg^+$ patients, but also the distribution of the antibody showed a close correlation with the type of disease, the antibody being infrequent in HCC patients but prevalent in AC cases. This is comparable with the tissue distribution of delta Ag, which was never present in HCC but positive in 10/14 subjects with AC.

A possible explanation of the different prevalence of markers of delta infection among patients with HCC and those with AC is that the delta agent could have been introduced only recently in our population, as documented in Sweden by *Hansson* et al. [9]. In this case patients with HCC, whose infection is usually due to exposure to HBV in an early period of life to a single carrier, are less likely to undergo superinfection with delta than subjects with AC, whose HBV infection was acquired more recently and in a more promiscuous context. An alternative explanation is that delta infection changes the sequence AC → IC → HCC, typical of the chronic HBsAg carrier state, by causing premature death of delta-infected subjects in the AC phase, while patients who are uninfected (or who clear the delta agent and possibly the HBV) proceed to HCC.

In summary, delta agent superinfection is positively correlated with active, advanced liver disease but not with inactive chronic liver disease or HCC. This finding suggests that delta infection has no role in the neoplastic transformation of hepatocytes, chronically infected with the HBV.

Summary

The prevalence of delta infection, as assessed by the presence of anti-delta in serum and delta Ag in the liver, was evaluated in 186 patients with hepatocellular carcinoma (53 $HBsAg^+$), 49 of whom has a liver biopsy available for tissue studies, in 52 patients with inactive cirrhosis (5 $HBsAg^+$), and in 53 patients with inactive cirrhosis (5 $HBsAG^+$). Anti-delta was present among $HBsAg^+$ cases in 5/53 patients with hepatocellular carcinoma, 2/5 with inactive cirrhosis, and 10/14 with active cirrhosis. Delta Ag was, however, found in the liver only in patients with inactive (1/5) or active (10/14) cirrhosis. Subjects who were $HBsAg^-$ but had other markers of HBV infection circulated low-titer anti-delta in 11% of cases, but delta Ag was never found in the liver. This was taken as evidence of resolved delta infection. No role of delta agent in oncogenic transformation was demonstrated.

References

1 Rizzetto, M.; Canese, M.G.; Aricò, S.; Crivelli, O.; Bonino, F.; Trepo, C.G.; Verme, G.: Immunofluorescence detection of a new antigen/antibody system (delta/anti-delta) associated with hepatitis B virus in liver and serum of HBsAg carriers. Gut *18:*997–1003 (1977).

2 Rizzetto, M.: Biology and characterization of the delta agent; in Viral hepatitis – 1981. Int. Symp. (Franklin Institute Press, Philadelphia 1982).

3 Rizzetto, M.; Purcell, R.H.; Gerin, J.L.: Epidemiology of HBV-associated delta agent: geographical distribution of anti-delta and prevalence in polytransfused HBsAg carriers. Lancet *i:*1215–1219 (1980).

4 Aricò, S.; Rizzetto, M.; Crivelli, O.; Canese, M.G.; Zanetti, A.; Ponzetto, A.; Ferrari, G.; Bonino, F.; Pera, A.; Verme, G.: The clinical and immunological significance of a new antigen/antibody system (delta/anti-delta) in chronic carriers of the HBsAg. Ital. J. Gastroenterol. *10:* 146–151 (1978).
5 Smedile, A.; Farci, P.; Verme, G.; Caredda, F.; Cargnel, A.; Caporaso, N.; Dentico, P.; Trepo, C.G.; Opolon, P.; Gimson, A.; Vergani, D.; Williams, R.; Rizzetto, M.: Influence of delta infection on severity of hepatitis B. Lancet *ii:* 945–947 (1982).
6 Smedile, A.; Lavarini, C.; Crivelli, O.; Raimondo, G.; Fassone, M.; Rizzetto, M.: Radioimmunoassay detection of IgM antibodies to the HBV-associated delta antigen: clinical significance in delta infection. J. med. Virol. *9:* 131–138 (1982).
7 Crivelli, O.; Rizzetto, M.; Lavarini, C.; Smedile, A.; Gerin, J.L.: Enzyme-linked immunosorbent assay for detection of antibody to the HBsAg-associated delta antigen. J. clin. Microbiol. *14:* 173–177 (1981).
8 Stöcklin, E.; Gudat, F.; Krey, G.; Dürmüller, U.; Gasser, M.; Schmid, M.; Stalder, G.; Bianchi, L.: Delta antigen in hepatitis B: immunohistology of frozen and paraffin-embedded liver biopsies and relation to HBV infection. Hepatology *1:* 238–242 (1981).
9 Hansson, B.G.; Moestrup, T.; Widell, A.; Nordenfeldt, E.: Infection with delta agent in Sweden: introduction of a new hepatitis agent. J. infect. Dis. *146:* 472–478 (1982).

Dr. A. Craxí, Divisione di Medicina Interna, Ospedale V. Cervello,
Via Trabucco 180, I-90100 Palermo (Italy)

Front. gastrointest. Res., vol. 8, pp. 197–202 (Karger, Basel 1984)

Drinking Habits in Students of Two Different Italian Regions

R. Naccarato, M. Salvagnini, D. Martines, M. Chiaramonte, R. Sciarrone, M. Cibin

Cattedra di Gastroenterologia, Clinica Medica I', Policlinico Universitario, Padova, Italia

Introduction

The Italian prevalence of alcohol-related diseases is among the highest in the World [1]. Moreover, when we consider, according to *Jellinek and Keller* [2], the liver cirrhosis mortality rate as a marker of alcoholism, we observe that its distribution changes from area to area [3]. In the northern regions the liver cirrhosis mortality rate is around 40/100,000 inhabitants annually, in southern Italy this rate is around 20/100,000. Furthermore, if we assume that alcoholism has a different diffusion in our country, we are still, however, far from having a clear explanation for this phenomenon.

It is commonly thought that several factors, passed on through the family and the social environment, may play an important role in inducing alcohol abuse [4]. On the other hand, the attitudes towards alcohol which form at an early age tend to continue into adult life [5]. In order to prevent the onset of alcoholism at an early age it appears necessary to obtain some information about the drinking habits of youngsters and the factors involved in the development of alcoholism in our country, where the knowledge of these aspects is as yet very limited.

Aim of the Study

This study was carried out in order to identify the drinking habits of teenage students in two different Italian areas, where the prevalence of alcoholism is quite different. Furthermore, we tried to identify some

Table I. Student population studied, divided according to sex

	Calabria	Veneto
Males	723	542
Females	796	222
Total	1,519	764

factors, which could account for the different diffusion of this phenomenon.

Methods

The study involved 2,283 high school students (table I); 764 were from Padua (a town in the Veneto region) and 1,519 from Cosence (a Calabria province). The age ranged between 14 and 19 years. Each student was asked to answer a questionnaire in a class, which contained several questions about personal drinking and smoking habits and about the drinking habits of the family. We classified the students in three groups, according their alcohol intake: (1) abstainers: who have never touched alcohol or who drink no more than once or twice per year; (2) moderate drinkers: i.e. students who drink only with meals and infrequently; (3) heavy drinkers: people who usually drink at meals and frequently at other times.

Data were coded and analysed by a computer. The χ^2 test was applied on the distribution of frequencies.

Results

Alcohol Intake. The abstainers in the Calabria sample were 23%; this percentage was more than double in comparison with the Veneto students (9%). The moderate drinkers were equally distributed in the two populations, with a small prevalence in the Southern Region (67 vs 59%). The heavy drinkers were significantly more in Padua (33 compared to 10%). The Padua data, in which male students predominated, did not change even when divided according to sex.

Age of Initiation to Drink. The students of Calabria (table II) showed an earlier first experience with alcohol than the Veneto teenagers. Indeed, 56% began to drink alcohol before the age of 12 in the Cosence sample, compared to 40% in Padua. However, we must consider

Table II. Age of Initiation to drink in the students

Age years	Calabria, %	Veneto, %
≤ 12	56	40
> 12	21	52
Abstainers	23	8

Table III. Fathers drinking habits of heavy drinking students

	Calabria, %	Veneto, %
Father moderate drinker	1.6	1.5
Father heavy drinker	4.3	10
	$\chi^2 = 4.4$	$\chi^2 = 13.1$
	$p < 0.05$	$p < 0.005$

that although the students of Veneto started to drink at an older age, they drank with a higher frequency (92 vs 77%).

Family Behavioural Patterns. We report data only from students fathers as the other family members did not supply sufficient information regarding their drinking habits (table III). A positive correlation was found between heavy alcohol intake in fathers and sons in both populations ($p < 0.001$). It is important to stress that the higher prevalence of heavy drinkers among the fathers observed in Padua reflects the higher prevalence of alcoholism in this area. If we express this finding calculating the difference between observed and expected frequencies in the students drinking habits correlated to those found in their fathers, we may represent this relationship better. Figure 1 demonstrates an excess of abstainers and moderate drinkers among the students whose parents were found to be in the same categories. Conversely, there is a similar excess of heavy drinkers in the children of heavy drinkers.

Alcohol and Smoking. A significant relationship was also found between smoking (more than 10 cigarettes per day) and a high alcohol intake in both populations (table IV).

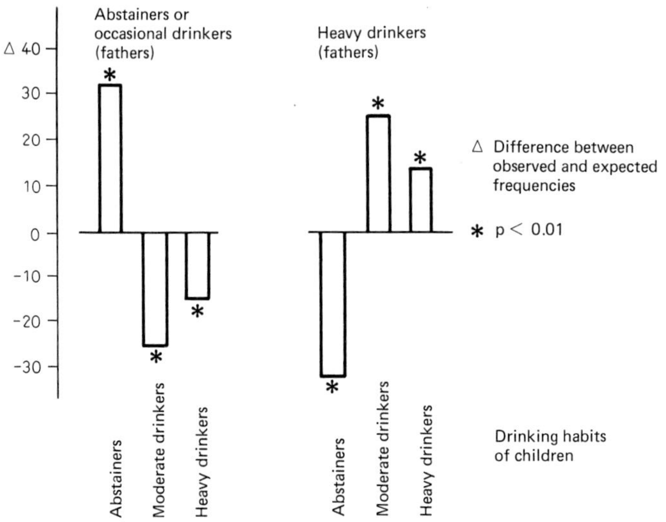

Fig. 1. Relationship between the drinking habits of fathers and their children.

Table IV. Correlation between alcohol and smoking habits

	Calabria: smoking		Veneto: smoking	
	yes	no	yes	no
Drinking: Yes	64 (4%)	98 (6%)	113 (27%)	58 (14%)
No	250 (16%)	1,129 (74%)	105 (25%)	141 (34%)
	$\chi^2 = 39$		$\chi^2 = 21.2$	
	$p < 0.001$	$p < 0.001$		

Discussion

In Veneto, a northern region of Italy with a high prevalence of alcoholism and a high alcohol production rate [6], more than 90% of the young population habitually drink alcohol. A third of them have drinking habits which can be defined as dangerous. Instead, in Calabria, a southern Italian region with a low prevalence of alcoholism, the abstainers represented 23% of the youngsters examined. In this latter population, less than 10% were heavy drinkers. These data stress that the drinking habits which characterize an entire population are acquired at a young age. We can furthermore confirm that the prevalence of alcoholism is basically linked to two mean factors: the availability of alcohol and the proportion of drinkers (including the normal drinkers). Moreover, a key point in the development of drinking habits must be recognized in the habits of the parents, which we found influenced the behaviour of the children. This influence develops as a transmission of a behavioural pattern which is a fundamental part of the family's culture and the social culture. The finding that the population at a lower prevalence of alcoholism has an earlier contact with alcohol drinking and develops less dangerous drinking habits does not contrast with this hypothesis. It has to be considered that Calabria is a Region with a prevalently rural culture where a moderate alcohol consumption is considered [7] 'magic, mystic and often therapeutic'. In this way the use of alcohol, which is 'intermittent but not toxic day by day', is passed on from generation to generation.

A further important observation regards the direct correlation between alcohol and tabacco abuse. This finding, already reported by others [8, 10], underlines that alcoholism cannot be considered a single pathological phenomenon, but especially in the case of youngsters it must be included in the wider field of addictions.

Summary

In order to clarify some factors involved in the development of alcohol abuse, the drinking habits of 2,283 students aged between 14 and 19, from two different Italian regions, were investigated. In areas with a high prevalence of alcohol-related disease (i.e. liver cirrhosis) and a high alcohol production, a surprisingly high proportion of students habitually drink an excessive amount of alcohol (33%).

No relationships between the age of initiation of drinking and drinking habits were found. A strong correlation was found between the drinking habits of the youngsters and their parents. The close correlation which was observed between alcohol drinking and cigarette smoking underlines the relationship between different addictions expecially in adolescence.

Acknowledgements

We are indebted to: Dr. *L. Gallimberti*, Dr. *D. Orlandini*, Dr. *G.P. Ricci*, Dr. *A. Del Borrello* (Alcohol and Drug Dependence Treatment Service, Dolo - Venezia), Dr. *G. Be-*

nussi (Istituto d'Igiene - Università di Trieste), Dr. *A. Piccoli* (Clinica Medica I' - Università di Padova), Dr. *A. Contino*, Dr. *F. Tassone* (Divisione Medica, Ospedale Civile di Praia a Mare - Cosenza), for their helpful collaboration.

References

1 Lint, J.; Schmidt, W.: The epidemiology of alcoholism, pp. 423-443, Israel, Mardones, (Wiley & Sons, New York 1971).
2 Jellinek, E.M.; Keller, M.: Rates of alcoholism in USA. Q. J Stud. Alcohol. *13:* 49-59 (1952).
3 Annuario di Statistiche Sanitarie, vol. 19, pp. 280-281 (Ist. Centr. Statistica, Roma 1978).
4 Akers, R.L.: Teenage drinking: a summary of research findings; Robinson, Alcohol problems, pp. 88-92 (MacMillan, London 1979).
5 Fillmore, K.M.; Bacon, S.D.; Himan, M.: Alcohol drinking behavior and attitudes: Rutges panel study. Report of N.I.A.A. (Rutges Center of Alcohol Studies, 1977).
6 Bollettino Mensile di Statistica, vol. 3, p. 245 (Ist. Centr. Statistica, Roma 1979).
7 Terzian H.: La cultura della vigna e del vino; in Patologia e problemi connessi all'uso inadeguato di alcolici. Quaderni di Educazione Sanitaria. Regione Veneto, Venezia 1982, pp. 17-33.
8 USA Department of Health Education and Welfare: National patterns of cigarettes smoking ages 12 through 18 in 1972 and 1974. (DHEW, Rockville 1976).
9 Kirk, S.R.: Use of alcohol and other drugs by rural teenagers; in Galanter, current in alcoholism, pp. 233-237 (Grune & Stratton, New York 1979).
10 National Commission of Marijhuana and Drug Abuse: Drug use in America. Problem in perspective (US Government Printing Office, Washington 1973).

Prof. R. Naccarato, Cattedra di Gastroenterologia, Clinica Medica I', Policlinico Universitario, Via N. Giustiniani, 2, I-35100 Padova (Italy)

Metabolic Aspects

Front. gastrointest. Res., vol. 8, pp. 203–207 (Karger, Basel 1984)

Pathometabolic Mechanism of Porphyria cutanea tarda

E. Ventura, E. Rocchi, P. Gibertini, T. Trenti, M. Cassanelli

Clinica Medica Generale e Terapia Medica III, Università di Modena, Modena, Italia

Porphyria cutanea tarda (PCT) is the full clinical and biochemical syndrome of chronic hepatic porphyria, which has aroused great interest and investigations in recent years. The aim of our studies is concerned with a better understanding of the basic enzyme disorder involved in the pathogenesis of PCT. PCT has long been considered an acquired disorder of porphyrin metabolism because most cases occur sporadically and late in life and seem to be induced or precipitated by liver disease, alcohol abuse, iron overload or estrogens taken in contraceptive drugs and in the treatment of breast and prostatic carcinoma. This hypothesis was supported by acquired toxic porphyria resembling PCT, which appeared in individuals accidentally exposed to polyalogenated aromatic hydrocarbons [1].

However, the low incidence of PCT as a complication of alcoholic liver disease, idiopathic or symptomatic hemochromatosis and estrogen administration must be stressed. Moreover, there is an increasing evidence of familial occurrence of PCT and abnormalities of porphyrin metabolism in apparently healthy relatives of PCT phenotypes. These findings suggest the occurrence of a genetic factor in the pathogenesis of PCT, possibly as an inherited enzyme defect having variable penetrance and expressivity.

The metabolic steps that lead from substrates glycine and succinyl-CoA to ALA, PBG, porphyrinogen and heme formation in a normal subject are outlined in figure 1a. The reactions are catalyzed by mitochondrial and cytosolic enzymes. In the liver, the rate of synthesis of ALA-S, and its uptake into mitochondria as well as the direct inhibition on its enzymatic activity are regulated by means of feedback repression exerted by heme, the end-product of the pathway. A decrease of the hepatic heme pool as well as increased requests for heme are met by the enhanced synthesis and activity of ALA-S induced by derepression of the negative feedback. Therefore, ALA-S plays the key-role of rate-limiting enzyme of heme formation [2].

Hepatic ALA-S can be induced: (1) by several lipid-soluble drugs, naturally occurring or synthetic steroids, alcohol and other chemicals that are substrates of cytochrome P_{450}, the terminal oxidase of microsomal drug metabolism; (2) by defects of enzymes of heme synthesis catalyzing the polymerization of PBG to uroporphyrinogen, and the steps of decarboxylation-oxidation of coproporphyrinogen to protoporphyrinogen and protoporphyrin, as happens in acute hepatic porphyrias.

Both pharmacometabolic mechanisms and enzyme deficiencies lead to a decrease of hepatic heme, that under physiological conditions would repress ALA-S. As a result, there is: (1) an enhanced ALA-S activity; (2) an increased production of precursors ALA and PBG, tending to compensate the partial block of heme synthesis; (3) a high urinary excretion of ALA and PBG, which becomes the most current biochemical pattern of acute hepatic porphyrias. The interaction between microsomal drug oxidation and heme synthesis explains how certain drugs may precipitate the clinical symptoms of acute intermittent porphyria, hereditary coproporphyria and variegate porphyria in patients who have latent disease or are in remission [2].

On the contrary, PCT taken as the full expression of chronic hepatic porphyria is characterized, apart from photosensitive cutaneous lesions, by the following features: (1) urinary excretion of ALA and PBG always within normal limits; (2) high urinary excretion of uroporphyrin (8-COOH) and eptacarboxyl porphyrin together with intermediate esa-, penta-, or tetracarboxyl porphyrin (coproporphyrin): their relative distribution is URO > EPTA > COPRO > PENTA > ESA (in acute intermittent porphyria the relative distribution of urinary porphyrin is URO > COPRO > PENTA > EPTA or ESA). These findings seem to exclude an overproduction of ALA and PBG and postulate an increased synthesis of uroporphyrinogen and a defect in its decarboxylation steps (fig. 1b).

In order to validate this hypothesis we investigated the enzyme activities involved in polymerization of PBG to uroporphyrinogen (URO-S) and in decarboxylation to coproporphyrin (URO-DEC). The

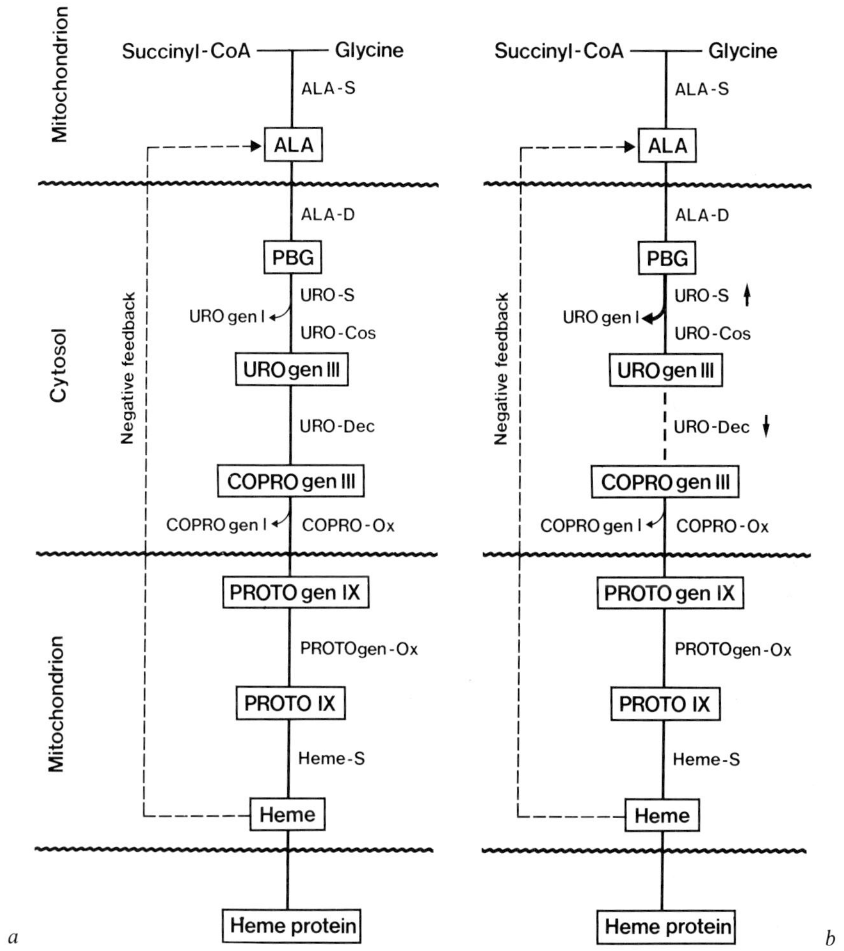

Fig. 1. Heme synthesis pathway in normal subject *(a)* and in PCT *(b)*.

study was carried out on liver tissue taken by biopsy during laparoscopy performed to define the morphological picture of liver disease. Liver specimens of nonalcoholic patients who underwent laparatomy for benign abdominal diseases were used as controls. Aiming to extend the investigation to other heme-forming tissues, erythrocytes were tested for the same enzymes [3].

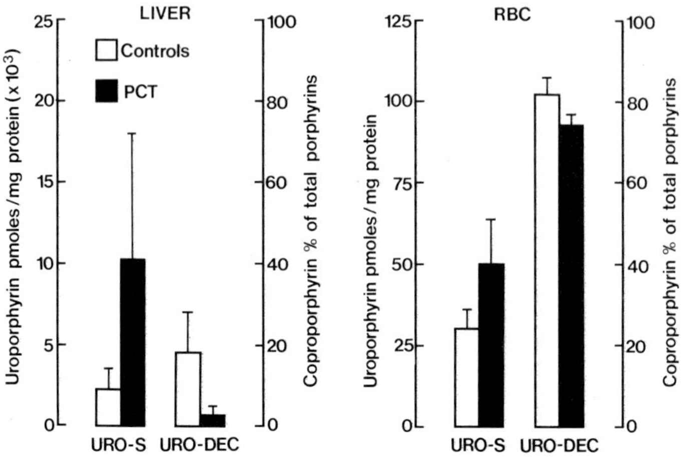

Fig. 2. Hepatic and RBC uroporphyrinogen synthetase (URO-S) and decarboxylase (URO-DEC) in normal subjects and in PCT. URO-S activity is expressed as uroporphyrin newly synthesized from porphobilinogen incubated with liver homogenate for 6 h and with RBC lysate for 30 min. URO-DEC activity is expressed as newly formed coproporphyrin, the end-product of decarboxylation process, as a percentage of total synthesized porphyrins identified as methyl-esters by thin-layer chromatography. Statistical analysis performed by a nonparametric test (rank-test). Hepatic and RBC URO-S, $p < 0.05$; hepatic URO-DEC, $p < 0.01$; RBC URO-DEC, $p < 0.001$.

As outlined in figure 2, PCT phenotypes showed a significant decrease in hepatic and red blood cell (RBC) URO-DEC activity. Moreover, the preceding metabolic step was characterized by a significant increase in both hepatic and RBC URO-S activity. The occurrence of the same enzymatic alterations in different heme-forming tissues could reflect a generalized and probably inherited metabolic disorder in PCT.

These results support the hypothesis that a deficiency of URO-DEC is the basic defect of PCT. The enhanced URO-S activity may be seen as a secondary effect directed to warrant the full utilization of PBG for the synthesis of octacarboxyl porphyrinogen, the first tetrapyrrole structure of the heme pathway. As a result, uroporphyrinogen I, principally responsible for the photocutaneous syndrome, accumulates; however, the rate of uroporphyrinogen I assembled to biologically active uroporphyrinogen III seems to be enough to provide for the physiological needs of hepatic heme. The normal excretion of precursors ALA and PBG excludes an induction of ALA-S that, on the contrary, should appear if the heme pool size were reduced. In other words, in spite of the defect of URO-DEC, heme synthesis seems to be unimpaired. We can assume that URO-DEC activity normally exceeds that required for physiological needs; then, hepatic heme synthesis may take place at a normal rate in spite of the decreased URO-DEC activity.

In conclusion, the defect of URO-DEC is the primary metabolic marker of PCT. This enzyme deficiency is compensated by an increased URO-S activity providing a continuous flux of substrate available for decarboxylation. Unlike acute hepatic porphyrias, the induction of ALA-S in PCT is lacking. This could mean: (1) a repressive-like effect on ALA-S exerted by increased end-products of URO-S activity; (2) a residual URO-DEC activity sufficient to prevent the block of the heme pathway, thus maintaining a normal heme pool size and avoiding any induction of ALA-S activity. For these reasons in the recent nosography PCT is considered a heme-compensated hepatic porphyria [4].

References

1 Ventura, E.; Rocchi, E.; Candini, R.: Studio del metabolismo porfirinico nelle epatopatie croniche. Il Fegato *21:*531 (1975).

2 Ventura, E.; Rocchi, E.; Gibertini, P.; Cremonini, C.; Zeneroli, M.L.: Erythrocyte uroporphyrinogen-I-synthetase activity in families with acute intermittent porphyria. Ital. J. Gastroenterol. *11:*163 (1979).

3 Ventura, E.; Rocchi, E.; Gibertini, P.: Hepatic and erythrocyte uroporphyrinogen synthetase and decarboxylase activities in porphyria cutanea tarda. Ital. J. Gastroenterol. *12:*71 (1980).

4 Ventura, E.; Manenti, F.; Di Marco, G.; Rocchi, E.; Gibertini, P.; Zeneroli, M.L.: Epatite cronica e cirrosi nella porfiria cutanea tarda; in Gentilini et al., Update on acute and chronic hepatitis. Soc. Ed. Universo, Roma, p. 263 (1981).

E. Ventura, MD, Clinica Medica Generale e Terapia Medica III, Università di Modena, I-41100 Modena (Italy)

Front. gastrointest. Res., vol. 8, pp. 208–216 (Karger, Basel 1984)

Volatile Hydrocarbons in the Breath of Patients with Chronic Alcoholic Liver Disease: A Possible Marker of Ethanol-Induced Lipoperoxidation[1]

Stefania Moscarella, G. Laffi, D. Coletta, U. Arena, A. Paola Cappellini, P. Gentilini[2]

Istituto di Clinica Medica IV, Università degli Studi di Firenze, Firenze, Italia

Introduction

Lipoperoxidation, the peroxidative degradation of polyunsaturated fatty acids (PUFA), as components of biological membranes, seems to have an important role in the cellular damage caused by some exogenous toxic agents and/or their metabolites (CCl_4, ethanol, ozone, paraquat, nitrogen dioxide, etc.).

Many different degradation products, such as short chain hydrocarbons, are formed from PUFA peroxidation.

In vitro experiments by *Dumelin and Tappel* [9] showed that these hydrocarbon gases evolve during iron, copper and/or hematin catalyzed decomposition of ω-6 and ω-3 fatty acid hydroperoxides and not during the peroxidation phase. The major hydrocarbon gas formed from ω-6 fatty acid hydroperoxides (linoleic acid and arachidonic acid) is pentane; other minor hydrocarbons (*n*-butane and propane) are also produced, but in smaller amounts. Ethane and ethylene, on the other hand, arise mainly from ω-3 fatty acid hydroperoxides (linolenic acid) but they too are produced in very small amounts from ω-6 fatty acid hydroperoxides.

The measurement of volatile hydrocarbons in the breath of laboratory animals has been used to study several lipoperoxidation-inducing

[1] Supported by the Ministero della Pubblica Istruzione, Rome, Italy, 1982.

[2] We would like to thank Prof. *G. Buzzelli* (Institute of Clinica Medica IV, University of Florence) for the statistic elaboration of data.

conditions such as carbon tetrachloride [16, 18], ozone [8, 24] and nitrogen dioxide [24, 25, 29] toxicity; or to evaluate the protective effect of vitamin E [4, 5, 14, 26] or catechols [17].

The problem whether or not lipoperoxidation has an important role in the pathogenic mechanism of acute and chronic ethanol toxicity, has been, for a long time, a subject of controversy. Several investigations [3, 6, 7, 31] offered evidence in animals for a peroxidative process, as measured by conjugated diene formation or malondialdehyde release, in liver homogenates after short-term ethanol administration. Other authors [15, 27] however, were unable to confirm these results.

Chronic ethanol consumption, also, led to conflicting results: some studies [22, 23, 28] showed the involvement of lipoperoxidation in ethanol toxicity and other studies the contrary [30]. The lipoperoxidative effect of ethanol has also been estimated by measuring volatile hydrocarbons in the breath of laboratory animals. The administration of an acute dose of ethanol (5 g/kg) to rats is reported to produce an increased release of ethane or pentane [1, 16, 20]. Chronic ethanol administration, on the contrary, did not influence exhaled basal pentane, but only an animal's susceptibility to the acute ethanol toxicity [19]. Experimental evidence regarding alcohol abuse and the monitoring of lipoperoxidation by breath analysis in humans, has not yet been reported in the literature.

In order to assess whether or not ethanol-induced liver disease, in humans, was accompanied by an enhancement of lipoperoxidative process, exhaled hydrocarbons were determined in chronic alcoholics with liver disease in comparison to non-alcoholic healthy subjects, as well as to non-alcoholic cirrhotic patients.

Materials and Methods

Subjects. 51 alcoholics (41 males, 10 females; mean age: 45.3 years) including 9 with alcoholic fatty liver (AFL), 23 with alcoholic hepatitis (AH) and 19 with alcoholic cirrhosis (AC) were studied. Their reported mean daily intake of ethanol was 256 g and reported mean duration of alcohol abuse was 15.3 years. 13 patients with post-hepatitis cirrhosis (PHC) (8 males, 5 females; mean age: 53 years) and 24 healthy subjects as controls (18 males, 6 females; mean age: 35 years) were also examined. All these subjects reported very moderate alcohol consumption.

The diagnosis of liver disease was based on serum biochemical tests (including total and conjugated bilirubin, glutamine-oxaloacetic transaminase (GOT), glutamic-pyruvic

Table I. Serum biochemical data in patients with chronic alcoholic liver disease and post-hepatitis cirrhosis (mean values ± SEM)

Diagnosis	Number of cases	Bilirubin, mg/dl		GOT IU/l (10–35)	GPT IU/l (10–40)	BSP %ret. 45 min (< 5)	AP IU/l (60–270)	GGT mU/ml (10–40)	Albumin g/dl (4–5.7)	GG g/dl (084–1.68)
		Total (0.2–1.0)	dir. (0–0.2)							
AFL	9	0.59 ± 0.1	0.02 ± 0.0	19 ± 1.5	16 ± 2.4	4 ± 0.8	170 ± 18.2	23 ± 2.9	3.9 ± 0.1	1.19 ± 01
AH	23	0.99 ± 0.1	0.17 ± 0.0	56 ± 2.5	53 ± 11.4	8 ± 0.8	200 ± 20.9	244 ± 34.2	3.9 ± 0.1	1.18 ± 0.0
AC	19	2.04 ± 0.6	0.71 ± 0.2	59 ± 9.4	56 ± 11.7	22 ± 2.1	278 ± 29.6	278 ± 50.0	3.5 ± 0.1	1.96 ± 1.8
PHC	13	1.17 ± 0.2	0.25 ± 0.1	47 ± 7.0	58 ± 12.1	20 ± 2.9	253 ± 44.5	60 ± 11.8	3.5 ± 0.2	2.20 ± 0.2

transaminase (GPT), bromsulphthalein (BSP), alkaline phosphatase (AP), γ-glutamyltranspeptidase (GGT), albumin, γ-globulins (GG), hepatitis B surface antigen) (table I), liver biopsy and ultrasonography. Mean daily intake of ethanol and mean duration of alcohol abuse were calculated using detailed interviews with each patient. The ethanol content of common beverages was calculated on the following basis: beer 3–5% v/v, wine 13% v/v, fortified wine 16–22% v/v, and spirits 42% v/v. The amount (ml) of alcohol consumption was then calculated considering the following: one glass of wine = 120 ml, one bottle of wine = 750 ml, one fiasco = 1,880 ml and one shot-glass = 40–60 ml. Blood alcohol levels were measured on admittance to hospital and repeated at frequent intervals during hospital stay. Informed consent was obtained from all subjects admitted to the study.

Breath Analysis. A Perkin-Elmer model 3920 gas chromatograph with a flame ionization detector was used for hydrocarbon gas analysis, according to *Frank* et al. [11]. A stainless steel column (3 mm × 5 m) filled with Porasil C (80–100 mesh) was used with a nitrogen carrier gas flow rate of 30 ml/min. The isothermal operating mode was used at a column temperature of 38 °C; the detector temperature was held constant at 170 °C. The identification of the sample peaks and the concentration of gases, expressed as pmol/ml, were determined using a standard mixture (ethane: 1.1 ppm; ethylene: 1.1 ppm; propane: 0.84 ppm; *n*-butane: 1.8 ppm; *n*-pentane: 1.8 ppm; 1-butene: 1.8 ppm). Each value was the average of three different determinations.

Hydrocarbon analysis was carried out at the highest instrumental sensitivity setting. Each breath sample was collected in a 2-liter gas sampling bag. 10 ml of the sample were withdrawn with a 10-ml glass syringe and injected into the gas chromatograph by means of a 5-ml gas sampling loop. A room air sample was analyzed at the same time for background contamination. Patients were fasted for at least 6 h before breath test. All smokers were asked not to smoke for at least 2 h before the test.

Statistical Analysis. All data are expressed as mean ± standard error of the mean. All groups were tested simultaneously with one-way analysis of variance; each group was then tested versus the other groups using Duncan's multiple range test with Kramer's modification for unequal class size.

The correlation coefficients between each hydrocarbon measured and some parameters of liver function - GOT, BSP, AP, GGT - were calculated in the 9 patients with AFL, in the 23 patients with AH and in the 19 patients with AC. In addition, the correlation coefficients between the amount of pentane and, respectively, the mean daily intake and the mean duration of alcohol abuse were also calculated. The same procedure was followed for ethane.

Results

Volatile hydrocarbon levels (pmol/ml) in the breath of patients with various types of chronic liver disease and healthy subjects are shown in table II. A statistically significant difference was found in the concentration of pentane ($F = 4.97$; $p < 0.01$), propane ($F = 5.71$; $p <$

0.01) and ethylene ($F = 3.25$; $p < 0.05$) in the various groups examined. The ethane concentrations of the groups were not significantly different.

Duncan's test (5%) showed that the mean concentration of exhaled pentane in patients with AC was significantly higher than in all the other groups examined. These patients, furthermore, exhaled greater amounts of butane than patients with PHC and AH, respectively. The mean concentration of exhaled propane was significantly higher in the control group than in all the other groups examined. Healthy subjects, moreover, showed greater amounts of exhaled ethylene as compared to patients with AC, AH and PHC.

The amounts of exhaled ethane and pentane detected in the 51 patients with chronic alcoholic liver disease, did not correlate either with individual liver function tests, the mean daily intake of ethanol or the mean duration of abuse. The same result was obtained with the other hydrocarbons (i.e. ethylene, propane, butane) examined in these patients. The mean values of hydrocarbon gases, relative to the fatty acid precursors, are shown in table III.

The analysis of variance showed a statistically significant difference ($F = 4.36$; $p < 0.01$) in the mean concentration of those hydrocarbons formed mainly from linoleic and arachidonic fatty acid hydroperoxides. The ω-6/ω-3 molar ratios and the mean concentrations of all five hydrocarbons were also significantly different in the same groups ($F = 3.58$; $p < 0.01$; $F = 3.26$, $p < 0.05$, respectively).

Table II. Exhalation of hydrocarbons by humans (mean values ± SEM)

	AFL [9]	AH [23]	AC [19]	PHC [13]	Controls [24]	
Ethane, pmol/ml	0.70 ± 0.10	0.62 ± 0.09	0.84 ± 0.12	0.75 ± 0.19	0.74 ± 0.15	F = 0.35 n.s.
Ethylene, pmol/ml	1.18 ± 0.26	1.20 ± 0.19	1.04 ± 0.14	1.28 ± 0.44	2.18 ± 0.32	F = 3.25 p < 0.05
Propane, pmol/ml	0.41 ± 0.10	0.40 ± 0.05	0.55 ± 0.07	0.37 ± 0.06	0.86 ± 0.12	F = 5.71 p < 0.01
n-Butane, pmol/ml	1.52 ± 0.25	1.70 ± 0.27	3.03 ± 0.77	1.20 ± 0.24	2.30 ± 0.31	F = 2.31 n.s.
n-Pentane, pmol/ml	0.51 ± 0.09	1.57 ± 0.57	3.35 ± 0.62	0.96 ± 0.32	1.11 ± 0.20	F = 4.97 p < 0.01

Table III. ω-3 and ω-6 fatty acid hydroperoxide-derived hydrocarbons (pml/ml) and their molar ratio (mean values ± SEM)

	AFL [9]	AH [23]	AC [19]	PHC [13]	Controls [24]	
Ethane + ethylene	1.90 ±	1.82 ±	1.87 ±	2.03 ±	2.92 ±	F = 1.99
(ω-3 FAHH)	0.30	0.27	0.20	0.50	0.43	n.s.
Propane + *n*-butane +	2.48 ±	3.52 ±	6.94 ±	2.54 ±	4.28 ±	F = 4.36
n-pentane (ω-6 FAHH)	0.38	0.75	1.25	0.44	0.53	$p < 0.01$
ω-6/ω-3 molar ratio	1.60 ±	2.57 ±	4.02 ±	1.95 ±	1.85 ±	F = 3.58
	0.35	0.50	0.72	0.54	0.19	$p < 0.01$
Total hydrocarb.	4.38 ±	5.34 ±	8.82 ±	4.57 ±	7.20 ±	F = 3.26
	0.52	0.87	1.31	0.72	0.86	$p < 0.05$

Duncan's test (5%) showed that the mean concentration of ω-6 fatty acid-derived hydrocarbons (propane, *n*-butane, *n*-pentane) and the ω-6/ω-3 molar ratio were significantly higher in patients with AC than in all the other groups, including controls. Moreover, these patients with AC showed a mean total hydrocarbon concentration significantly greater than in patients with AFL, PHC and AH.

Discussion

The significantly elevated levels of exhaled pentane observed in chronic alcoholics with liver cirrhosis, in this study, may indicate an enhancement of lipoperoxidative process. No significant increase of expired ethane, on the contrary, was detected. The presence of pentane in the breath, however, seems to be a more sensitive index, than ethane, of an ethanol-induced peroxidative damage. In fact ω-6 PUFA, especially arachidonic acid, are the most abundant types in the membrane lipids of mammals.

Other authors [2], however, measured volatile alkanes in red blood cells and reported a decreased release of pentane and an elevated release of ethane from erythrocytes in patients with alcoholic cirrhosis. These apparently contrasting findings might be due to a different ω-6/ω-3 PUFA ratio in biological membranes of human red cells and liver organelles, which are both susceptible to injury by ethanol alone and/or its metabolites.

However, no significant change in exhaled pentane levels was detected in chronic alcoholics with either steatosis or AH as compared to controls. A possible relationship between the extent of lipoperoxidation, evaluated by pentane release, and the degree of histological ethanol-induced liver damage, might be suggested.

In patients with steatosis, *Poulsen* et al. [21] showed no reduction in hepatic glutathione, the most important sulfhydryl free radical scavenger involved in cellular protective mechanisms. This relationship between reduced glutathione (GSH) content, which is believed to correlate with peroxidative damage, and the degree of liver impairment, seems to confirm the present results.

In baboons, an experimental model in which not only fatty liver but also cirrhosis develops, chronic alcohol administration is accompanied by evidence of lipid peroxidation, measured by diene conjugates, in association with GSH reduction [28].

Endogenous hydrocarbons are metabolized, for the most part, to corresponding alcohols through hydroxylation by hepatic monoxygenases [12]. Enhanced exhalation of volatile hydrocarbons by rats following administration of 5 g/kg ethanol (an inhibitor of the hydroxylation pathway) may not show an increased lipid peroxidation, but rather an inhibited hydrocarbon metabolism [10].

The measurement of these hydrocarbons in breath probably does not allow us to specify that the liver is the organ where ethanol-induced lipoperoxidation takes place. However, several investigations in vivo showed evidence for a peroxidative injury in hepatic subcellular structures. Therefore, a possible extrahepatic source of pentane, as reported by *Gelmond* et al. [13] in anaesthetized rats, should be considered.

The in vivo source of the other hydrocarbons (ethylene, propane and *n*-butane), examined with this procedure, still remains to be clarified, as only in vitro experiments have so far been reported.

Summary

Lipid peroxidation products (ethane, ethylene, propane, *n*-butane, *n*-pentane) were measured by a gas-chromatographic method in respiratory gases of 51 alcoholics with chronic liver disease (19 with cirrhosis, 23 with alcoholic hepatitis and 9 with fatty liver). 13 patients with post-hepatitis cirrhosis and 24 healthy subjects were also examined.

The greatest increase of exhaled pentane (3.35 pmol/ml) was found in the 19 patients with alcoholic cirrhosis compared with all the other groups examined. No significant difference was detected in the release of exhaled ethane among the five groups. These results may suggest that pentane could be considered a more sensitive index than ethane, for ethanol-induced lipoperoxidation in humans. Furthermore, chronic alcohol abuse in this study seems to induce a significant enhancement of lipoperoxidation, detectable by expired pentane measurement, only in conditions of maximal liver impairment.

References

1 Burk, R.F.; Lane, J.M.: Ethane production and liver necrosis in rats after administration of drugs and other chemicals. Toxicol. appl. Pharmacol. *50:*467–478 (1979).

2 Clemens, M.R.; Remmer, H.: Volatile alkanes produced by erythrocytes: an assay for in vitro studies on lipid peroxidation. Blut *45:*329–335 (1982).

3 Comporti, M.; Hartman, A.; Di Luzio, N.R.: Effect of in vivo and in vitro ethanol administration on liver lipid peroxidation. Lab. Invest. *16:*616–624 (1967).

4 Dillard, C.J.; Dumelin, E.E.; Tappel, A.L.: Effect of dietary vitamin E on expiration of pentane and ethane by the rat. Lipids *12:*109–114 (1977).

5 Dillard, C.J.; Litov, R.E.; Tappel, A.L.: Effect of dietary vitamin E, selenium and polyunsaturated fats on in vivo lipid peroxidation in the rats as measured by pentane production. Lipids *13:*396–402 (1978).

6 Di Luzio, N.R.: The role of lipid peroxidation and antioxidants in ethanol-induced lipid alterations. Exp. molec. Path. *8:*394–402 (1968).

7 Di Luzio, N.R.; Hartman, A.D.: The effect of ethanol and carbon tetrachloride administration on hepatic lipid soluble antioxidant activity. Exp. molec. Path. *11:* 38–52 (1969).

8 Dumelin, E.E.; Dillard, C.J.; Tappel, A.L.: Effect of vitamin E and ozone on pentane and ethane expired by rats. Archs envir. Hlth *33:*129–135 (1978).

9 Dumelin, E.E.; Tappel, A.L.: Hydrocarbon gases produced during in vitro peroxidation of polyunsaturated fatty acids and decomposition of preformed hydroperoxides. Lipids *12:*894–900 (1977).

10 Frank, H.; Hintze, T.; Bimboes, D.; Remmer, H.: Monitoring lipid peroxidation by breath analysis: endogenous hydrocarbons and their metabolic elimination. Toxicol. app. Pharmacol. *56:*337–344 (1980).

11 Frank, H.; Hintze, T.; Remmer, H.: Volatile hydrocarbons in breath, an indicator for peroxidative degradation of lipids; in Kolb, Applied head-space gas chromatography (Heyden, London 1980).

12 Frommer, U.; Ullerich, V.; Staundinger, H.J.: Hydroxylation of aliphatic compounds by liver microsomes. Hoppe-Seyler's Z. physiol. Chem. *351:* 903–912 (1970).

13 Gelmond, D.; Stein, R.A.; Mead, J.F.: The bacterial origin of rat breath pentane. Biochem. biophys. Res. Commun. *102:*932–936 (1981).

14 Hafeman, D.G.; Hoekstra, W.G.: Lipid peroxidation in vivo during vitamin E and selenium deficiency in the rat as monitored by ethane evolution. J. Nutr. *107:* 666–672 (1977).

15 Hashimoto, S.; Recknagel, R.O.: No chemical evidence of hepatic lipid peroxidation in acute ethanol toxicity. Exp. molec. Path. *8:*225–242 (1968).

16 Köster, U.; Albrecht, D.; Kappus, H.: Evidence for carbon tetrachloride and ethanol-induced lipoperoxidation in vivo demonstrated by ethane production in mice and rats. Toxicol appl. Pharmacol. *41:*639–648 (1977).
17 Köster-Albrecht, D.; Köster, U.; Kappus, H.; Remmer, H.: Inhibition of $CBrCl_3$-induced lipid peroxidation in rats in vivo by (+)-cyanidanol-3. Toxicol. Lett. *3:* 363–368 (1979).
18 Lindstrom, T.D.; Anders, M.W.: Effect of agents known to alter carbon tetrachloride hepatotoxicity and cytochrome P-450 levels on carbon tetrachloride-stimulated lipid peroxidation and ethane expiration in the intact rat. Biochem. Pharmacol. *27:*563–567 (1978).
19 Litov, R.E.; Gee, D.L.; Downey, J.E.; Tappel, A.L.: The role of lipid peroxidation during chronic and acute exposure to ethanol as determined by pentane expiration in the rat. Lipids *16:*52–57 (1981).
20 Litov, R.E.; Irving, D.H.; Downey, J.E.; Tappel, A.L.: Lipid peroxidation: a mechanism involved in acute ethanol toxicity as demonstrated in vivo pentane production in the rat. Lipids *13:*305–307 (1978).
21 Poulsen, H.E.; Ranek, L; Andreasen, P.B.: The hepatic glutathione content in liver disease. Scand. J. clin. Lab. Invest. *41:*573–576 (1981).
22 Reitz, R.C.: A possible mechanism for peroxidation of lipids due to chronic ethanol ingestion. Biochim. biophys. Acta *380:*145–154 (1975).
23 Remmer, H.; Albrecht, D.; Kappus, H.: Lipid peroxidation in isolated hepatocytes from rats ingesting chronically. Naunyn-Schmiederberg's Arch. Pharmacol. *298:* 107–113 (1977).
24 Roehm, J.M.; Hadley, J.G.; Menzel, D.B.: Oxidation of unsaturated fatty acids by ozone and nitrogen dioxide. Common mechanism of action. Archs envir. Hlth *23:* 142–148 (1971).
25 Sagai, M.; Ichinose, H.O.; Kubota, K.: Studies on biochemical effects on nitrogen dioxide. I. Lipid peroxidation as measured by ethane exhalation of rats exposed to nitrogen dioxide. Lipids *16:*64–67 (1981).
26 Sagai, M.; Tappel, A.L.: Effect of vitamin E on carbon tetrachloride-induced lipid peroxidation as demonstrated by in vivo pentane production. Toxicol. Lett. *2:* 149–155 (1978).
27 Scheig, R.; Klatskin, G.: Some effects of ethanol and carbon tetrachloride on lipoperoxidation in rat liver. Life Sci. *8:*855–865 (1969).
28 Shaw, S.; Jayatilleke, E.; Ross, W.A.; Gordon, E.R.; Lieber, C.S.: Ethanol-induced lipid peroxidation: potentially by long-term alcohol feeding and attenuation by methionine. J. Lab. clin. Med. *98:*417–424 (1981).
29 Thomas, H.V.; Meuller, P.K.; Lyman, R.L.: Lipoperoxidation of lung lipids in rats exposed to nitrogen dioxide. Science *159:*532–534 (1968).
30 Torrielli, M.V.; Gabriel, L.; Dianzani, M.; Ethanol-induced hepatotoxicity: experimental observations on the role of lipid peroxidation. J. Path. *126:* 11–25 (1978).
31 Videla, L.A.; Fernandez, V.; Ugarte, G.; Valenzuela, A.: Effect of acute ethanol intoxication on the content of reduced glutathione of the liver in relation to its lipoperoxidative capacity in the rat. FEBS Lett. *111:*6–10 (1980).

Dr. Stefania Moscarella, Istituto di Clinica Medica IV, Viale Morgagni 85,
I-50100 Firenze (Italy)

Front. gastrointest. Res., vol. 8, pp. 217–228 (Karger, Basel 1984)

Sterol Metabolism in Liver Cirrhosis

N. Carulli, M. Ponz de Leon, P. Loria, F. Zironi, R. Iori
Clinica Medica I-Università degli Studi di Modena, Modena, Italy

Introduction

The liver plays a key role in the overall regulation of sterol metabolism in man: it is the major site of lipoprotein synthesis and catabolism, contributes a large portion of the body cholesterol pool and it has the unique role of secreting cholesterol into the bile either as intact molecule or after its degradation to bile acids. All these processes are interlocked in such a way that in physiological conditions cholesterol balance is kept in equilibrium. From the above considerations it could be anticipated that impairment of liver function may be associated with alterations of cholesterol metabolism. However, detailed studies on this subject are few.

Several authors have reported that plasma cholesterol levels are somewhat reduced in patients with cirrhosis [1–3]. A recent study from our laboratory has extended these findings in a large population of subjects with chronic liver disease [4]. This study has clearly shown that the reduction of plasma cholesterol correlates with the severity of the disease and that the cholesterol content of the three lipoprotein fractions is similarly affected. Hypocholesterolemia in cirrhosis could represent the net result of a complex and multifactorial alteration of sterol metabolism involving the different processes contributing to the input as well as to the output of sterol in the body. This paper will deal with some steps of cholesterol and bile acid metabolism abnormalities observed in patients with cirrhosis.

Cholesterol Absorption

Two reports have dealt specifically with the absorption of cholesterol in patients with chronic liver disease. In the first study, *von Bergmann* et al. [5] using a combination of intestinal perfusion and sterol balance technique found that patients with liver cirrhosis show an absorption of cholesterol which is within the range of normal values. Such conclusions, as admitted by the authors, are somewhat limited by the fact that the investigated group of liver patients had a well-compensated form of cirrhosis.

The second study [6], performed in our laboratory using Borgstrom's technique, was carried out on patients with different degrees of liver cirrhosis and demonstrated that the absorption of cholesterol was impaired in 50% of the subjects examined. The mean value found in the cirrhotic group was 26.4% of the administered dose as compared to 38.6% of a group of normal subjects (fig. 1).

Furthermore, a close relationship was observed between reduction of the fractional absorption of cholesterol and the severity of the disease. Although the regulation of cholesterol absorption is not completely defined in its details, evidence so far accumulated indicates that bile acids play an important role in every step of this complex process [7, 8].

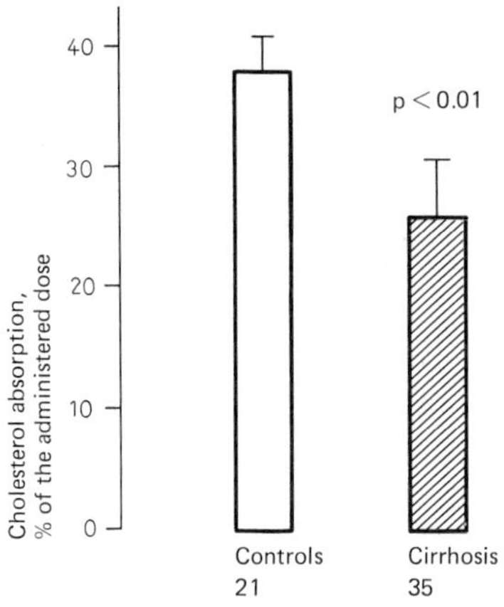

Fig. 1. Cholesterol absorption in patients with cirrhosis. Results are expressed as mean ± SEM [adapted from ref. 6].

In our study a linear relationship was noted between cholesterol absorption and colic acid (CA) pool size, whereas no relationship was found with the other bile acid pool size, thus suggesting that this trihydroxylated bile acid might play a role in promoting cholesterol absorption, at least in cirrhosis. Such an implication is supported by the observation, made on laboratory animals, that CA stimulates the esterification, within the intestinal mucosa, of the absorbed cholesterol [9], a necessary step in order for cholesterol to be transported into the lymph [10].

Whether or not the same holds true for the absorption of cholesterol in man remains to be elucidated and it is actually under investigation in our laboratory [11]. On the other hand, it might be that the reduction of both cholesterol absorption and CA pool size observed in patients with advanced liver cirrhosis are not linked by a cause-effect relationship but represent two associated events. In theory both alterations could be produced by an other mechanism such as a general decrease of the absorptive processes caused by portal hypertension. Clearly more work needs to be done to clarify such a complex problem.

Cholesterol Synthesis

In patients with moderately advanced but compensated cirrhosis, *von Bergmann* et al. [5] have reported that cholesterol synthesis, evaluated by means of the sterol balance technique, does not show an appreciable difference as compared to that of normal subjects. *Miettinen* [12], using the same technique but studying patients with more severe grades of liver disease, showed cholesterol synthesis to be depressed in cirrhosis.

In the complex pathway leading to cholesterol synthesis, the transformation of hydroxymethylglutaryl-CoA into mevalonate seems to be the major rate-limiting step [13]. In vitro estimation of hydroxymethylglutaryl-CoA reductase (HMG-CoA-R) activity provides relative estimates of the rate of cholesterol synthesis and, under appropriate conditions, it could be used for comparative purposes when studying the effect of a disease or of treatment on hepatic cholesterol synthesis [14].

Using this approach, we evaluated the hepatic HMG-CoA-R activity in 6 cirrhotic patients. Liver specimens were obtained at surgery for portocaval anastomosis. Procedures and conditions of the assay were

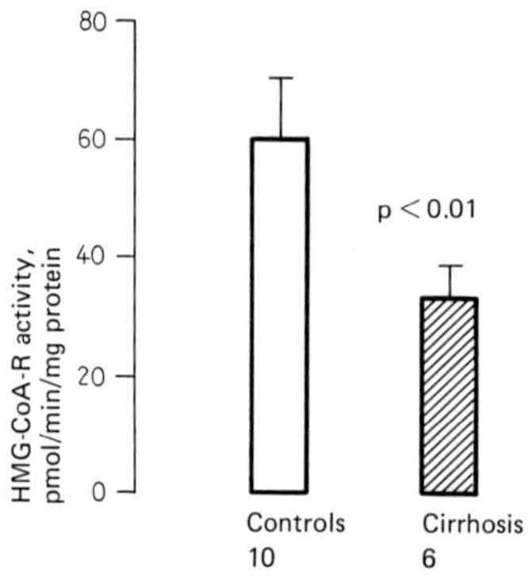

Fig. 2. Specific activity of HMG-CoA-R of microsomal preparation from liver specimen of cirrhotic patients and normal controls. Results are expressed as mean ± SEM.

as described in our previous reports [15, 16]. Enzyme activity in the cirrhotic liver was found to be decreased, the mean value being nearly 50% that of a control group [17] (fig. 2).

The few data available in the literature do not allow any conclusion on the mechanism(s) responsible for the changes of cholesterol synthesis in cirrhosis. The results obtained with the sterol balance technique seem to suggest that the size of the functional hepatic mass could represent a critical determinant in the sense that its reduction would be paralleled by a global decrease of the amount of synthesized cholesterol.

On the other hand, our finding of a reduced HMG-CoA-R activity points to an altered regulation of cholesterol synthesis in patients with cirrhosis. The ultimate mechanism regulating hepatic cholesterol synthesis is the balance between the amount of cholesterol entering the hepatocyte and the amount leaving it. A positive balance will depress cholesterol synthesis whereas a negative balance will lead to an increased synthesis.

If the observation on the reduced absorption of cholesterol in patients with advanced cirrhosis is correct, then an increased synthesis would be expected unless the disposal of cholesterol by the hepatocyte is even more depressed. In this respect it is worthy to note that the major ways for the liver cell to dispose of cholesterol, i.e. secretion into the bile, bile acid synthesis and lipoprotein secretion, are affected in patients with cirrhosis [18–21].

Clearly the quantitation of the input as well as the output processes of cholesterol in the liver cell would be necessary in order to define the alteration of synthesis.

Hepatic Cholesterol Degradation

Breakdown of cholesterol in the hepatocyte consists of a series of hydroxylations of the steroid nucleus and cleavage of the side chain of the molecule to give rise to 2 primary bile acids: cholic acid (3α,7α,12α-5β-trihydroxy-cholanoic acid) and chenodeoxycholic acid (3α,7α-dihydroxy-5βcholanoic acid, CDCA). From the quantitative point of view, the most relevant pathway is the one involving the hydroxylation of cholesterol at the 7α-position [22] but other pathways as well have been demonstrated to be possible although their contribution in physiological conditions remains to be established [23–25].

Synthetic rate of CA is almost double that of CDCA; however, their relative proportion into the bile acid pool is approximately the same, owing to the better preservation of CDCA in the enterohepatic circuit [26].

The main abnormality of bile acid metabolism exhibited by patients with advanced cirrhosis is the decrease of total bile acid pool size down to 50% of the normal value (fig. 3).

Kinetic studies have provided evidence that the reduction of the bile acid pool can be accounted for mainly by a preferential decrease of CA synthesis as compared to that of CDCA which seems to be less affected [27, 28]. In fact, whereas CA synthesis is reduced to 20–30% of the normal, CDCA synthesis is nearly halved [29].

A simple explanation of these abnormalities was put forward by earlier investigations suggesting that a relative deficiency of 12α-hydroxylase, a step necessary for the synthesis of CA, could be responsible for the different synthetic rate of the two bile acids observed in patients with cirrhosis [28, 30].

However, extensive work done with labelled precursors in order to elucidate the possible sites of blockage in the pathway have not led to such a clear-cut conclusion [29, 31, 32]. From these investigations it appears that the main factors potentially responsible for the impairment

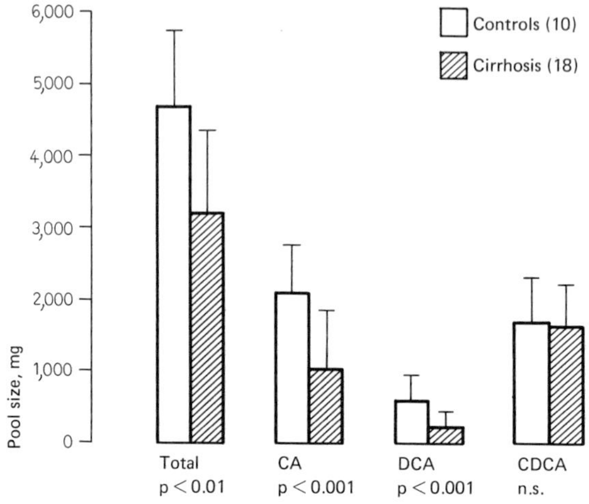

Fig. 3. Total and individual bile acid pool size in cirrhosis and normal controls. Results are given as mean ± SEM.

of bile acid synthesis could possibly include a defective regulation of the feedback mechanism, an impairment of the 7α-hydroxylation of cholesterol, the first reaction common to the synthesis of both CA and CDCA, and, in addition, other, as yet undefined, blockages in the pathway to CA.

To gain further insight into this problem we attempted to evaluate the 7α-hydroxylation of cholesterol in patients with advanced cirrhosis, compared to subjects without liver desease. The first approach we used consisted of an in vitro assay of the 7α-hydroxylase activity measured in liver specimens obtained during operation of portocaval anastomosis. The method, first described by *van Cantfort* et al. [33], is based on the use of [7α-^{3}H]cholesterol as a substrate. During the reaction, the tritium is released into the incubation medium and its quantitation gives a measure of the enzyme activity. The procedure has been validated and details have been previously reported by us [15, 16].

The obtained results, published in a preliminary form [17], showed that the specific activity of cholesterol-7α-hydroxylase in the liver of patients with cirrhosis was not impaired (fig.4). These data suggest that in the artificial conditions, such as those of an in vitro system, the cirrhotic liver hydroxylates cholesterol at the 7α-position at the same rate as in the healthy liver.

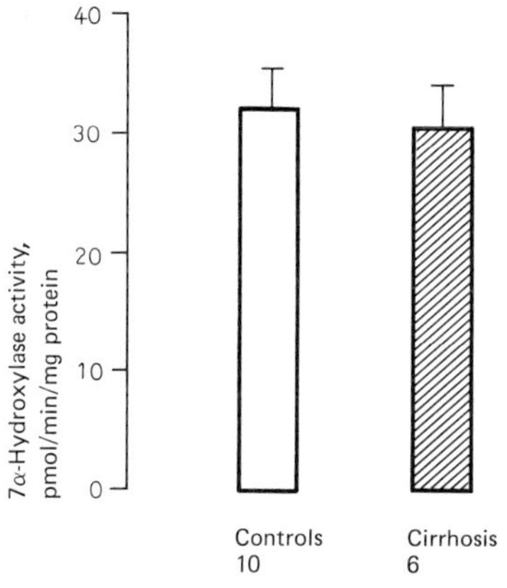

Fig. 4. Specific activity of cholesterol-7α-hydroxylase of microsomal preparation from liver specimen of cirrhotic patients and normal controls. Results are given as mean ± SEM.

If these in vitro results could be extrapolated to the in vivo situation then we should assume that the initial step of cholesterol degradation is not defective and that a reduction of liver cell mass should also be taken into account to explain the global decrease of bile acid synthesis observed in cirrhosis. With this rationale we tried to quantitate in vivo the 7α-hydroxylation of cholesterol in patients with and without cirrhosis.

The principle used was the same as that followed for the in vitro assay. Using cholesterol specifically labelled with tritium at the 7α-position as a precursor, the tritium, relased during 7α-hydroxylation, labels body water and the specific activity of water gives a measure of cholesterol undergoing degradation. Such a principle has been validated for the study of cholesterol degradation in vivo using [24,25-^{3}H]cholesterol [34].

In our study [7α-^{3}H]cholesterol was dissolved in plasma in order to label lipoprotein cholesterol and then administered i.v. to the investigated subjects. Blood samples were taken at fixed intervals for several days after administration. The specific activity of both free and esterified cholesterol was estimated on each sample. Sublimate of red blood cells was used to measure tritium body water. Cholesterol undergoing 7α-hydroxylation at any given time interval could be calculated from the equation: increase of specific activity of body water (as a percentage of injected cholesterol radioactivity)/specific activity of plasma cholesterol.

Preliminary data [35] indicate that the amount of cholesterol undergoing 7α-hydroxylation/24 h in patients with cirrhosis is significantly lower than in subjects without liver disease.

The discrepancy between the results obtained in vitro and those observed in vivo on the hydroxylation of cholesterol could suggest that in patients with cirrhosis, notwithstanding an intact cellular mechanism for cholesterol degradation, a critical role in the rate of bile acid synthesis is played by the size of functional hepatic mass. Two observations lend support to this interpretation. First, if we assume that the functional hepatic mass decreases as the severity of the disease progresses, a close relationship has been demonstrated between reduction of bile acid synthesis and the severity of cirrhosis [36]. Secondly, cirrhotic liver has the capacity to increase severalfold the synthetic rate in response to appropriate stimuli that trigger the feedback mechanism such as bile fistula [32] and treatment with cholestyramine [12, 37]. In such situations the synthesis of CA becomes greater than that of CDCA. If our findings suggest that the first committed step of the major pathway leading to bile acids is depressed in patients with cirrhosis of the liver, they do not give information on the preferential reduction of CA synthesis. 7α-Hydroxycholesterol is a common precursor of CA as well CDCA. A reduced formation of this intermediate should affect the synthesis of both bile acids to the same extent. This does not seem to be the case so that other factors should be invoked that might affect in a different way the synthesis of the two primary bile acids.

Bile Acid Pool Composition and Sterol Metabolism

In patients with advanced cirrhosis the bile acid pool composition differs from that of normal controls. CDCA constitutes more than 50% of the pool, CA proportion is decreased and deoxycholic acid (DCA) is present nearly in trace amounts (fig. 5). Abnormalities of both the synthesis and the intestinal metabolism of primary bile acids are believed to be the cause of these changes [27]. Whatever the responsible mechanism, it is tempting to speculate that the altered composition of bile acid pool could bear relevance on the regulation of sterol metabolism in

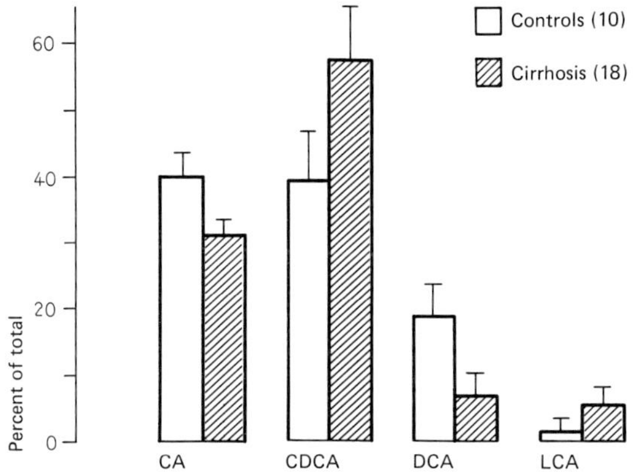

Fig. 5. Bile acid pool composition in cirrhotic patients and normal controls. Results are given as mean ± SEM.

cirrhosis. This suggestion arises from the observation that manipulations of bile acid pool composition affect several steps of cholesterol metabolism.

In this respect the most convincing evidence comes from the studies of subjects treated with CDCA. The administration of this bile acid in appropriate dosage leads to a bile acid pool made up to 75% by the administered bile acid. In such instances, cholesterol secretion into the bile has been shown to be depressed [38, 39] and this constitutes the rationale for bile acid therapy for gallstone dissolution. In addition, several authors have reported that during CDCA feeding the synthesis of cholesterol, as measured by evaluating HMG-CoA-R activity in the liver, is reduced to about 50% of the controls [15, 40–42].

Finally cholesterol absorption seems to be impaired [43] although the issue is still controversial [44, 45].

The mechanism by which a bile acid pool predominantly CDCA brings about changes of cholesterol metabolism is poorly understood. Nevertheless, the similarity between cirrhotics and subjects given CDCA, as far bile acid pool composition and sterol metabolism are concerned, might suggest that bile acid pool composition could be a contributing factor in the observed abnormalities of cirrhosis.

References

1 Eder, H.A.; Russ, E.M.; Rees Protchett, R.A.; Wilber, M.M.; Barr, D.P.: Protein lipid relationship in human plasma: in biliary cirrhosis, obstructive jaundice and acute hepatitis. J. clin. Invest. *34:* 1147–1162 (1955).

2 Switzer, S.: Plasma lipoprotein in the differential diagnosis of liver disease. Gastroenterology *53:* 790–793 (1967).

3 Simon, J.B.; Kepray, D.L.; Poon, R.: Serum cholesterol esterification in human liver disease: role of lecithin-cholesterol acyltransferase and cholesterol esters hydrolase. Gastroenterology *66:* 539–547 (1974).

4 Ponz de Leon, M.; Paterlini, P.; Caroleo, A.; Carulli, N.: Serum cholesterol levels and lipoprotein cholesterol in liver cirrhosis. Ital. J. Gastroenterol. *13:* 167–172 (1981).

5 von Bergmann, K.; Mok, H.Y.; Hardison, W.G.M.; Grundy, S.M.: Cholesterol and bile acid metabolism in moderately advanced, stable cirrhosis of the liver. Gastroenterology *77:* 1183–1192 (1979).

6 Ponz de Leon, M.; Loria, P.; Iori, R.; Carulli, N.: Cholesterol absorption in cirrhosis: the role of total and individual bile acid pool size. Gastroenterology *80:* 1428–1437 (1981).

7 Wilson, J.D.: The role of bile acids in the overall regulation of steroid metabolism. Archs. intern. Med. *130:* 493–505 (1975).

8 Holt, P.R.: The role of bile acids during the process of normal fat and cholesterol absorption. Archs intern. Med. *130:* 574–583 (1972).

9 Treadwell, C.R.; Vahouny, G.V.: Cholesterol absorption; in Code, Heidel, Handbook of physiology, vol. V, pp. 2391–2407 (Am. Physiol. Soc., Washington 1968).

10 Ito, M.; Connor, W.E.; Blanchette, E.J.; Treadwell, C.R.; Vahouny, C.R.: Inhibition of lymphatic absorption of cholesterol by cholestane 3-5-6-Triol. J. Lipid Res. *10:* 694–702 (1969).

11 Carubbi, F.; Ponz de Leon, M.; Di Donato, P.; Carulli, N.: Cholesterol esterase activity of human intestinal mucosa. Properties of the enzyme and effect of different bile acids (Abstract). Gut *23:* A887 (1982).

12 Miettinen, T.A.: Lipid absorption, bile acids, and cholesterol metabolism in patients with chronic liver disease. Gut *13:* 682–689 (1972).

13 Dietschy, J.M.; Wilson, J.D.: Regulation of cholesterol metabolism. New Engl. J. Med. *282:* 1128, 1179, 1241 (1970).

14 Carulli, N.; Ponz de Leon, M.; Zironi, F.: Evaluation of hepatic cholesterol synthesis in man: results with the HMG-CoA reductase activity assay; in Bile acids in gastroenterology (MTP Press, Lancaster 1983).

15 Carulli, N.; Ponz de Leon, M.; Zironi, F.; Pinetti, A.; Smerieri, A.; Iori, R.; Loria, P.: Hepatic cholesterol and bile acid metabolism in subjects with gallstones: comparative effects of short-term feeding of chenodeoxycholic and ursodeoxycholic acid. J. Lipid Res. *21:* 35–43 (1980).

16 Carulli, N.; Ponz de Leon, M.; Zironi, F.; Iori, R.; Loria, P.: Bile acid feeding and hepatic sterol metabolism: effect of deoxycholic acid. Gastroenterology *79:* 637–641 (1980).

17 Carulli, N.; Zironi, F.; Bosco, F.; Zambarda, E.: Relationship between rate-limiting

enzymes of hepatic cholesterol and bile acid metabolism and bile lipid composition. Digestion *16:*325–326 (1977).
18 Vlahcevic, Z.R.; Prugh, M.F.; Gregory, D.H.; Swell, L.: Distrubances of bile acid metabolism in parenchymal liver cell disease. Clin. Gastroenterol. *6:*25–43 (1977).
19 Schwartz, C.C.; Almond, H.R.; Vlahcevic, Z.R.; Swell, L.: Bile acid metabolism in cirrhosis. V. Determination of biliary lipid secretion rates in patients with advanced cirrhosis. Gastroenterology *77:*1177–1182 (1979).
20 Day, R.C.; Harry, D.S.; McIntyre, N.: Plasma lecithin-cholesterol acyltransferase activity and the lipoprotein abnormalities of parenchymal liver disease. Clin. Sci. mol. Med. *54:*36P (1978).
21 McIntyre, N.: Plasma lipids and lipoproteins in liver disease. Gut *19:* 526–530 (1978).
22 Mosbach, E.M.: Hepatic synthesis of bile acids. Biochemical steps and mechanism of rate control. Archs. intern. Med. *130:*478–487 (1972).
23 Anderson, K.E.; Kok, E.; Javitt, N.B.: Bile acid synthesis in man: metabolism of 7α-hydroxycholesterol ^{14}C and 26-hydroxycholesterol. J. clin. Invest. *51:* 112–117 (1972).
24 Vlahcevic, Z.R.; Schwartz, C.C.; Gustafsson, J.; Halloran, L.G.; Danielsson, H.; Swell, L.: Biosynthesis of bile acids in man. Multiple pathways to cholic acid and chenodeoxycholic acid. J. biol. Chem. *255:*2925–2933 (1980).
25 Swell, L.; Gustafsson, J.; Schwartz, C.C.; Halloran, L.G.; Danielsson, H.; Vlahcevic, Z.R.: An in vivo evaluation of the quantitative significance of several potential pathways to cholic and chenodeoxycholic acids from cholesterol in man. J. Lipid Res. *21:*455–466 (1980).
26 Carey, M.C.: The enterohepatic circulation; in Arias, Popper, Schachter, Shafritz, The liver: biology and pathology, pp. 429–465 (Raven Press, New York 1982).
27 Vlahcevic, Z.R.; Buhac, I.; Farrar, J.T.; Bell, C.C.; Swell, L.: Bile acid metabolism in patients with cirrhosis. I. Kinetic aspects of cholic acid metabolism. Gastroenterology *62:*491–498 (1971).
28 Vlahcevic, Z.R.; Juttijudata, P.; Bell, C.C., Jr.; Swell, L.: Bile acid metabolism in patients with cirrhosis. II. Cholic and chenodeoxycholic acid metabolism. Gastroenterology *62:*1174–1181 (1972).
29 Patteson, T.E., III; Vlahcevic, Z.R.; Schwartz, C.C.; Gustafsson, J.; Danielsson, H.; Swell, L.: Bile acid metabolism in cirrhosis. VI. Sites of blockage in the bile acid pathways to primary bile acids. Gastroenterology *79:*620–628 (1980).
30 Carey, J.B., Jr.: The serum trihydroxy-dihydroxy bile and ratio in liver and biliary tract disease. J. clin. Invest. *37:*1494–1503 (1958).
31 Goldman, M.; Vlahcevic, Z.R.; Schwartz, C.C.; Gustafsson, J.; Swell, L.: Bile acid metabolism in cirrhosis. VIII. Quantitative evaluation of bile acid synthesis from [7β-^{3}H]7α-hydroxycholesterol and [G-^{3}H]26-hydroxycholesterol. Hepatology *2:* 59–66 (1982).
32 Vlahcevic, Z.R.; Goldman, M.; Schwartz, C.C.; Gustafsson, J.; Swell, L.: Bile acid metabolism in cirrhosis. VII. Evidence for defective feedback control of bile acid synthesis. Hepatology *2:*146–198 (1981).
33 van Cantfort, J.; Renson, J.; Gielen, J.: Rat-liver cholesterol 7α-hydroxylase. I. Development of a new assay based on the enzymic exchange of the tritium located on the 7α position of the substrate. Eur. J. Biochem. *55:*23–31 (1975).

34 Rosenfeld, R.S.; Bradlow, H.L.; Levin, J.; Zumoff, B.: Preparation of [24,25-^{3}H]cholesterol. Oxidation in man as a measure of bile acid formation. J. Lipid Res. *19:*850–855 (1978).

35 Bertolotti, M.; Zironi, F.; Ponz de Leon, M.; Carulli, N.: In vivo evaluation of 7α-hydroxylation of cholesterol in man. VIIth Int. Bile Acid Meet. Bile Acids and Cholesterol in Health and Disease, Basel 1982.

36 McCormick, W.C.; Bell, C.C.; Swell, L.; Vlahcevic, Z.R.: Cholic acid synthesis as an index of the severity of liver disease in man. Gut *14:*895–902 (1973).

37 Angelin, B.; Einarsson, K.; Hellstrom, K.: Effect of cholestyramine on bile acid kinetics in patients with portal cirrhosis of the liver. Evidence of a selective defect in the formation of cholic acid. Am. J. dig. Dis. *23:*1115–1120 (1978).

38 Northfield, T.C.; LaRusso, N.F.; Hofmann, A.F.; Thistle, J.L., II: Effect of chenodeoxycholic acid treatment in gallstone subjects. Gut *16:*12–17 (1975).

39 Iser, J.H.; Dowling, R.H.; Mok, H.Y.I.; Bell, G.D.: Chenodeoxycholic acid treatment of gallstones. A follow-up report and analysis of factors influencing response to therapy. New Engl. J. Med. *293:*378–383 (1975).

40 Coyne, M.J.; Bonorris, G.G.; Goldstein, L.I.; Schoenfield, L.J.: Effect of chenodeoxycholic acid and phenobarbital on the rate-limiting enzymes of hepatic cholesterol and bile acid synthesis in patients with gallstones. J. Lab. clin. Med. *87:* 281–291 (1976).

41 Maton, P.N.; Ellis, H.J.; Higgins, M.J.P.; Dowling, R.H.: Hepatic HMG-CoA reductase in human cholelithiasis: effects of chenodeoxycholic and ursodeoxycholic acids. Eur. J. clin. Invest. *10:*325–332 (1980).

42 Ahlberg, J.; Angelin, B.; Einarsson, K.: Hepatic 3-hydroxy-3-methylglutaryl coenzyme a reductase activity and biliary lipid composition in man: ralation to cholesterol gallstone disease and effects of cholic and chenodeoxycholic acid treatment. J. Lipid Res. *22:*410–422 (1981).

43 Ponz de Leon, M.; Carulli, N.; Loria, P.; Iori, R.; Zironi, F.: The effect of chenodeoxycholic acid (CDCA) on cholesterol absorption. Gastroenterology *77:* 223–230 (1979).

44 Tangedahl, T.N.; Thistle, J.L.; Hofmann, A.F.; Matseshe, J.W.: Effect of β-sitosterol alone or in combination with chenic acid on cholesterol saturation of bile and cholesterol absorption in gallstone patients. Gastroenterology *76:* 1941–1946 (1979).

45 Einarsson, K.; Grundy, S.M.: Effects of feeding cholic acid and chenodeoxycholic acid on cholesterol absorption and hepatic secretion of biliary lipids in man. J. Lipid Res. *21:*23–34 (1980).

N. Carulli, MD, Clinica Medica I, Policlinico, Via del Pozzo 71,
I-41100 Modena (Italy)

Front. gastrointest. Res., vol. 8, pp. 229–234 (Karger, Basel 1984)

Pharmacokinetic Studies in Liver Cirrhosis

L. Okolicsanyi[a], *M. Venuti*[a], *R. Orlando*[a], *T. Berti*[b]

[a] Inst. Clinica Medica I', University of Padua, Padua; [b] Institute of Pharmacology, University of Padua, Padua, Italy

Introduction

The liver plays the key role in the metabolism and elimination of water-insoluble, relatively non-polar drugs, which are bound principally to albumin. Hepatic biotransformation is a two-step process: (1) molecular modification to introduce substituent groups (hydroxylation, sulfoxidation, dealkylation, reduction and oxidation) which can be conjugated, and (2) covalent conjugation with highly polar molecules (uronic acids, sugars, sulfate, glutathione, taurine and glycine). The result is the formation of metabolites with increased polarity (and therefore increased water solubility) which are then excreted into bile or by the kidney.

In liver disease, particularly in cirrhosis, hepatic drug elimination may be altered, either by changes in the activity of drug-metabolizing enzymes or by changes in hepatic blood flow, resulting in modifications of the intensity of therapeutic and toxic effects [1]. These considcrations are well reflected by the behavior of the kinetics of different compounds, particularly antipyrine, bromosulphthalein and cardiogreen (ICG) [2, 3].

However, patients with liver cirrhosis need, among others, medical therapy against common infections and frequently specific drugs, in order to treat lesions of the gastrointestinal tract, such as erosions and ulcerations, often complicated with acute bleeding. It seemed interesting, therefore, to investigate the pharmacokinetics of antimicrobial and H_2-receptor-blocking agents in such a condition, in which any drug must be used with particular care. Recently, we studied the pharmacokinetics of a series of drugs in patients with compensated liver cirrhosis, without ascites whose renal function was normal. In this report we discuss the most relevant methods and results obtained.

Chemiotherapeutic Agents

In patients with liver cirrhosis such drugs must be used not only for the treatment of common infectious diseases, but also for more severe infections occurring after abdominal surgery (on the portal tract: porta-systemic shunting, on the biliary tree, etc.). In the first case, many antibiotics including penicillines and other penicilline-derivative pro-drugs [4, 5], may be administered orally. Their bioavailability does not seem to be impaired in such conditions. Moreover, for more severe cases higher doses and/or parenteral administration, particularly of cephalosporins and aminoglucosides are needed when Gram-negative strains are involved.

There have so far only been a few studies, dealing with the pharmacokinetics of antibiotics in liver cirrhosis [6–9], while a great number of publications have reported the bioavailability of antibiotics in patients with a variety of liver diseases [9–15].

Among the more commonly used antibiotics, we studied the kinetics of cefuroxime in cirrhotics [16]. This antibiotic does not seem to be metabolized to any extent and loses none of its activity in the body. The whole dose administered, therefore, remains fully active in the body until it is excreted, in active form, by the kidney. In patients with liver cirrhosis, liver cell degeneration and necrosis occur, followed by the proliferation of connective tissue, which causes secondary changes in intrahepatic blood circulation. As the liver has an important role in blood circulation, it is open to question whether the morphological changes mentioned here might be connected with an altered bioavailability of foreign compounds, even if the renal function is well preserved. In this study the serum concentrations and the urinary excretion of cefuroxime, following i.m. and i.v. injection, were estimated. As expected (table I) normal pharmacokinetics of this antibiotic was found in cirrhotic patients, thus indicating its safety in such pathological conditions. In the same pathological condition, we studied the pharmacokinetics of josamycin, a macrolide, which is a so-called 'second choice' antibiotic [17]. It must be administered orally and, after ab-

Table I. Pharmacokinetic parameters of cefuroxime, josamycin and cimetidine

	Dose	Route	Vd liters	T/2 min	Cl ml/min	AUC g/mg/h	Peak concentration g/ml
Cefuroxime							
Normals (n = 6)	750 mg	i.m.	23	94.6	210.1	58.1	23.7
	1 g	i.v.	18.1	109.5	174.8	98.4	-
Cirrhotics (n = 6)	750 mg	i.m.	22.5	74	213.5	61.1	24.3
	1 g	i.v.	15.3	62.1	226.5	79.0	-
Josamycin							
Normals (n = 6)	1 g	o.s.	-	-	-	3.64	1.28
Cirrhotics (n = 6)	1 g	o.s.	-	-	-	12.92	4.17
Cimetidine							
Cirrhotics (n = 6)	200 mg	o.s.	210	175.2	51.8	4.69	0.9
	400 mg	o.s.	164.5	145.8	46.9	8.82	1.93
	200 mg	i.v.	159.6	166.8	39.9	6.04	-
	400 mg	i.v.	177.8	142.2	52.5	8.68	-

Vd = volume of distribution; T/2 = half-life; Cl = clearance; AUC = area under the curve

sorption, it is extensively metabolized by the liver, mainly by hydroxydation [18]. Blood and urine samples were collected. Results (table I) revealed altered kinetics in cirrhotics with significantly higher plasma levels and increased renal excretion.

H_2-Receptor Blocking Agents

Peptic ulcer disease and upper gastrointestinal hemorrhage from varices and/or gastric erosions are frequent complications of liver cirrhosis [19–22]. A series of H_2-receptor-blocking agents has been proved useful in the management of these conditions and is widely used in such patients [23, 24].

Cimetidine is the first drug used for these purposes. It may be administered orally and intravenously, it is mainly excreted unchanged by the kidney but it also undergoes different metabolic transformations (principally sulfoxydation) in the liver, particularly after oral administration [25, 26]. Moreover, it has been shown in healthy subjects to interfere, at different stages with the hepatic elimination of other drugs [27, 28]. Therefore, we studied the bioavailability of p.o. and i.v. cimetidine in 12 cirrhotics in a balanced incomplete block designed crossover study [29]. Single doses of 200 and 400 mg were used for both administration routes. Plasma concentrations and urinary recovery were determined by the high-pressure liquid chromatography (HPLC) method. Results (table I), showed no kinetic differences from healthy subjects.

Ranitidine, a new H_2-receptor blocking agent, is now becoming extensively used. This drug also undergoes some hepatic biotransformation (even if it is extensively eliminated unchanged by the kidney) and there is recent evidence [30] of impaired pharmacokinetics in patients with cirrhosis. However, these data need further confirmation because most of the patients studied by *Young* et al. [30] presented free ascitic fluid.

Conclusions

In this report we discuss the pharmacokinetic data of different drugs: (1) a newly synthesized first-choice antibiotic, cefuroxime; (2) a relatively recent macrolide, josamycin, and (3) the well-known H_2-receptor blocking agent, cimetidine. As already mentioned, the kinetics of these drugs is fundamentally different even in well compensated cirrhotics: certainly, their structure and metabolic fate play a determinant role in influencing bioavailability. In fact, drugs such as cefuroxime which are excreted by the kidney, completely unchanged and in active form, are normally handled by cirrhotics.

This is not the case of extensively metabolized drugs, like josamycin, and its pharmacokinetics in chronic liver disease patients are significantly altered. This observation is also of practical interest because drug monitoring in such pathological conditions could be requested. Cimetidine, which represents a metabolically intermediate compound between cefuroxime and josamycin, seems to be handled normally by patients with compensated liver cirrhosis. Our observations suggest the further need of such pharmacological studies in order to give more detailed informations about the degree of impairment of the various metabolic processes of the cirrhotic liver.

References

1 Wilkinson, G.R.; Shand, D.G.: A physiological approach to hepatic drug clearance. Clin. Pharmacol. Ther. *18:*377–390 (1975).

2 Venuti, M.; Nassuato, G.; Iemmolo, R.M.; Orlando, R.; Okolicsanyi, L.: Cinetica degli anioni organici in epatologia; in Furlanut, Segre, Farmacocinetica clinica e monitoraggio dei livelli ematici dei farmaci, pp. 283–291 (1981).

3 Okolicsanyi, L.; Venuti, M.; Orlando, R.; Lirussi, F.; Bussolon, R.; Dal Brun, G.: Antipyrin-Kinetik bei Patienten mit Leberzirrhose nach Behandlung mit Silymarin: in De Ritis, Csomos, Braatz, Legalon Symp. XI Int. Kongr. für Gastroenterologie, Hamburg 1980, pp. 83–86.

4 Fracasso, M.E.; Leone, R.; Dall'Antonia, F.: Cinetica dell'ampicillina in pazienti cirrotici prima e dopo intervento di anastomosi porto-cavale. Fracastoro *5/6:* 224–229 (1976).

5 Okolicsanyi, L.; Venuti, M.; Orlando, R.; Benoni, G.; Dorigo, M.T.; Berti, T.; Serembe, M.; Porro, A.; Vettori, G.: Pharmacokinetics of bacampicillin and amoxicillin in humans: a comparative multicentre study. Drugs exp. clin. Res. *VII:*725–730 (1981).

6 Azzolini, F.; Gazzaniga, A.; Lodola, E.; Natangelo, R.: Elimination of chloramphenicol and thiamphenicol in subjects with cirrhosis of the liver. Int. J. clin. Pharm. Ter. Tox. *6:*130–134 (1972).

7 Kunin, C.; Glazko, A.I.; Finland, M.: Persistence of antibiotics in blood of patients with acute renal failure. II. Chloramphenicol and its metabolic products in the blood of patients with severe renal disease or hepatic cirrhosis. J. clin. Invest. *38:* 1498–1508 (1959).

8 Avant, G.R.; Schenker, S.; Alford R.h.: The effect of cirrhosis on the disposition and elimination of clindamycin. Am. J. dig. Dis. *20:*223–230 (1975).

9 Narang, A.P.S.; Datta, D.V.; Nath, N.; Mathur, V.S.: Pharmacokinetic study of chloramphenicol in patients with liver disease. Eur. J. clin. Pharmacol *20:*479–483 (1981).

10 Neuman, M.; Kazmierczak, A.; Cachin, M.: The serum, biliary and urinary antibacterial levels of trimethoprim-sulfamethoxazole in relation to the liver function. Proc. 7th Int. Congr. of Chemotherapy, vol. 1, pp. 85–87 (1971).

11 Rieder, J.; Schwartz, D.E.: Pharmakokinetik der Wirkstoffkombination Trimethoprim + Sulfamethoxazol bei Leberkranken im Vergleich zu Gesunden. Arzneimittel-Forsch. *25:*656–666 (1975).

12 Hoffman, T.A.; Cestero, R.; Bullock, W.E.: Pharmacodynamics of carbenicillin in hepatic and renal failure. Ann. intern. Med. *73:*173–178 (1970).

13 Williams, D.N.; Crossley, K.; Hoffman, F.; Sabath, L.D.: Parenteral clindamycin phosphate: pharmacology with normal and abnormal liver function and effect on nasal staphylococci. Antimicrob. Agents Chemother. *7:*153–158 (1975).

14 Medina, A.; Fiske, N.; Hjelt-Harvey, I.; Brown, C.D.; Prigot, A.: Absorption, diffusion, excretion of a new antibiotic, lincomycin. Antimicrob. Agents Chemother. *4:* 189–196 (1963).

15 Bellamy, H., Jr.; Bates, B.B.; Reinarz, J.A.: Lincomycin metabolism in patients with hepatic insufficiency: effect of liver disease on lincomycin serum concentration; in

Hobby, Antimicrob. Agents Chemother, pp. 36–41 (Am. Society for Microbiology, Washington 1966).
16 Okolicsanyi, L.; Venuti, M.; Orlando, R.; Xerri, L.; Pugina, M.: Pharmacokinetic studies of cefuroxime in patients with liver cirrhosis. Arzneimittel-Forsch. *32:* 777–782 (1982).
17 Venuti, M.; Okolicsanyi, L.; Orlando, R.; Iemmolo, R.M.; Nassuato, G.; Miglioli, P.A.; Festa, A.; Grollo, A.; Berti, T.: The pharmacokinetics of josamycin in patients with liver cirrhosis. Drugs exp. clin. Res. (in press).
18 Umezawa, H.: Discovery of josamycin and the basis of its clinical study. Int. Symp. Josamycin, Gaz. Med. suppl., pp. 6–12 (1979).
19 McGray, R.S.; Martin, F.; Amir-Ahmadi, H.; Sheaken, D.G.; Zamcheck, N.: Erroneous diagnosis of hemorrhage from oesophageal varices. Am. J. dig. Dis. *14:* 755–760 (1969).
20 Dagradi, A.E.; Mehler, R.; Ton, D.T.: Sources of upper gastrointestinal bleeding in patients with liver cirrhosis and large esophagogastric varices. Am. J. Gastroenterol. *54:*458–463 (1970).
21 Waldram, R.; Davis, M.; Nunnerley, H.; Williams, R.: Emergency endoscopy after gastrointestinal haemorrhage in 50 patients with portal hypertension. Br. med. J. *iv:* 94–96 (1974).
22 Villeneuve, J.P.; Huet, P.M.; Joly, J.G.; Marleau, D.; Cote, G.; Legare, A.; Lafortune, M.; Lovoie, P.; Vaillet, A.: Idiopathic portal hypertension. Am. J. Med. *61:* 459–464 (1976).
23 MacDougall, B.R.D.; Bailey, R.J.; Williams, R.: H_2-Receptor antagonist in prophylaxis and control of bleeding in liver disease. Digestion *14:*465 (1976).
24 Pagliaro, L.; Amuso, M.; Cottone, M.; D'Amico, G.; Gatto, G.; Maringhini, A.; Pinzello, G.: L'impiego della cimetidina nelle emorragie del tratto gastroenterico superiore negli epatopazienti; in Lucchelli, Cimetidina-Farmacologia e clinica, pp. 199–207 (Smith, Kline & French, Milan 1978).
25 Burland, W.L.; Duncan, W.A.M.; Hesselbo, D.; Mills, J.G.; Sharpe, P.C.; Haggie, S.J.; Wyllie, J.H.: Pharmacological evaluation of cimetidine, a new H_2-receptor antagonist, in healthy man. Br. J. clin. Pharmacol. *2:*481–486 (1975).
26 Taylor, D.G.; Cresswell, P.R.; Bartlett, D.C.: The metabolism and elimination of cimetidine, a histamine H_2-receptor antagonist in the rat, dog and man. Drug. Metab. Disposit. *6:*21–30 (1978).
27 Roberts, R.K.; Grice, J.; Wood, L.; Petroff, V.; McGuffie, C.: Cimetidine impairs the elimination of theophylline and antipyrine. Gastroenterology *81:*19–21 (1981).
28 Feely, J.; Wilkinson, G.R.; Wood, A.J.J.: Reduction of liver blood flow and propranolol metabolism by cimetidine. New Engl. J. Med. *304:*692–695 (1981).
29 Okolicsanyi, L.; Venuti, M.; Orlando, R.; Lirussi, F.; Nassuato, G.; Benvenuti, C.: Oral and intravenous pharmacokinetics of cimetidine in liver cirrhosis. Int. J. clin. Pharm. Ter. Tox. *20:*482–487 (1982).
30 Young, C.J.; Daneshmend, T.K.; Roberts, C.J.C.: Effects of cirrhosis and ageing on the elimination and bioavailability of ranitidine. Gut *23:*819–832 (1982).

L. Okolicsanyi, MD, Institute Clinica Medica I', University of Padua,
I-35100 Padua (Italy)

Complications

Front. gastrointest. Res., vol. 8, pp. 235–245 (Karger, Basel 1984)

Pathogenetic Mechanisms in Renal Functional Impairment in Cirrhotics[1]

P. Gentilini, G. Laffi, G. La Villa, M. Pinzani, G. Buzzelli, A. Pampana, C. Smorlesi, Stefania Moscarella, Anna Birardi

Clinica Medica IV dell'Università di Firenze, Firenze, Italia

Introduction

The term hepatorenal syndrome (HRS) has been used to describe many pathological states showing alterations of both the liver and kidney. HRS was first described by *Budd* [1] in 1845 and then by *Flint* [2] in 1863. Other authors in this century described an acute renal failure in patients undergoing surgical treatment for cholestasis, using the terms 'liver death' [3, 4] and 'liver-kidney syndrome' [5]. *Lichtman and Sohval* [6] used the term HRS to indicate various pathological situations characterized by simultaneous involvement of the liver and kidney.

The inclusion of all types of impairment in both these organs in the same definition, without knowing the precise mechanisms of correlation, can be misleading. For this reason, *Conn* [7] in 1973 and *Baldus and Summerskill* [8] in 1975 proposed a classification of these pathological pictures. Table I shows a re-evaluation of these two classifications modified by *Gentilini* et al. [9] in 1981. The true HRS was considered only in the third part of table I. According to *Conn* [7], the other situations should be defined as 'pseudo-HRS'. This definition is similar in some aspects to that formulated by *Vesin* [10] who proposed the term of 'functional renal failure in cirrhotics'. A specific definition of the syndrome, however, has not yet been made. In 1978, The Sassari meeting [11] drew up the following diagnostic criteria for HRS:

[1] This work was supported by grants from the Ministero della Pubblica Istruzione, Progetto Nazionale «Cirrosi Epatica» and from Consiglio Nazionale delle Ricerche (No. 82.0274.04).

Table I. Classification of pathological situations involving both the liver and the kidney

A	Disorders with simultaneous liver and kidney involvement
1	Congenital Polycystic disease Caroli's syndrome Sickle cell anaemia
2	Toxic agents Vegetable toxins (poisonous toadstools) Drugs (tetracyclin, halothane, sulfamides, streptomycin, paracetamol, etc.) Industrial toxins (carbon tetrachloride, copper sulphate, chromium)
3	Systemic Infections (yellow fever, sepsis, miliary TBC, hepatitis B, etc.) Connective tissue diseases Circulatory alterations (heart failure, shock) Unknown aetiology (amyloidosis, sarcoidosis, Reye's syndrome)
B	Disorders with primary renal and secondary hepatic involvement Hypernephroma Sarcoma Xanthogranulomatous pyelonephritis Acute renal failure
C	Disorders with primary hepatic and secondary renal involvement Intra- and extra-hepatic cholestasis with secondary renal involvement Chronic liver disease with glomerulonephritis Renal tubular acidosis Renal functional alterations in chronic liver disease, leading to true hepatorenal syndrome

(1) Renal insufficiency (plasma creatinine > 1.5 mg/dl) which progresses over days or weeks in the presence of severe liver disease and in the absence of recognized nephrotoxic agents.

(2) Tubular function initially intact as measured by: (a) $U/P_{osm} > 1.0$; (b) U/P creatinine > 30; (c) U Na remarkably low: < 10, often < 5 m Eq/l.

(3) The above findings undergo no sustained improvement with expansion of the intravascular space to achieve a central venous pressure up to 10 cm H_2O.

According to some authors [12, and *Zipser,* personal communication], the following elements should also be considered: (1) creatininaemia and azotaemia usually increase suddenly and rapidly with creatinine values rising to more than 4 mg/dl; (2) urinary sodium values must be expressed in mEq/24 h (less than 10); (3) the lack of response to plasma volume expansion should be recognized only after both acute and chronic expansion, the latter being effective in many cases, as reported by *Gentilini and Migone* [12]; (4) plasma expansion should be performed after evaluation of central venous pressure, since only with low pressure values is it possible to perform acute expansion without dangerous side effects.

Plasma expansion is useful in distinguishing between true and pseudo-HRS. In fact, after administration of strong diuretics, profuse diarrhoea or bleeding, plasma volume can suddenly decrease and blood urea nitrogen (BUN) increases in a way similar to that observed in kidneys in shock. This situation is defined 'pre-renal azotaemia' and can rapidly recover with plasma expansion.

Taking into account the above considerations, and also the practical aspects of this syndrome, it is necessary to distinguish between the *clinically latent phase* of HRS, which corresponds to the pre-azotaemic phase [8] or the less severe, latent HRS [13] in which impairment of renal clearances and modifications in renal haemodynamics can be demonstrated, and the *open or clinically overt stage* of HRS which may characterize the final stage of chronic liver disease, especially cirrhosis (table II).

In both the latent and overt stages of HRS, there is complete preservation of renal tubular function and this explains the almost total reabsorption of sodium in the advanced stage of the disease. In fact, HRS is characterized by an almost complete lack of histological modifications in the kidney [14, 15] and the organ can function normally after transplantation [16] or when the cirrhotic liver is substituted with that of a healthy donor [17].

For these reasons, it can be concluded that the HRS is an 'unexplainable renal failure occurring in patients with cirrhosis in the absence of clinical or anatomical evidence of other known causes of renal failure' [18].

Clinical and Diagnostic Aspects

The preclinical stage of HRS may also be seen in patients affected by chronic active hepatitis (CAH) and alcoholic hepatitis with or with-

Table II. The hepatorenal syndrome

Stage I:	Latent or preclinical stage
	Renal plasma flow and glomerular filtration rate decreased by more than 50%, with vascular alterations and diversion of blood from outer cortex to inner cortex and medulla
	Opsiuria
	Reduction in water and sodium excretion
Stage II:	Open or clinically overt stage
	Azotaemia > 1 f/l
	Creatininaemia > 4 mg/dl
	Decrease in diuresis
	Water and sodium excretion < 5 mEq/24 h
	U_{osm}/P_{osm} normal or > 1
	U/P creatinine > 30
	Lack of response to plasma volume expansion

out cirrhosis or clinical evidence of portal hypertension. In these patients, careful observation may demonstrate a significant decrease of renal clearances and opsiuria, or delayed urinary excretion of water loading. Clinically, in cirrhotics we can observe water and salt retention with formation of oedema in the lower limbs and abdomen.

In the open stage of HRS, the patient is affected by advanced cirrhosis with severe portal hypertension and refractory ascites. In these cases, liver failure and a worsening in renal function is seen. Clinically, we observe deterioration of mental capacity and awareness, tense ascites, peripheral oedema and haemorrhagic tendency. Even if the 'emergent' coma can be considered both hyperammoniemic and hyperazotaemic, death is caused by hepatic failure. Some authors think that the transition between latent and overt phases of HRS can occur almost unapparently [19]. Others [18], however, state that the real HRS can appear suddenly and sometimes dramatically. When this happens, it may be difficult to make an accurate diagnosis, as it is necessary to exclude pre-renal azotaemia due to hypovolaemic states, acute renal failure due to tubular necrosis and urinary infections which are frequent in patients with hepatic cirrhosis [7, 18, 20]. An early and correct diagnosis is very important in some cases such as pre-renal azotaemia and pyelonephritis which respond well to therapy. On the contrary, medical and surgical treatment have so far been found to have little effect in the open HRS [18, 21]. Administration of plasma expanders is therefore almost always recommended as an aid for correct diagnosis. Patients with tubular necrosis can be recognized because these subjects are usually affected by a deep cholestatic jaundice or treated with nephrotoxic drugs, or have septic shock. In these cases, urine excretion is poor with a high sodium and protein content and sediment rich in red cells and cylinders [8, 20].

Pathogenetic Hypotheses

Numerous pathogenetic hypotheses have been put forward. The most important are reported in table III. The underlined hypotheses are more probable.

Patients with HRS show a reduction of renal blood flow and glomerular filtration rate with redistribution of blood from the outer to inner cortex and medulla, as demonstrated by selective angiography and xenon renal scan. These haemodynamic alterations also occur in cirrhotic patients without overt HRS [22], and may be observed in patients with CAH without cirrhosis and sodium retention.

In these patients, the opening of intrarenal shunts may sometimes occur as demonstrated by *Gentilini* et al. by the infusion of ^{99}Tc labelled microspheres of human albumin, 30 μm in diameter, directly into the renal artery. Because of these shunts, the tracer gets to the lungs, bypassing the kidney filter [23–25].

The hypoperfusion of the renal outer cortex and the alterations of interlobar renal arteries seen by angiography were interpreted by *Epstein* et al. [26] as an effect of active arterial vasoconstriction. But this concept is difficult to share. Active vasoconstriction, in fact, should last for months or years, since the previously described haemodynamic alterations were also observed in CAH without portal hypertension [23, 24].

This particular behaviour of the renal vascular tree has been attributed to a generalized circulatory failure and based essentially on dynamic deficiency of the heart. This complex situation was called 'hepato-circulatory failure' [13, 27], but this is contradicted by the observations that acute plasma expansion in cirrhotics with ascites and sodium retention provokes a normal response of ventricular function and peripheral resistences in the majority of patients [28, 29]. It can therefore be concluded that systemic haemodynamics are not substan-

Table III. Pathogenetic mechanisms in the hepatorenol syndrome

1	Active vasoconstriction in the renal cortex
2	Decrease of effective plasma volume
3	Opening of intrarenal arteriovenous shunts
4	Compression of abdominal blood vessels by ascites
5	False neurotransmitters
6	Endotoxaemia
7	Altered balance of vasoactive substances

tially impaired and that HRS is due to a particular response of the kidney in these patients.

Taking into consideration the hormones involved in the pathogenesis of salt and water retention and renal functional impairment in cirrhosis, the natriuretic hormone and other substances such as VIP probably do not play an important role [30]. On the contrary, the renin-angiotensin-aldosterone system has been proved to be activated, as shown by the effective action of spironolactone on diuresis. This, however, is not constant and occurs only in late stages as the renal response to both hyponatraemia and hypoperfusion [23].

Antidiuretic hormone (ADH) is also increased but only in patients with severe water and salt retention, probably due to a reduction in plasma volume [31]. However, circulating levels of ADH have not been found significantly decreased in ascitic patients with a LeVeen shunt who excreted less concentrated urines with correction of hyponatraemia [32]. The improvement of renal function observed in those cirrhotic patients with LeVeen shunt was probably due to plasma volume expansion. In fact, a similar improvement in renal function follows plasma expansion achieved by head-out water immersion [33] or by infusion of plasma expanders [27].

Considering renal haemodynamics, the outer cortex hypoperfusion may be due to an initial short-lasting active arterial vasoconstriction. This circulatory modification can be considered a response to a reflex coming from the portal vein [34] or to a decrease of effective plasma volume [22]. A release of vasoactive substances with vasodilating action (PGE_2, PGI_2) may follow vasoconstriction and might induce the opening of intrarenal arteriovenous shunts. This situation may have the following effects: (1) It allows a sufficient blood supply to the inner cortex and medulla even during the course of decreased plasma volume, with a faster return of the blood to the heart. (2) It increases the oxygenation in the medulla and especially in the inner cortex in which the long-loop nephrons, efficient in sodium excretion, are located. (3) It increases the release of renin, which in turn may induce an increased synthesis of vasodilating prostaglandins.

Personal Data

As far as the role of the prostaglandin system in the pathogenesis of renal functional impairment in cirrhotics is concerned, an increase in PGE_2 concentration has been found in the renal vein of cirrhotic patients [22]. *Zipser* et al. [35] showed an increased urinary excretion of PGE_2. Parenteral administration of 100 mg indomethacin to cirrhotic patients causes a significant decrease in glomerular filtration rate and renal plasma flow [36].

A method to measure the main metabolites of arachidonic acid in the urine is now available [37, 38]. Urinary concentrations of PGE_2, $PGF_{2\alpha}$, thromboxane B_2 and 6-keto $PGF_{1\alpha}$ were determined in a group of 6 female patients compared with 13 healthy women. Thromboxane B_2 and 6-keto $PGF_{1\alpha}$ are non-enzymatic metabolites of thromboxane A_2 and PGI_2, respectively. These products were extracted with chloroform, separated by column chromatography and measured by a radioimmunoassay method [38]. As it can be seen in table IV, there is a significant increase of $T \times B_2$ ($p < 0.001$) and PGE_2 ($p < 0.005$), a significant increase in the $T \times B_2$/6-keto $PGF_{1\alpha}$ ratio ($p < 0.001$) and a significant inversion of the $PGE_2/T \times B_2$ ratio ($p < 0.001$). These data showed an evident prevalence of the vasoconstrictive $T \times A_2$ when compared to the behaviour of the vasodilatory substances, PGE_2, which increased slightly, and 6 keto $PGF_{1\alpha}$, which remained unchanged.

Conclusions

It is known that the functional impairment of the kidney may last many months in some patients suffering from cirrhosis and other chronic liver diseases. This long period of renal impairment may be defined as 'functional renal impairment in the cirrhotic' or the latent stage of HRS and, in a few cases, is followed by the open HRS, with a

Table IV. Urinary excretion (pg/ml) of main arachidonic acid metabolites (mean ± SEM) compared with control subjects and ratios between vasodilatory and vasoconstrictory substances

	Sex	n	$T\times B_2$	6-keto $PGF_{1\alpha}$	PGE_2	PGF_2
Cirrhotics	F	6	8.87 ± 2.16	4.18 ± 1.07	11.87 ± 2.90	16.90 ± 2.93
Controls	F	13	1.80 ± 0.17	4.28 ± 0.27	6.70 ± 0.96	17.86 ± 2.18
Significance, p			< 0.001	n.s.	< 0.005	n.s.
			$T\times B_2$/6-keto $PGF_{1\alpha}$		$PGE_2/T\times B_2$	
Cirrhotics	F	6	2.52 ± 0.54		1.43 ± 0.36	
Controls	F	13	0.43 ± 0.03		3.78 ± 0.45	
Significance, p			< 0.001		< 0.01	

strong decrease in sodium excretion and a significant increase of creatininaemia and BUN. The open stage of HRS corresponds to the worsening of liver cirrhosis with refractory ascites and a lack of response to acute and chronic plasma volume expansion.

From a pathogenetic point of view, it is possible to confirm the hypothesis that in the renal functional impairment the hypoperfusion of the outer renal cortex may be due firstly to active vasoconstriction with release of renin and vasodilating prostaglandins. This phenomenon can be followed by the opening of arteriovenous shunts. There is also an increased activity of the renin-angiotensin-aldosterone system on renal sodium handling, probably with other hormonal influences. We showed an increase of prostaglandins in patients affected by cirrhosis, both in the renal venous blood and in the urine, and the worsening of renal function, together with a significant decrease of PGE after infusion of indomethacin solution.

On the basis of these data the hypothesis that arachidonic acid metabolites are somehow involved in regulating renal haemodynamics is then even more suggestive. The increase of vasodilatory PGEs (perhaps secondary to enhanced intrarenal synthesis of renin) may maintain a renal haemodynamic homeostasis. This tends to be progressively impaired, at least in some cases, by increasing thromboxane which has a vasoconstrictive action. When this prevails, the cirrhotic patient with severe water and salt retention enters the untreatable stage of HRS.

Summary

After pointing out the nosological concepts which led to the present definition of the hepatorenal syndrome in two different stages (latent and open or terminal), clinical and diagnostic aspects are reviewed. Pathogenetically, the main points are: (1) the decrease of effective blood volume; (2) the hypoperfusion of the outer cortex of the kidney with redistribution of the blood; (3) the opening of intrarenal arteriovenous shunts; (4) hormonal modifications (in particular, renin-angiotensin-aldosterone system); (5) renal prostaglandin system alterations. As far as the latter is concerned, the authors report personal data showing a significant increase of PGE_2 (11.87 ± 2.9 vs 6.7 ± 0.96 pg/ml) and an even greater increase of $T \times B_2$ (8.87 ± 2.16 vs 1.8 ± 0.17 pg/ml) in the urine of 6 female patients with cirrhosis compared to 13 female control subjects. The 6-keto $PGF_{1\alpha}$ (metabolite of PGI_2) and PGF_2 remained unmodified. Consequently, the $T \times B_2$/6-keto $PGF_{1\alpha}$ and $PGE_2/T \times B_2$ ratios were significantly modified (2.52 ± 0.54 vs 0.43 ± 0.03 and 1.43 ± 0.36 vs 3.78 ± 0.45, respectively). These data suggest that arachidonic acid metabolites are involved in regulating renal haemodynamics and that the vasodilatory PGEs are counteracted by the stronger increase of vasoconstricting thromboxane, a situation in which the cirrhotic patient with severe water and sodium retention becomes untreatable.

References

1 Budd, G.: On the diseases of the Liver (1845).

2 Flint, A.: Clinical report on hydro-peritoneum based on an analysis of 46 cases. Am. J. med. Sci. *45:*306–339 (1863).

3 Clairmont, P.; Von Habeler, H.: On the anuria following operation for gallstones. Mitt. Grenzgeb. Med. Clin. *1:*159 (1911).

4 Heyd, C.G.: Liver deaths in surgery of the gall bladder. J. Am. med. Ass. *97:*1847–1850 (1931).

5 Helvig, F.; Schultz, C.B.: A liver-kidney syndrome: clinical, pathological and experimental studies. Surgery Gynec. Obstet. *55:*570–580 (1932).

6 Lichtman, S.S.; Sohval, A.R.: Clinical disorders with associated hepatic and renal manifestations with special reference to the so-called hepatorenal syndrome. Am. J. dig. Dis. *4:*26-32 (1937).

7 Conn, H.O.: A rational approach to the hepatorenal syndrome. Gastroenterology *65:*321–340 (1973).

8 Baldus, W.P.; Summerskill, W.H.J.: Liver-kidney interrelationships; in Schiff, Diseases of the liver (Lippincott, Philadelphia 1975).

9 Gentilini, P.; Laffi, G.; Buzzelli, G.; Smorlesi, C.; La Villa, G.: Chronic liver disease and renal functional impairment. Ital. J. Gastroenterol. *13:*201–210 (1981).

10 Vesin, P.: Late functional renal failure in cirrhosis with ascites. Pathophysiology, diagnosis and treatment; in Martini, Aktuelle Probleme der Hepatologie (Theme, Stuttgart 1962).

11 Earley, L.E.: Conclusive remarks and presentation of diagnostic criteria of the hep-

atorenal syndrome; in Bartoli, Chiandussi, Hepato-renal syndrome (Piccin, Padova 1979).

12 Gentilini, P.; Migone, L.: La sindrome epato-renale. L'insufficienza funzionale renale dell'epatopatico cronico; in Atti dell'82° Congr. della Società Italiana di Medicina Interna (Pozzi, Roma 1981).

13 Cohn, J.N.: Hepatocirculatory failure. Med. Clins N. Am. *59:*955–962 (1975).

14 Hecker, R.; Sherlock, S.: Electrolyte and circulatory changes in terminal liver fail-ure. Lancet *ii:*1121–1125 (1956).

15 Papper, S.; Belsky, J.L.; Bleifer, K.H.: Renal failure in Laennec's cirrhosis of the liver. I. Description of clinical and laboratory features. Ann. intern. Med. *51:* 759–763 (1959).

16 Koppel, M.H.; Coburn, J.W.; Mims, M.M.; et al.: Transplantation of cadaveric kidneys from patients with hepatorenal syndrome. Evidence for the functional nature of renal failure in advanced liver disease. New Engl. J. Med. *280:* 1367–1371 (1969).

17 Iwatsuki, S.; Popovtzer, M.M.; Corman, J.L.; et al.: Recovery from hepatorenal syndrome after orthotopic liver transplantation. New Engl. J. Med. *289:*1155–1160 (1973).

18 Vaamonde, C.A.; Papper, S.: The kidney in liver disease; in Earley, Gottschalk, Strauss and Welt's diseases of the kidney (Little, Brown, Boston 1979).

19 Vesin, P.: Water, electrolyte and acid-base disorders in liver disease; in Maxwell, Kleeman, Clinical disorders of fluid and electrolyte metabolism (McGraw, New York 1972).

20 Wong, P.; McCloy, G.C.; Spielberg, A.; Milofa, R.V.; Balint, J.A.: Clinical conference: the hepatorenal syndrome. Gastroenterology *77:*1326–1330 (1979).

21 Epstein, M.: Peritoneovenous shunt in the management of ascites and the hepatorenal syndrome. Gastroenterology *82:*790–799 (1982).

22 Epstein, M.; Levison, R.; Sancho, J.; Haber, E.; Re, R.: Characterisation of the renin-aldosterone system in decompensated cirrhosis. Circulation Res. *41:* 818–822 (1977).

23 Gentilini, P.; Laffi, G.; Buzzelli, G.; Stefani, R.; Scarpelli, P.; Paladini, S.; Smorlesi, C.; La Villa, G.; Forti, G.: Functional renal alterations in chronic liver disease. Digestion *20:*73–78 (1980).

24 Gentilini, P.; Laffi, G.; Buzzelli, G.; Stefani, P.; Masi, R.: Early renal impairment in chronic hepatitis. Rendic. Gastroenterol. *7:*194–196 (1975).

25 Gentilini, P.; Laffi, G.; Stefani, P.; Masi, R.; Buzzelli, G.; Salvadori, M.; Mazzanti, R.; Moggi, A.: Early functional modifications of the kidney in chronic active hepatitis; in Gentilini, Popper, Teodori, Chronic hepatitis (Karger, Basel 1976).

26 Epstein, M.; Berk, D.P.; Hollenberg, N.K.; Adams, D.F.; Chalmers, T.C.; Abrams, H.L.; Merrill, J.P.: Renal failure in the patient with cirrhosis. The role of active vasoconstriction. Am. J. Med. *49:*175–185 (1970).

27 Tristani, F.E.; Cohn, J.N.: Systemic and renal haemodynamics in oliguric renal failure: effect of volume expansion. J. clin. Invest. *46:*1894–1906 (1967).

28 Fantini, F.; Di Donato, M.; Dabizzi, R.P.; Laffi, G.; Pinzani, M.; Buzzelli, G.; Alterazioni emodinamiche a livello sistemico. Atti dell'82° Congresso, Soc. Ital. Med. Int. (Pozzi, Roma 1981).

29 Gentilini, P.; Laffi, G.; Fantini, F.; Di Donato, M.; Smorlesi, C.; La Villa, G.; Pin-

zani, M.; Buzzelli, G.; Alterini, B.: Systemic haemodynamics and renal function in cirrhotic patients during plasma volume expansion. Digestion (in press).
30 Epstein, M.: Deranged sodium homeostasis in cirrhosis. Gastroenterology *76:* 622-635 (1979).
31 Bichet, D.; Szatalowicz, V.; Chaimovitz, C.; Schrier, R.W.: Role of vasopressin in abnormal water excretion in cirrhotic patients. Ann. intern. Med. *96:* 413-417 (1982).
32 Blendis, L.M.; Reznick, R.K.; Langer, B.; Taylor, B.R.; Seif, S.: Hyponatraemia and ADH secretion in hepatic ascites. Gut *23:* A449 (1982).
33 Epstein, M.: Renal sodium handling in cirrhosis; in Epstein, The kidney in liver disease (Elsevier, New York 1978).
34 Bazzato, G.; Gordon, S.; Onesti, G.: Contributo sperimentale alla conoscenza della fisiopatologia della sindrome epatorenale. Min. Nefrol. *20:* 74 (1973).
35 Zipser, R.; Hoefs, J.; Speckart, P.; Zia, P.; Horton, R.: Prostaglandins: modulators of renal function and pressor resistance in chronic liver disease. J. clin. Endocr. Metab. *48:* 895-900 (1979).
36 Laffi, G.; Smorlesi, C.; Salvadori, M.; Chiarantini, E.; La Villa, G.; Pampana, A.; Magnolfi, P.; Betti, D.; Buzzelli, G.: Alterazioni funzionali renali precoci. Atti dell'82° Congresso Soc. Ital. Med. Int. (Pozzi, Roma 1981).
37 Ciabattoni, G.; Pugliese, F.; Spaldi, M.; Cinotti, G.A.; Patrono, C.: Radioimmunoassay measurement of prostaglandin E_2 and $F_{2\alpha}$ in human urine. J. Endocr. Invest. *2:* 173-182 (1979).
38 Patrono, C.; Pugliese, F.; Ciabattoni, G.; Patrignani, P.; Maseri, A.; Chierchia, S.; Perkar, B.A.; Cinotti, G.A.; Simonetti, B.M.; Pierucci, A.: Evidence for a direct stimulatory effect of prostacyclin on renin release in men. J. clin. Invest. *69:* 231-239 (1982).

Prof. P. Gentilini, Clinica Medica IV, Università di Firenze,
Viale Morgagni 85, I-50100 Firenze (Italy)

Front. gastrointest. Res., vol. 8, pp. 246–253 (Karger, Basel 1984)

Effects of a Protein Meal in Patients with Liver Cirrhosis: Relation to Decreased Liver Function[1]

E. Pisi, G. Marchesini, G.P. Bianchi, Cristina Dondi, M. Zoli

Istituto di Patologia Medica 1, Università di Bologna, Bologna, Italia

Several changes in plasma amino acid (AA) profile were repeatedly demonstrated in patients with liver cirrhosis [1–3], and given as the cause for hepatic encephalopathy [4–6]. They possibly reflect an altered AA metabolism due to decreased liver function and to relevant changes in hormonal balance and enzyme levels. Increased levels of aromatic amines are generally ascribed to impaired liver function [7], since the liver is the sole site of their metabolism [8]. Decreased levels of branched-chain AAs (BCAAs = valine, isoleucine and leucine) are usually related to hyperinsulinaemia [9, 10], but the exact relationship which links BCAAs to insulin has never been established in patients with cirrhosis [11].

Following a protein meal, amino acids absorbed from the gastrointestinal tract reach the liver to enter the systemic circulation. Part of them are immediately cleared up by the liver; part, mainly BCAAs, reach the general circulation where insulin promotes their uptake in muscle [12].

The present study was aimed at evaluating the changes in peripheral plasma AAs which occur in cirrhotic patients after a standard protein meal. The results were correlated to the degree of hepatocellular dysfunction and to the extent of portal-systemic shunting.

[1] Supported by Ministero della Pubblica Istruzione, Rome, Italy, 1982.

Materials and Methods

Patients

10 control subjects (5 males and 5 females), aged 32–65 years (median 56), and 12 cirrhotic patients (6 males and 6 females), aged 23–66 years (median 57), were studied. All had normal or near normal basal glucose (< 100 mg/dl in controls and < 120 mg/dl in cirrhotics), and their body weight was within ± 10% of ideal body weight.

The cirrhotic group consisted of 5 patients with alcoholic cirrhosis, 4 with $HBsAg^{+}$ postnecrotic cirrhosis, and 3 with cryptogenic cirrhosis. The diagnosis was based on histological findings (liver biopsy and laparoscopy). Their liver function tests were as follows: albumin, 3.6 ± SE 0.4 g/dl; γ-globulin, 1.9 ± 0.2 g/dl; cholesterol, 152 ± 11 mg/dl; prothrombin activity, 70 ± 6%; bilirubin, 1.8 ± 0.7 mg/dl; galactose elimination capacity, 5.1 ± 0.3 mg/kg · min; alanine transaminase, 19 ± 5 U/l; aspartate transaminase, 24 ± 3 U/l; alkaline phosphatase, 30 ± 6 U/l. Most patients had evidence of portal-systemic shunting (oesophageal varices grade 1–2 in 5 cases and grade 3–4 in 4 cases). All had fasting venous ammonia < 40 μmol/l and no evidence of hepatic encephalopathy at the time of the study. All subjects were on a weight-maintaining diet of about 30 kcal/kg of body weight.

Methods

After an overnight fast, patients and controls consumed 200 g of lean boiled beef (22% protein, 5% fat) within 15 min. Venous blood samples for the determination of hormones and substrates were collected in the basal state and at 60, 120, and 180 min.

Blood glucose was determined by a Technicon autoanalyzer. Insulin and glucagon were determined by radio-immunoassay [11]. Plasma free AAs were measured by a single-column 3 A 28 Carlo-Erba amino acid analyzer or by fluorometry [11]. AA response to protein ingestion was expressed as the area under the plasma increment curve of AA level above basal, calculated by the trapezoidal rule, divided by 180 min.

Statistical analyses were carried out by means of t-test for paired and unpaired data. The r coefficients of linear regression analysis between AA responses and parameters of liver function were also calculated.

Results

Blood glucose did not vary significantly during the test. Both basal insulin and glucagon were higher in the cirrhotics, and more extensively increased following protein ingestion (table I).

Plasma AAs are reported in table II. In the basal state BCAAs and total tryptophan were significantly reduced in cirrhotics, while aromatic AAs (phenylalanine and tyrosine), methionine and free trypthophan were increased. Following the protein meal, all AAs increased in controls. BCAAs showed the greatest and most prolonged increments, while free tryptophan increased little and, in some cases, did not vary.

Table I. Blood glucose, insulin and glucagon levels in the basal state and following protein ingestion in control subjects and in patients with liver cirrhosis

		Basal	Protein meal		
			60 min	120 min	180 min
Blood glucose mg/dl	control	85 ± 6	88 ± 3	85 ± 3	84 ± 3
	cirrhosis	94 ± 6	99 ± 5	96 ± 5	95 ± 5
Insulin μU/ml	control	11 ± 1	19 ± 3[d]	20 ± 2[c]	15 ± 1[d]
	cirrhosis	21 ± 2[b]	33 ± 3[b,c]	47 ± 9[a,c]	41 ± 9[a,c]
Glucagon pg/ml	control	101 ± 14	182 ± 40[c]	208 ± 51[c]	178 ± 22[d]
	cirrhosis	232 ± 37[a]	316 ± 46[a,d]	415 ± 42[b,d]	469 ± 59[b,d]

The results are expressed as means ± SEM.
[a] Significantly different from the corresponding value of controls: $p < 0.05$.
[b] Significantly different from the corresponding value of controls: $p < 0.005$.
[c] Significantly different from basal value (paired t test): $p < 0.05$.
[d] Significantly different from basal value (paired t test): $p < 0.005$.

The maximal increments were usually observed at 120 min. In patients with cirrhosis, the maximal increments of most AAs were comparable to those observed in controls, although they occurred more frequently at 180 min. In contrast, valine, isoleucine, leucine, phenylalanine and total tryptophan exaggeratedly increased, so that their total 3-hour responses were significantly higher as compared to controls. The sum of BCAAs increased by 248 ± 42 nmol/ml·min in cirrhotics and by 153 ± 14 in controls ($p<0.05$). The sum of phenylalanine and tyrosine increased by 56 ± 8 nmol/ml·min in patients and by 37 ± 2 in controls ($p < 0.05$).

Aromatic AA response significantly correlated with tests of liver function and portal-systemic shunting (prothrombin activity, r = 0.601; basal ammonia, r = 0.683; galactose elimination capacity, r = -0.788 (fig. 1); and the degree of oesophageal varices, r = 0.678 - Spearman's rank correlation). No correlation was present with tests of liver cell necrosis (aspartate and alanine transaminases), or cholestasis (bilirubin, alkaline phosphatase). All these parameters of liver dysfunction failed to correlate with BCAA and total tryptophan response to protein ingestion.

Table II. Basal plasma amino acids and amino acid response to protein ingestion in control subjects and in patients with cirrhosis

Amino acid	Control subjects			Cirrhotic patients		
	basal level nmol/ml	maximal increment nmol/ml	3-hour response, nmol/ml·min	basal level nmol/ml	maximal increment, nmol/ml	3-hour response nmol/ml·min
Threonine	124 ± 14	56 ± 11	41 ± 11	132 ± 10	89 ± 15	45 ± 7
Serine	114 ± 12	41 ± 12	30 ± 8	142 ± 13	43 ± 14	23 ± 8
Glutamate	56 ± 5	45 ± 18	28 ± 9	68 ± 7	42 ± 13	63 ± 36
Glutamine	668 ± 64	82 ± 75	32 ± 55	711 ± 43	106 ± 56	77 ± 25
Proline	228 ± 26	48 ± 25	24 ± 15	231 ± 16	59 ± 24	42 ± 15
Glycine	206 ± 27	30 ± 21	22 ± 19	194 ± 16	53 ± 21	39 ± 12
Alanine	374 ± 33	97 ± 40	22 ± 19	362 ± 46	65 ± 60	24 ± 16
Valine	249 ± 11	63 ± 16	38 ± 10	174 ± 11[b]	170 ± 43[a]	89 ± 21[a]
Cysteine	44 ± 14	13 ± 11	7 ± 7	59 ± 12	16 ± 7	10 ± 5
Methionine	22 ± 2	25 ± 3	17 ± 3	39 ± 6[a]	35 ± 7	20 ± 3
Isoleucine	75 ± 5	75 ± 8	46 ± 6	59 ± 2[a]	115 ± 13[a]	72 ± 10[a]
Leucine	162 ± 10	135 ± 11	84 ± 8	121 ± 7[b]	175 ± 23	110 ± 9[a]
Tyrosine	70 ± 5	49 ± 9	25 ± 5	128 ± 14[b]	57 ± 10	30 ± 7
Phenylalanine	50 ± 3	22 ± 4	13 ± 4	72 ± 9[a]	44 ± 8[a]	27 ± 5[a]
Ornithine	64 ± 5	59 ± 8	33 ± 5	84 ± 9	64 ± 9	30 ± 9
Lysine	188 ± 11	159 ± 25	110 ± 18	213 ± 18	191 ± 32	106 ± 16
Histidine	66 ± 3	27 ± 7	20 ± 4	73 ± 4	26 ± 8	20 ± 5
Arginine	78 ± 9	67 ± 20	50 ± 16	103 ± 13	74 ± 14	46 ± 6
Free tryptophan	5.0 ± 0.5	1.1 ± 0.6	0.4 ± 0.3	7.1 ± 0.3[b]	1.6 ± 0.8	0.2 ± 0.4
Total tryptophan	56 ± 3	18 ± 3	12 ± 2	47 ± 3[a]	30 ± 5[a]	19 ± 3a

The results are expressed as means ± SEM. [a] Significantly different from control value; $p < 0.05$. [b] $p < 0.005$.

Discussion

The plasma AA response to protein ingestion was first studied by *Frame* [13] in normal man. He proved that plasma levels of different AAs increase after protein ingestion, and that BCAAs show the

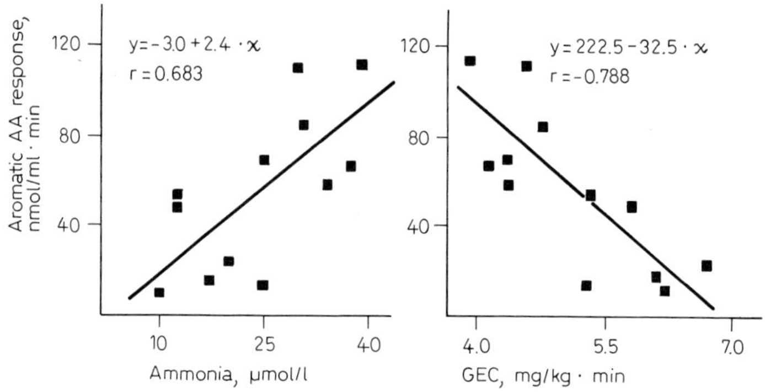

Fig. 1. Correlation between liver function (basal ammonia levels and galactose elimination capacity, GEC) and aromatic AA response to protein ingestion (sum of phenylalanine and tyrosine) in 12 patients with liver cirrhosis.

greatest and most prolonged increments. These data are in keeping with the unimportant role of the liver in the metabolism of BCAAs [8].

In patients with liver cirrhosis, we observed an exaggerated increase in plasma levels of aromatic AAs, total trypthophan, and BCAAs after a protein-rich meal. Shunting of blood around the liver and hepatocellular dysfunction are likely to reduce the uptake of substrates arriving from the splanchnic bed, thus affecting the levels of amines which are primarily metabolized in the liver. In agreement with this hypothesis, we found a close correlation between decreased liver function and aromatic AA response to protein ingestion. The same mechanism is possibly responsible for increased basal levels of aromatic AAs, which also correlate with decreased liver function.

BCAA response to protein was also increased in cirrhotic patients, without any correlation with decreased liver function, and in the presence of an increased insulin response. BCAAs have a unique role in providing nitrogen repletion to muscle after feeding [14]. In type 1 diabetes mellitus, the lack of circulating insulin is responsible for an exaggerated BCAA increase after protein ingestion, and for a failure to turn from a net muscular BCAA output to a net BCAA uptake [15]. In hyperinsulinaemic states, BCAA response to protein, as well as basal BCAA levels, should be reduced, due to a more rapid AA uptake in muscle. In accordance with this hypothesis, *Job* et al. [16] demonstrated an apparently more rapid clearance of BCAA from blood in cirrhosis after a protein-rich meal, although the heterogeneous composition of the test meal, which also included large amounts of fats and carbohydrates, questions the validity of their results. In patients with cirrhosis insulin resistance is present together with hyperinsulinaemia [17]. Increased insulin fails to maintain the basal plasma levels of different insulin-dependent substrates within the normal range. Enhanced BCAA response to protein should be due to insulin resistance, which might also prevent the clearance of ingested BCAAs.

This hypothesis is in keeping with studies of insulin resistance carried out by means of the glucose 'clamp' technique in cirrhosis. The clearance of infused BCAAs [18], and the fall in plasma BCAA levels [19] during euglycaemic hyperinsulinism, is reduced in cirrhotic patients in strict correlation to insulin resistance.

These data contrast the hypothesis that decreased BCAA levels in patients with cirrhosis are merely effects of hyperinsulinaemia. Insulin resistance removes the effects of hyperinsulinaemia on AA metabolism, as well as on glucose metabolism, where it fails to induce hypoglycaemia. Non-insulin-dependent metabolic pathways of BCAA degradation in organs different from skeletal muscle require further consideration.

Summary

The effects of a protein meal (200 g of lean boiled beef) on hormone and substrate levels were studied in 10 controls and 12 subjects with liver cirrhosis. The results were correlated with the extent of liver dysfunction. In the period 0–180 min, an exaggerated response of insulin and glucagon was observed in cirrhotics. The total response of valine, isoleucine, leucine, phenylalanine, and trypthophan was also significantly higher in patients with cirrhosis. Aromatic amines are exclusively metabolized by the liver. Their exaggerated increase following protein ingestion is likely to be due to liver failure, as supported by the strict correlation between aromatic amino acid response and impaired liver function tests or portal-systemic shunting. Increased branched-chain amino acid response did not correlate with any parameter of hepatic dysfunction, in agreement with the well-known unimportance of the liver in branched-chain amino acid metabolism. Insulin resistance might alter branched-chain amino acid uptake in muscle, leading to their abnormal response to protein.

References

1 Wu, C.; Bollman, J.C.; Butt, H.R.: Changes in free amino acids in the plasma during hepatic encephalopathy. J. clin. Invest. *34:*845–849 (1955).

2 Iber, F.I.; Rosen, H.M.; Levenson, S.H.; Chalmers, T.C.: The plasma amino acids in patients with cirrhosis of the liver. J. Lab. clin. Med. *50:*417–425 (1957).

3 Cascino, A.; Cangiano, C.; Calcaterra, V.; Rossi-Fanelli, F.; Capocaccia, L.: Plasma amino acid imbalance in patients with liver disease. Dig. Dis. Sci. *23:* 591–598 (1978).

4 Fischer, J.E.; Funovics, J.M.; Aguirre, A.; James, J.H.; Keane, J.M.; Wesdorp, R.I.C.; Yoshimura, N.; Westman, T.: The role of plasma amino acids in hepatic encephalopathy. Surgery, St Louis *78:*276–290 (1975).

5 Rosen, H.M.; Yoshimura, N.; Hodgman, J.M.; Fischer, J.E.: Plasma amino acid pattern in hepatic encephalopathy of differing etiology. Gastroenterology *78:* 677–683 (1980).

6 Marchesini, G.; Zoli, M.; Dondi, C.; Cecchini, L.; Angiolini, A.; Bianchi, F.B.; Pisi, E.: Prevalence of subclinical hepatic encephalopathy in cirrhotics and relationship to plasma amino acid imbalance. Dig. Dis. Sci. *25:*763–768 (1980).

7 Zoli, M.; Marchesini, G.; Angiolini, A.; Dondi, C.; Bianchi, F.B.; Pisi, E.: Plasma amino acids as markers of liver dysfunction in cirrhotics. Scand. J. Gastroent. *16:* 689–692 (1981).

8 Miller, L.L.: The role of the liver and the non-hepatic tissues in the regulation of free amino acid levels in the blood; in Holten, Amino acid pools, pp. 708–721, (Elsevier, 1962).

9 Munro, H.N.; Fernstrom, J.D.; Wurtman, R.J.: Insulin, plasma amino acid imbalance, and hepatic coma. Lancet *i:*722–724 (1975).

10 Soeters, P.B.; Fischer, J.E.: Insulin, glucagon, amino acid imbalance, and hepatic encephalopathy. Lancet *ii:*880–882 (1976).

11 Marchesini, G.; Forlani, G.; Zoli, M.; Angiolini, A.; Scolari, M.P.; Bianchi, F.B.; Pisi, E.: Insulin and glucagon levels in liver cirrhosis: relationship with the plasma amino acid imbalance of chronic hepatic encephalopathy. Dig. Dis. Sci. *24:* 594–601 (1979).

12 Pozefsky, T.; Felig, P.; Tobin, J.D.; Soeldner, S.; Cahill, G.F.: Amino acid balance across tissues of the forearm in postabsorptive man. Effect of insulin at two dose levels. J. clin. Invest. *48:*2273–2282 (1969).

13 Frame, E.G.: The levels of individual free amino acids in the plasma of normal man at various intervals after a high protein meal. J.clin. Invest. *37:*1710–1723 (1958).

14 Adibi, S.A.: Roles of branched-chain amino acids in metabolic regulations. J. Lab. clin. Med. *95:*475–484 (1980).

15 Wahren, J.; Felig, P.; Hagenfeldt, L.: Effect of protein ingestion on splanchnic and leg metabolism in normal man and in patients with diabetes mellitus. J. clin. Invest. *57:*987–999 (1976).

16 Job, V.; Coon, W.W.; Sloan, M.: Altered clearance of free amino acids from plasma in patients with cirrhosis of the liver. J. surg. Res. *6:*233–239 (1966).

17 Collins, J.R.; Crofford, O.B.: Glucose intolerance and insulin resistance in patients with liver disease. Archs. intern. Med. *124:*142–148 (1969).

18 Ferenci, P.; Brautsch-Marrain, P.; Waldhausl, W.; Korn, A.; Novotni, P.: Insulin resistance in patients with cirrhosis-associated defects of plasma clearance of amino acids (Abstract). Gastroenterology *80:* 1332 (1981).

19 Marchesini, G.; Forlani, G.; Zoli, M.; Dondi, C.; Bianchi, G.P.; Bua, V.; Vannini, P.; Pisi, E.: Effect of euglycaemic insulin infusion on plasma levels of branched-chain amino acids in patients with liver cirrhosis. Hepatology *3:* 184–187 (1983).

Prof. E. Pisi, Istituto di Patologia Medica 1, Università di Bologna, Policlinico S. Orsola, 9, Via Massarenti, I-40138 Bologna (Italy)

Front. gastrointest. Res., vol. 8, pp. 254–265 (Karger, Basel 1984)

Different Nutritional Approaches in Liver Cirrhosis

Effects on Nitrogen Balance
Preliminary Report

F. Fiaccadori, F. Ghinelli, G. Pedretti, G. Pelosi, D. Sacchini

Cattedra di Malattie Infettive – Università di Parma, Parma, Italia

Introduction

In its advanced stages, hepatic cirrhosis is characterized by complex alterations in the liver's metabolic capacity; among them is a reduction in urea synthesis [1], the most important pathway for the removal of excess nitrogen. Caused by a decrease in the activity of the enzymes of the Krebs-Henseleit cycle [2], this reduction leads to an accumulation of ammonia which, directly or indirectly, plays a central role in the pathogenesis of hepatic encephalopathy (HE). The resulting nitrogen metabolism imbalance is expressed by low urea levels and hyperammoniaemia, together with an altered amino acid (AA) profile – high aromatic and low branched-chain AA (BCAA) levels. This condition is a consequence of the use of endogenous body protein as a source of energy, the cirrhotic liver's incapacity to metabolize aromatic AA and increased consumption of BCAA by the muscle, kidney and fat.

The observation of alterations in plasma AAs together with changes in blood-brain barrier permeability led to the hypothesis that the synthesis of false neurotransmitters may play an essential role in the pathogenesis of HE [3]. Whatever factors are implicated in HE (blood ammonia, plasma AAs, etc.) it is generally agreed that HE is closely related to a nitrogen metabolism imbalance and particularly to a catabolic state [4]. Treatments based on the use of AA mixtures rich in BCAA and low in aromatic AA (AAA) [5–7] give a protein intake which can correct the quantitative and qualitative imbalance of nitrogen metabolism and, at the same time, positively affect neurotransmission in the central nervous system (CNS). This therapeutic approach guarantees to the cirrhotic an adequate quantity of the proteins which are essential for synthetic processes and so prevents the onset of a catabolic state.

In the present study we examined the changes in nitrogen metabolism and in neuroclinical state induced by dietetic and/or parenteral treatment that favours BCAA. The subjects were a group of cirrhotic patients in different stages of functional impairment. 5 patients affected simultaneously by hepatic cirrhosis with recurrent HE and by renal failure were also studied. The accumulation of nitrogen was particularly evident in them and the diet was supplemented with the α-keto acids of the BCAA; since it is thought that the keto acids are converted into their respective AAs using the excess nitrogen [8].

Material and Methods

74 patients with decompensated liver cirrhosis (LC) were studied. All had negative nitrogen balance (NB) and their hepatic decompensation was caused by the following factors: HE (39 cases; 52.7%); jaundice (11 cases; 14.9%); ascites (9 cases; 12.2%); gastrointestinal bleeding (11 cases; 14.9%); hepatorenal syndrome (4 cases; 5.3%).

The diagnosis of LC was based on clinical and laboratory data and confirmed in 70 cases (94.5%) by liver biopsy. The presence of HE and its grade (evaluated by neuroclinical and electro-encephalographic parameters) were the criteria used to assign the patients to one of the following therapeutic protocols: (1) 34 patients, without HE, were given a 'metabolic' diet prepared according to their tastes with foods selected on the basis of their high BCAA content. The protein intake was 0.8–1 g/kg body weight/day. (2) In 15 patients with grade 1 HE, the 'metabolic' diet was supplemented by the parenteral administration (through an antecubital vein) of an AA solution high in BCAA and low in AAA, supplied by Boehringer-Biochemia-Robin under the trade name of BS 666 (table I). The total protein intake (i.e. dietary proteins + those given parenterally) was again 0.8–1 g/kg body weight/day. (3) 24 patients with grade 2, 3 or 4 HE were put on total parenteral nutrition (TPN), through the subclavian vein, with hypertonic glucose (30%) plus the special AA solution BS 666. The protein intake was 0.8–1 g/kg body weight/day. (4) 5 patients (table II) affected by both LC with recurrent HE and chronic renal failure (CRF) (glomerular filtration rate < 20 ml/min) were followed for 4 months: during the first 2 months, they were kept on a protein-restricted diet (0.5 g/kg body weight/day); during the following 2 months the same diet was supplemented by the oral administration of branched-chain keto analogues (BCKA) in an amount equivalent to 20 g proteins (table III).

The caloric intake ranged between 30 and 35 kcal/kg body weight/day for all patients, independently of the protocol used. Treatment with lactulose (administered orally, through nasogastric tube or enema, at a dose varying from 150 to 300 ml/day) was

Table I. Comparison between a commercial AA mixture and the selective one (BS 666) used in this study

Components	Commercial solution, g/l	BS 666; g/l
1 *L*-Essentials		
Isoleucine	2.95	4.50
Leucine	3.85	5.50
Valine	2.80	4.20
Methionine	2.25	0.50
Phenylalanine	2.40	0.50
Tryptophan	0.65	0.38
Threonine	1.70	2.25
Lysine-HCl	3.85	3.80
2 *L*-Non-essentials		
Alanine	3.00	3.75
Arginine	1.55	3.00
Glycine	9.00	4.50
Histidine	1.20	1.20
Proline	4.75	4.00
Serine	2.50	2.50
Cysteine-HCl	0.02	0.25

Table II. Clinical and morphological details of 5 patients with both kidney and liver chronic disease

Patient	Sex	Age, years	Diagnosis (liver and kidney biopsy)
1	M	42	post-necrotic LC (ascites, HE) mesangioproliferative glomerulonephritis
2	M	55	post-necrotic LC (ascites, HE) mesangioproliferative glomerulonephritis
3	F	66	alcoholic LC (HE) mesangial glomerulonephritis
4	F	58	alcoholic LC (HE) mesangioproliferative glomerulonephritis
5	M	55	alcoholic LC (HE) chronic pyelonephritis

Table III. Diet schedules used for the treatment of 5 patients with both liver cirrhosis with recurrent HE and chronic renal failure

	Range
Period 1 (2 months)	
Diet Schedule	
Intake kg body weight/day	
kcal 30–35	1,800–2,685
Proteins (g) 0.5	30–45
Sodium m*M* 1.0–1.5	60–112
Potassium m*M* 0.5–1.5	30–112
Phosphorus m*M* 0.3–0.6	18–45
Magnesium m*M* 0.1–0.2	6–15
kcal/N ratio 300 (m.v.)	
Period 2 (2 months) The same diet schedule as period 1 plus oral supplementation with:	
BCKA 20 g protein equivalents:	
KIVA 40%	
KICA 30%	
KMVA 30%	
BS 704 Boehringer-Biochemia-Robin	

combined with each of the therapeutic protocols listed above. Biochemical tests were performed regularly in all patients to measure nitrogen metabolism: blood urea (enzymatic method), blood ammonia (enzymatic method), and nitrogen balance. Nitrogen intake was calculated from the weight of the food taken and the content of the solution administered. Urinary (24-hour pool) and faecal (3-day pool) losses of nitrogen were determined by Kieldhal's method. Cutaneous losses were taken to be 0.5 g/day. The following liver function tests were also performed: albuminaemia, prothrombin time, *p*-cholinesterase.

Hepatic encephalopathy was assessed by two independent observers through the evaluation of mental state, asterixis and number connection test following the system proposed by *Conn* [9]. EEGs were also performed. The plasma AA profile was determined (Technicon Multisample Aminoanalyzer) in 8 of the 24 cases with grade 2–4 HE on the 1st, 3rd, 7th, 10th day of treatment.

In the 5 patients with both LC and CRF, a biopsy of the femoris quadriceps muscle was performed, as described by *Bergstrom* [10], in order to determine the following parameters: total (TW), extracellular (ECW) and intracellular water (ICW) expressed in kg/kg free fat dry solid (FFDS), sodium, chloride, potassium, magnesium, phosphorus, expressed in m*M*/kg FFDS; alkali-soluble protein nitrogen (ASPN), as g/kg/FFDS; intra- and extracellular K concentration [11–14]. Renal function was also evaluated regularly by means of the following parameters: blood urea (enzymatic method), blood creatinine (enzymatic method) and creatinine clearance.

Results

All 74 patients with decompensated LC reached a positive NB with simultaneous improvement of their clinical-neurological conditions. The NB became positive, on average, at the 5th day of treatment both in the 24 cases put on TPN and the 15 treated with oral diet supplemented by parenteral administration of BS 666; the 35 patients who received the oral diet alone achieved a positive NB, on average, at the 8th day of treatment. An inverse correlation was observed between NB and blood ammonia levels which showed a statistically significant decrease (fig. 1) in each group of patients independently of the type of treatment. In the same groups, blood urea showed a highly significant increase ($p < 0.001$) (fig. 2).

As to the tests indicative of protein synthesis, significant increases were found in serum albumin and *p*-cholinesterase levels at the end of a 3-week observation period (fig. 3). The AA profile was determined for 8 patients with HE: when the NB was negative, a negative correlation was found between it and the BCAA/AAA ratio; when the NB became positive as a result of treatment, its values showed a positive correlation with the BCAA/AAA ratio (fig. 4). The incidence of mortality after 6 months was markedly inferior (31%) in the 74 patients studied and

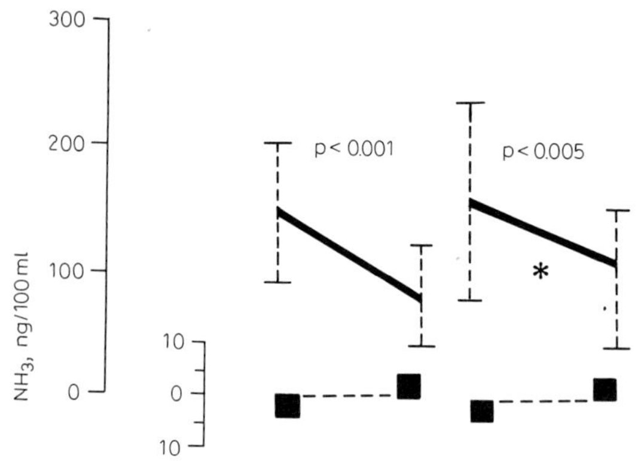

Fig. 1. Comparison of NH_3 and NB changes, before and after treatment, in patients with (▬*) and without (▬) HE.

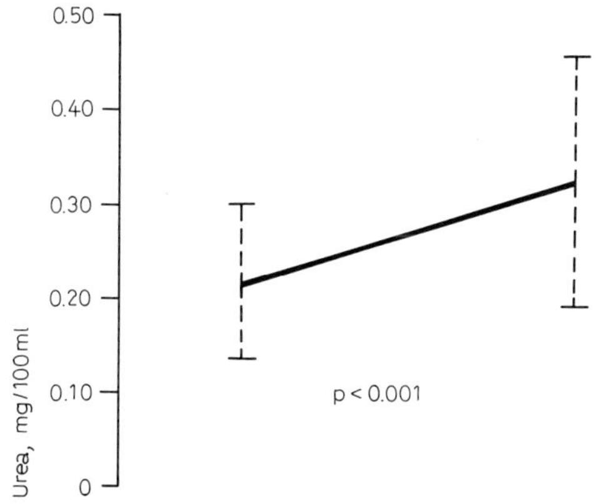

Fig. 2. Comparison between blood urea values before and after treatment in the subjects with a NB at first negative and afterwards positive.

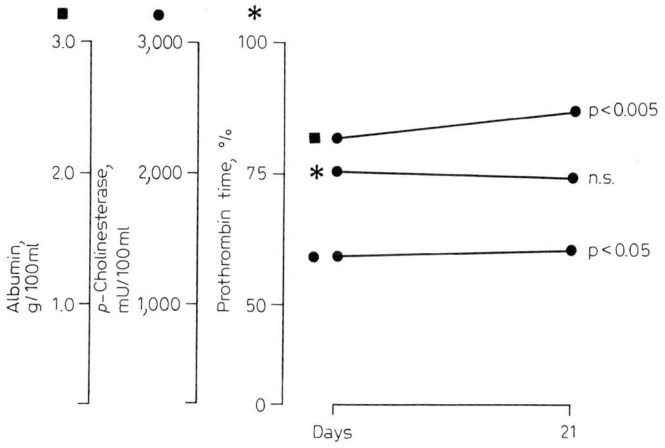

Fig. 3. Changes in some parameters of hepatie function before and after 21 days of treatment.

in 69 cirrhotics who showed a continuously positive NB to that of a control group of 10 patients who, for lack of compliance, always had a negative NB (fig. 5).

The 5 patients also affected by CRF showed no significant changes in the indices of hepatic and renal function during the treatment periods. But the use of BCKA as a diet supplement led to a return to posi-

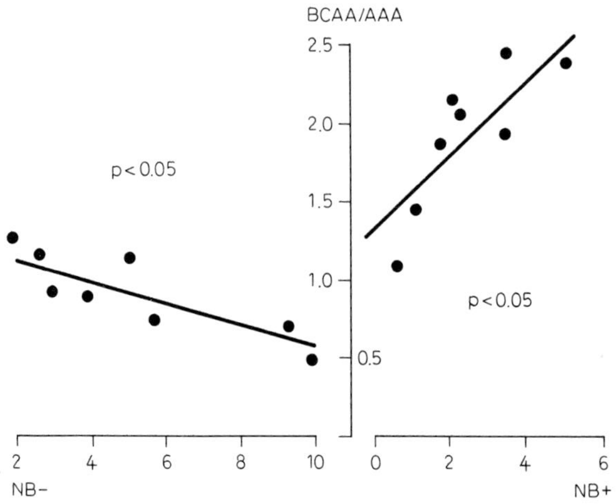

Fig. 4. Correlation between the NB and the BCAA/AAA ratio under basal conditions and after 7 days of treatment with TPN.

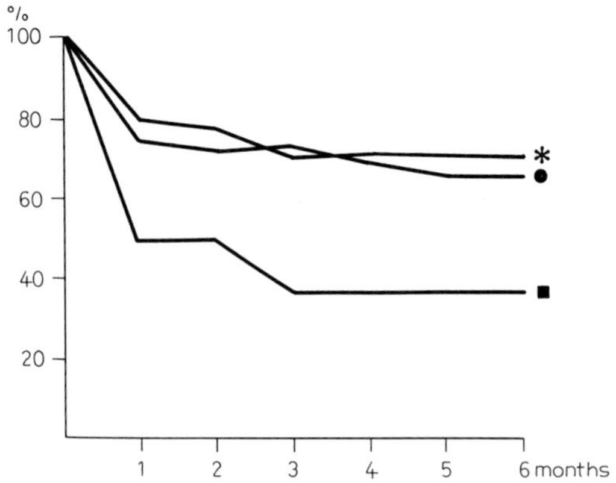

Fig. 5. Comparison, at a distance of 6 months, of the actuarial curves in patients with persistent negative NB (■), in those with positive NB (●), in those with NB at first negative and afterwards positive (★).

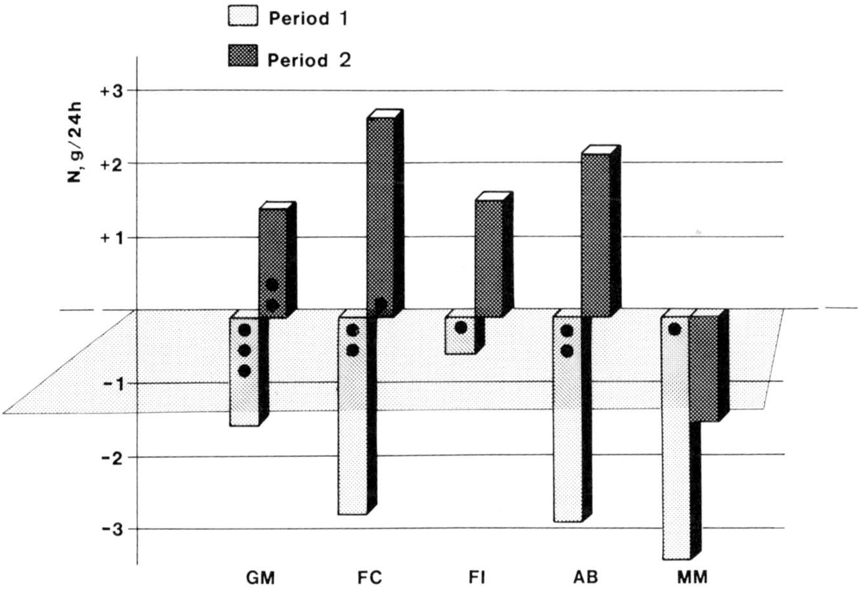

Fig. 6. N balance in period 1 (2 months without keto acid supplementation) and in period 2 (2 months with keto acid supplementation), (mean of 10 determinations for each period). ● indicates HE episodes.

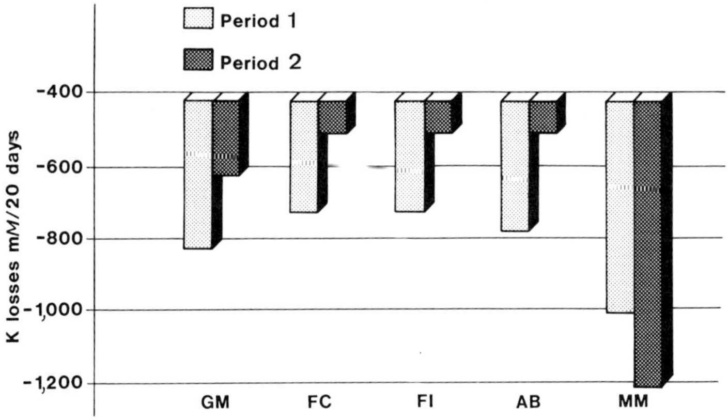

Fig. 7. Cumulative K losses (20 days) in period 1 (2 months without keto acid supplementation) and period 2 (2 months with keto acid supplementation).

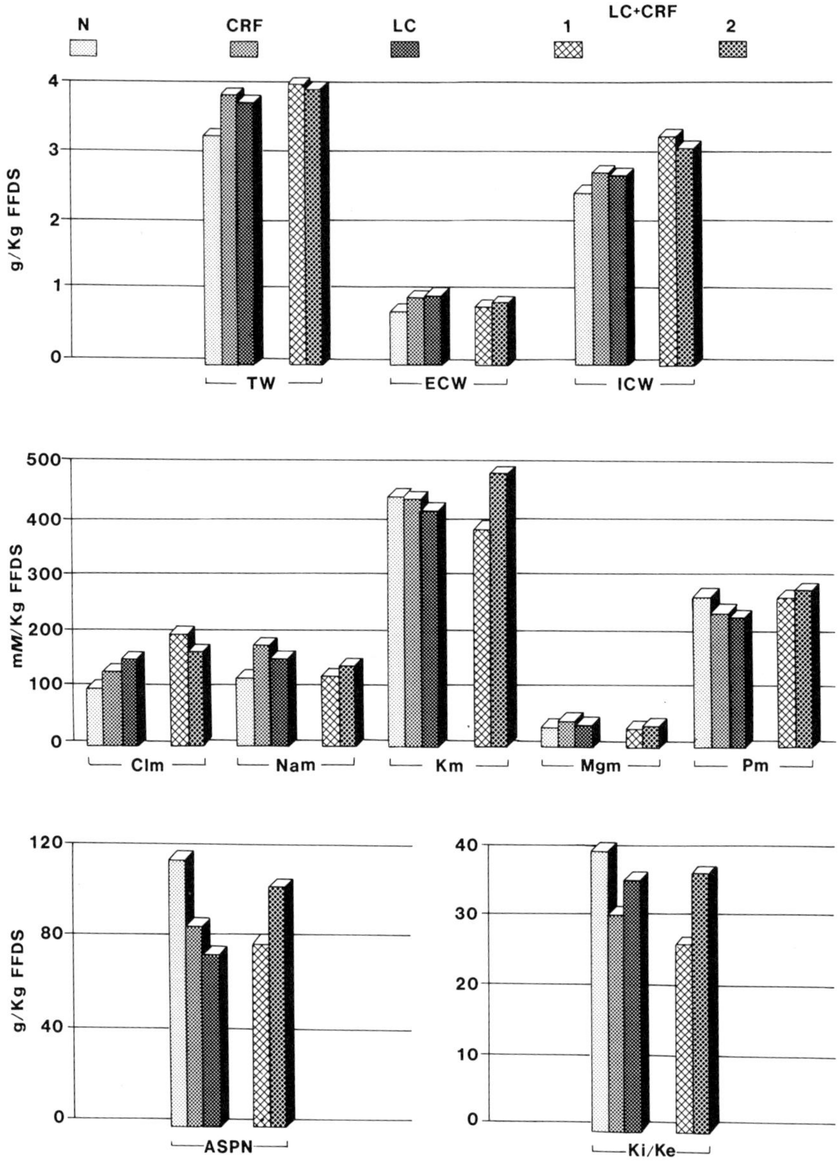

Fig. 8. Muscle composition (needle biopsy) in normal subjects (N) and in 24 patients with CRF, 21 with LC and 5 with LC+CRF. In this last group the changes induced by a protein-restricted diet alone (period 1) are compared with those from a BCKA dietary supplementation (period 2).

tive of the NB in 4 cases, a marked decrease in the incidence of HE episodes (from 9 to 3) (fig. 6) and a reduction in the total urinary potassium losses (4 cases) (fig. 7). Muscle composition at the end of the first period was similar to that seen in a larger number of previously studied cirrhotic or CRF patients. In particular, potassium and ASPN were lower than normal. After treatment with BCKA, a significant improvement was seen with values within the normal limits for potassium, ASPN and the intra- and extracellular potassium (Ki/Ke) ratio (fig. 8).

Conclusions

Patients with advanced hepatic cirrhosis tend to develop protein intolerance. A protein intake which in other circumstances would be normal (1 g/kg body weight/day) may be a precipitating cause of encephalopathy. A drastic reduction in dietary protein intake unfortunately favours the onset of a catabolic state in subjects already consuming its scarse reserves of structural proteins as a source of energy.

Vegetable and milk proteins are better tolerated than those in meat [15, 16]. This fact can be explained by the differences in AA content: vegetable proteins contain less methionine and milk proteins have a lower concentration of AAA. The unappealing nature of a diet based on these two kinds of food make it difficult to use as a treatment for cirrhotics. Best adapted to their needs seems to be a diet with a mixed animal and vegetable protein intake, designed to favour foods rich in BCAA and low in AAA. When we supplied such a diet to our patients there was a marked improvement in the parameters used for nitrogen metabolism (NB, NH_3, blood urea). When the patients' oral diet is inadequate because of encephalopathy or other causes, BCAA-enriched solutions used as a supplement or total parenteral nutrition seem to fully meet their needs. New synthetic products, currently under trial, offer another possibility. They are for oral intake, have a high BCAA, low AAA content and can be used as a supplementary source of protein.

During this study, we achieved very positive results with the α-keto analogues of the three BCAA. They were used under conditions characterized by a very high retention of nitrogen (simultaneous hepatic and renal failure). In such circumstances, the exogenous protein intake becomes critical and it seems to us that the three α-keto acids met the need both to reduce the damage deriving from the accumulation of nitrogenous metabolites and to prevent the onset of a catabolic state provoked by protein depletion. This is indicated by the return to positive of NB and the significant decrease in urinary loss of K (4 cases) and, above all, by the significant increase in potassium and ASPN. The method for estimating muscle nitrogen content excludes nitrogen bound to the structural proteins, i.e. collagen, elastin, and not involved in cellular function: so it can be assumed as a valid parameter of metabolism.

The ion balance across the cell membrane is maintained at a high ATPase energy cost and so an improvement in the Ki/Ke ratio can be considered as favouring a real improvement in the energy state of the cell. In conclusion, protein malnutrition in cirrhotics can be successfully corrected in most cases by a diet favouring foods rich in BCAA. AA solutions rich in BCAA have shown themselves to be at least a temporary answer to the nutritional requirements of the cirrhotic patient. Cases of especially high protein intolerance can also be treated with a diet supplement of synthetic products that are high in or exclusively made of BCAA or, better, their α-keto analogues.

Summary

The authors examined the changes in nitrogen metabolism and neurological state of patients with cirrhosis, induced by dietetic and/or parenteral treatment favouring AA mixtures rich in BCAA. 74 patients with decompensated LC were studied. All had negative NB. They were classified into four groups on the basis of different grades of HE. One of these groups also had chronic renal impairment. Different treatment was carried out in the various groups: (1) dietary measures; (2) an infusion of an amino acid mixture rich in BCAA and poor in AAA (BS 666); (3) parenteral infusion of glucose and an AA solution of BS 666; (4) oral administration of keto analogues.

All 74 patients with decompensated cirrhosis acquired a positive NB with simultaneous improvement of their clinical and neurological state. An inverse correlation was observed between NB and blood ammonia levels, which showed a statistically significant decrease in each group of patients independent of the type of treatment. In the same groups, blood urea showed a highly significant increase ($p < 0.001$).

In conclusion, protein malnutrition in cirrhotics can be successfully corrected in most cases by a diet favouring foods rich in BCAA. AA solutions rich in BCAA have proved to be at least a temporary answer to the nutritional requirements of the cirrhotic patient. Cases of high protein intolerance can be treated with a diet supplement of synthetic products which are high in or are exclusively made of BCAA, or even better, their α-keto analogues.

References

1 Rudman, D.; Di Fulco, T.J.; Galambos, J.T.; Smith, R.B.; Salam, A.A.; Warren, W.D.: Maximal rates of excretion and synthesis of urea in normal and cirrhotic subjects. J. clin. Ivest. *52:*2241 (1973).
2 Khatra, B.S.; Smith, R.B.; Millikan, W.J.; Sewell, C.W.; Warren, W.D.; Rudman, D.: Activities of Krebs-Henseleit enzymes in normal and cirrhotic human liver. J. Lab. clin. Med. *84:*709 (1974).
3 Fischer, J.E.; Funovics, J.M.; Aguirre, A.; James, J.H.; Keane, J.M.; Wesdorf, R.I.C.; Yoshimura, H.; Westman, T.: The role of plasma amino acids in hepatic encephalopathy. Surgery, St Louis *78:*276 (1975).
4 Rikkers, L.; Jenko, P.; Rudman, D.; Freides, D.: Subclinical hepatic encephalopathy detection, prevalence and relationship to nitrogen metabolism. Gastroenterology *75:*462 (1978).
5 Fischer, J.E.; Rosen, H.M.; Ebeid, A.M.; James, J.H.; Keane, J.M.; Soeters, P.B.: The effect of normalization of plasma amino acids on hepatic encephalopathy in man. Surgery, St Louis *80:*77 (1976).
6 Fiaccadori, F.; Ghinelli, F.; Pelosi, G.; Sacchini, D.; Vaona, G.L.; Zeneroli, M.L.; Rocchi, E.; Santunione, V.; Gibertini, P.; Ventura, E.: Selective amino acid solutions in hepatic encephalopathy treatment. Ric. Clin. Lab. *10:*377 (1980).
7 Fiaccadori, F.; Ghinelli, F.; Pedretti, G.; Pelosi, G.; Sacchini, D.; Spadini, G.: Negative nitrogen balance in cirrhosis. A correct therapeutic approach. Ric. Clin. Lab. *11:*259 (1981).
8 Walser, M.; Lund, P.; Riderman, B.; Coulter, A.W.: Synthesis of essential amino acids from their alpha-ketoanalogues by perfused rat liver and muscle. J. clin. Invest. *52:*2865 (1973).
9 Conn, H.O.: Assessment of mental state; in The hepatic coma syndrome and lactulose (Williams & Wilkins, Baltimore 1979).
10 Bergstrom, J.: Muscle electrolytes in man. Scand. J. clin. Lab. Invest. *14:*suppl. 68, 1 (1962).
11 Cotlove, E.: Determination of the true chloride content of biological fluids and tissues. II. Analysis by simple, non-isotopic methods. Analyt. Chem. *35:*101 (1963).
12 Bradbury, M.W.; Cleeman, C.R.; Bagdoyan, H.; Berberian, A.: The calcium and magnesium content of skeletal muscle, brain an cerebrospinal fluid as determined by atomic absorption plane photometry. J. Lab. clin. Med. *71:*884 (1968).
13 Chen, P.S.; Toribara, T.Y.; Warner, H.: Microdetermination of phosphorus. Analyt. Chem. *28:*1756 (1956).
14 Lowry, O.H.; Rosebrough, N.J.; Farr, L.; Randall, R.J.: Protein measurement with the folin phenol reagent. J. biol. Chem. *193:*265 (1951).
15 Fenton, J.C.B.; Knight, E.J.; Humpherson, P.L.: Milk and cheese diet in portal-systemic encephalopathy. Lancet *i:*164 (1966).
16 Greenberger, N.J.; Carley, J.; Shenker, S.; Bettinger, I.; Stamnes, C.; Beyer, P.: Effect of vegetable protein diets in chronic hepatic encephalopathy. Am. J. dig. Dis. *22:*845 (1977).

F. Fiaccadori, MD, Cattedra di Malattie Infettive, Università di Parma, I-43100 Parma (Italy)

Front. gastrointest. Res., vol. 8, pp. 266–277 (Karger, Basel 1984)

Portal Systemic Shunts in the Treatment of Bleeding Esophageal Varices

A 15-Year Experience

G. Pezzuoli, G.P. Spina, F. Galeotti, E. Opocher, R. Santambrogio

Università degli Studi di Milano, Istituto di Clinica Chirurgica II, Milano, Italia

The surgical treatment of bleeding esophageal varices may follow two main courses of action: (1) portal pressure decompression by means of portal-systemic shunt, or (2) direct operation for esophageal varices by means of ligature or transection. Previous experience shows that portal-systemic shunts, removing portal hypertension, represent the only definitive treatment for the prevention of variceal relapse. However, suppression of hepatopetal flow by portal-systemic shunts may cause the appearance of encephalopathy and liver function deterioration.

Direct operations, maintaining portal hypertension, preserve the hepatopetal portal flow, but they are penalized by a greater risk of variceal relapse. We prefer portal-systemic shunts, as we believe them to be the best choice in the management of esophageal varices and our experience is based almost exclusively on results drawn from such operations [1, 2]. However, they are not the ideal solution from a therapeutic point of view, and their 'anti'-physiological nature may cause serious complications. In view of this, we have always tried to improve surgical techniques and patient selection. Initially, we only performed traditional shunts – porta caval (PC), splenorenal (SR), mesocaval (MC) – and then we tried selective SR shunts (Warren's operation), which theoretically appear to ensure a definitive decompression of esophageal varices preserving the hepatopetal portal flow [3].

On the other hand, to obtain a better patient selection for surgery, we have tried, using statistical tests, to identify preoperative risk parameters influencing survival rate and incidence of encephalopathy [4, 5]. The knowledge we have gained from shunt surgery for portal hypertension is the object of this study.

Materials and Methods

From 1967 to 1982, we performed 584 portal-systemic shunts: they consisted of 367 side-to-side portacaval and 37 end-to-side shunts when the anatomical positions did not permit the correct approach to the two veins. 94 were SR shunts performed as a second-ary-choice operation when the portal vein was not utilizable, or as a first choice when splenectomy was necessary. 10 were MC shunts, of which 9 were performed using a Dacron graft. Since 1977, we have performed 76 selective SR shunts, complete with a venous portamesentric and gastrosplenic (PM-GS) disconnection, carried out with ligation of the left gastric and the gastroepiploic veins. The main surgical indication was portal hypertension of intrahepatic origin, complicated by one or more previous hemorrhages from esophageal varices.

506 patients underwent elective operation, and 43 emergency shunts were performed to control bleeding. This relatively limited number is explained by our tendency to treat hemorrhages initially with balloon tamponade tubes and also by the elective character of our admissions.

28 patients who required splenectomy, but did not present complicated esophageal varices, underwent prophylactic shunt. 6 side-to-side PC shunts were performed to treat Budd-Chiari syndrome, and one shunt for refractory ascites. However, these last three indications were excluded from the study due to particular problems they caused.

Patients underwent the usual preoperative liver function tests, and were divided into risk sections: mild (19%), moderate (77%) and poor (4%), according to *Childs's* [6] classification. Preoperative investigations were completed with EEG, esophagogastric X-ray and/or endoscopy.

Since 1975 celiac-mesenteric arteriography with venous phase has substituted splenoportography and ileoportography in the morphologic and hemodynamic evaluation of the portal system. The portal pressure was measured intraoperatively using a water manometer both before and after shunt.

Long-term follow-up of the patients was carried out with periodic controls at the Outpatient Specialist Division, where approximately 2500 examinations were carried out. The main objective of this follow-up was to prevent alcoholism, encephalopathy, following a rigid hypoproteic diet and lactulose treatment, and periodically control the esophago-gastric tract by means of X-ray and/or endoscopy.

Results and Discussion

Hemodynamic Findings in Traditional and Selective Shunts

Intraoperative portal pressure showed that portal-systemic shunts efficiently reduce hypertension (table I).

Traditional shunts (PC, SR, MC) permitted reduction of the mesenteric pressure to almost physiological values, and no significant difference was observed in the results of the different shunts. The drop in

Table I. Venous pressure fall after shunt

Shunt	Venous pressure	
	pre-shunt	pressure fall
Portacaval	37.83 ± 5.50	16.27 ± 4.90
Splenorenal	33.39 ± 5.54	13.11 ± 5.14
Mesocaval	37.30 ± 5.52	19.87 ± 6.64
Warren's operation	38.40 ± 4.64	17.00 ± 4.01[1]

[1] Splenic stump

blood pressure occurred throughout the splanchnic area and caused the total and sudden suppression of the hepatopetal flow.

This event has already been reported in the literature [7, 8], and was confirmed in our experience where none of 29 postoperative angiographic controls showed the persistence of a hepatic portal perfusion. In those shunts where the interruption of the portal vein is not performed, the diversion of the portal flow occurs, as the portal vein transforms into an outflow tract at low resistence.

The only hepatic artery, although it can double its flow [9], sometimes does not ensure an adequate parenchymal perfusion. When this occurs, an acute liver failure may cause death during the early postoperative period, with a clinical and histological pattern of diffused hepatocellular necrosis.

The selective SR shunt also efficiently decreases the splanchnic pressure, maintaining a high mesenteric pressure however, which is capable of forcing the portal flow in a hepatopetal direction, and probably responsible for the gradual decompression of varices, seen in the postoperative endoscopy (table II).

Despite this limit, selective shunt appears 'more' physiological than traditional shunt, as it maintained the hepatopetal flow in 80% of 45 angiographic controls performed during the early postoperative period [10, 11]. Consequently, the hepatic artery did not develop a rapid hypertrophy and there was no significant difference in pre- and postoperative artery diameters (pre: 7.57 + 2.14 mm; post: 8.36 + 1.95 mm, p = NS). This favorable hemodynamic pattern is, however, unstable and the portal flow is bound to decrease due to the progressive deve-

Table II. Decompression of varices after shunt

	Selective shunts,%		Non-selective shunts,%	
	61 cases post-op.	32 cases long-term	386 cases post-op.	248 cases long-term
Disappearance	16	50	53	80
Regression	75	43	43	18
Unchanged	9	7	4	2

lopment of collateral veins between the portomesenteric high pressure area and the gastrosplenic low pressure area, which the disconnection manages to separate only temporarily. Angiographic checks, carried out within 1 month of shunt, showed the early development of the collateral veins. In fact, only 8% of cases showed an ideal disconnection. In 50% the portal vein diameter appeared reduced and the PM-GS collaterals were evident. Finally, in 42%, collateral veins were highly developed and the portal flow remarkably reduced, or the portal vein showed partial or total thrombosis, in 4 cases and 8 cases, respectively.

Although, from a hemodynamic point of view, the final results of these changes do not differ from those obtained from traditional shunts (that is suppression of the hepatopetal portal flow), this result is obtained more gradually with major respect towards physiopathology of the hepatic cirrhosis. [11–13].

If the long-term hemodynamic and clinical advantages of selective shunts appear controversial (see below), in the early postoperative period this shunt seems to give better liver protection from sudden flow changes [14].

Selective shunts are, however, more difficult to perform than the traditional shunts, and are not always suitable for all patients. The fact that many patients with cirrhosis who undergo traditional shunts seem to tolerate sudden suppression of the portal flow, with no apparent consequences, suggests that the use of selective shunts could be more justified only in those patients whose hepatic artery has a limited capacity of compensation and where the progressive reduction of portal flow could be better tolerated.

However, these hypotheses still have be confirmed, and are presently only the basis of a clinical and hemodynamic study, which our team is carrying out using an intraoperative electromagnetic flowmeter.

Operative Mortality

Cumulative operative mortality of elective shunts was 7% and appeared highly influenced by the degree of liver function (risk A: 4%; B: 9%; C: 18%). Operative mortality, however, has been reduced to 5% in the last 3 years. We believe this to be due to a more accurate patient selection with elimination of poor C risks which, apart from an excessive operative mortality, showed no improvement in long-term survival after shunt (fig. 3). In these patients, we actually perform endoscopic sclerosis of the varices and, when the varices appear enlarged and extended, esophageal transection with EEA stapler. This may be followed by additional sessions of sclerotherapy, at the first reappearance of varices, during long-term follow-up.

The type of shunt used does not influence operative mortality, and in particular the selective shunt showed identical values to those shown by traditional shunts (7%). We would like to stress, however, that in the two groups of patients, the causes of death were different. Hepatic failure as a primary cause of death in the traditional shunt group (80%) was substituted by postoperative variceal hemorrhagic relapse (60%) in the selective shunt group.This could be a further confirmation of more limited hemodynamic changes caused by the selective shunt, thanks to the preservation, at least initially, of a certain degree of portal hypertension.

In emergency shunts, the operative mortality was obviously higher (28%), but this value may be strongly influenced by the strategy of treatment applied. If an initial non-operative management with the use of balloon tamponade tubes does not control bleeding, as in one third of cases [15], and the patient presents a mild-moderate operative risk, we presently prefer a more aggressive surgical approach. This consists of a portal-systemic shunt before the patient's general state of health contro indicates the performance of the operation.

In poor-risk patients, on the other hand, we have substituted the shunt with the esophageal transection using an EEA stapler, which is effective in the immediate control of esophageal variceal bleeding, without disrupting the hemodynamics of the liver, already affected by hemorrhagic shock. Using this procedure, the last 7 emergency shunts, which were reserved for selected patients operated on at the right time, were all successful.

Hemorrhagic Relapses

Hemorrhagic relapses in the early and long-term postoperative period are caused both by varices and gastroduodenal ulcers. As far as the varices are concerned, our experience shows that the main objective of portal-systemic shunts is largely reached, as only 2% of 485 patients rebled from varices in long-term control. Traditional and selective shunts, giving identical values, confirmed an equally efficient detention.

However, during the early postoperative period of selective shunts, variceal hemorrhagic relapses represented a major problem (12%). This was also confirmed by endoscopic visualization of a 'more' progressive variceal detention following selective shunt, with respect to the traditional shunt. This characteristic leads us to believe that this shunt is unadvisable in the emergency management of variceal bleeding (table II).

The use of the Dacron graft proved rather discouraging, as 6 of the 8 patients who underwent long-term controls developed shunt thrombosis. When a by-pass is required in the performance of a shunt, we actually prefer the use of an internal jugular vein allograft [16].

Hemorrhagic digestive relapses were more often caused by gastroduodenal ulcers (7%). Moreover, if we consider those patients who did not present with a bleeding ulcer, the percentage rose to 16% and this is a real problem in shunted cirrhotic patients. As these ulcers are often asymptomatic and often bleed, an accurate follow-up with periodic radiologic and/or endoscopic checks and a prophylactic management in risk patients are justified.

Portal-Systemic Encephalopathy

The most dreaded complication of the portal-systemic shunt showed, in our experience of 472 cases, an overall percentage of 42%, excluding terminal signs of hepatic coma. This alone appears striking enough to discourage the shunt; however, it takes on another value if its components are analyzed. In fact, in these patients, portal-systemic encephalopathy (PSE) appeared as a multiform neurologic syndrome with either acute or chronic patterns and a different degree of seriousness such as to influence the quality of survival.

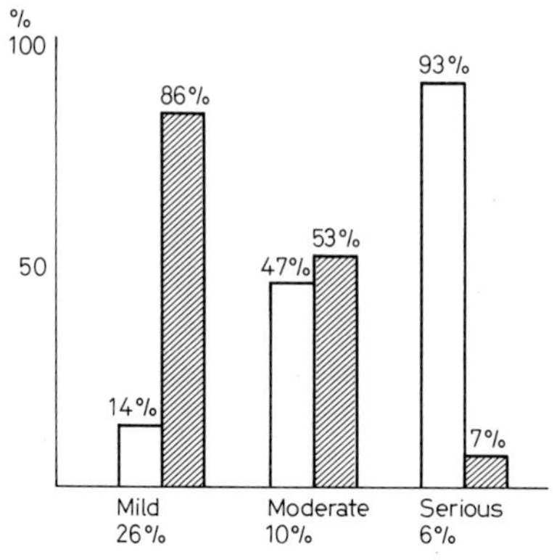

Fig. 1. Portal-systemic encephalopathy and degree of seriousness. Open columns = acute; hatched columns = chronic.

Although acute PSE usually presents serious symptoms, it is of brief duration and easily controllable with elimination of dietary and pharmacologic imbalances. On the other hand, symptoms of chronic PSE more often cause invalidity and are more difficult to manage. However, patients affected by the serious form, so as to endanger their life, are a very small percentage (6%) (fig. 1).

These data must not underestimate the importance of this complication, which is a constant and serious problem in this type of surgery. On the other hand, they help to put this complication in its right perspective. In fact patient selection and above all PSE prevention, with frequent follow-up controls to warn the patient against the danger of meat intoxication and constipation, can reduce this complication to tolerable percentages [4, 17].

Traditional shunts (PC, SR) have shown the same percentage of PSE. Furthermore, our expectations in selective shunts were, in part, disappointed, although they had shown a lower percentage of PSE in randomized studies [18–21]. This shunt seems to present a lower tendency to develop PSE, but this is not sufficient to be statistically significant (table III).

At present, we believe that better results can be obtained only through an accurate selection of patients who are less exposed to PSE by means of preoperative risk factors. In a multifactorial statistical study based on 350 shunted patients, the factors shown in table IV

Table III. Percentage of encephalopathy after shunt

Shunt	Number	Encephalopathy		
		total,%	acute,%	chronic,%
Portacaval*	341	44	15	29
Splenorenal**	69	39	12	27
Warren's operation***	62	32	19	13

* Mean follow-up 46 months; ** mean follow-up 44 months; *** mean follow-up 14 months.

Table IV. Encephalopathy high-risk patient characteristics

PSE preoperative episodes
Over 55 years
Serious hepatic fibrosis or cellular degeneration
Seriously affected tests of cellular necrosis

show a significant correlation with postoperative encephalopathy percentages [17].

Future prospects of controlling this complication seem to depend on a greater knowledge of its pathogenesis and new pharmacological managements.

Long-Term Survival

The survival actuarial curves are shown in figures 2 and 3. The mean 5-year survival rate was 60% and the comparison between traditional (PC, SR) and selective shunts showed no significant difference.

On the other hand, survival appeared highly influenced by preoperative liver function. All patients from *Child's* [6] group C died within 4 years of shunt, thus showing that anastomosis might perhaps be substituted with better results by alternative techniques, such as sclerotherapy either done or preceded by esophageal transection with an EEA stapler. A multifactorial statistical study on 350 shunted patients, analyzing 20 preoperative clinical and laboratory parameters, confirmed the importance of liver function. A significant correlation with survival was evident only in some of these parameters, which on the whole virtually represent *Child's* [6] classification grading (table V).

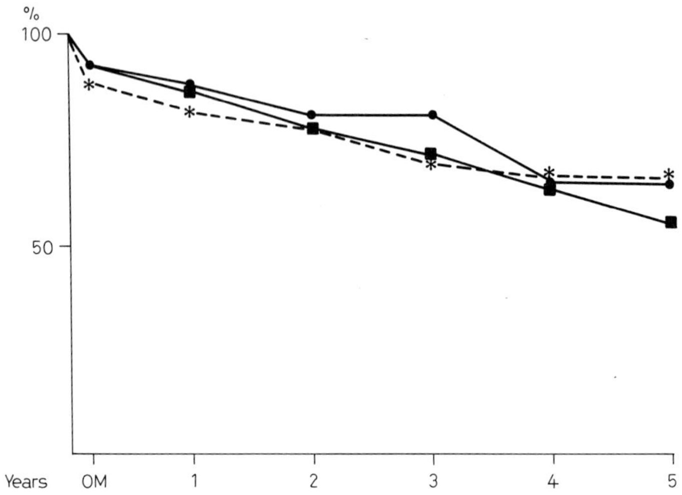

Fig. 2. Survival curves: Portacaval shunt (■), splenorenal shunt (★), Warren's shunt (●).

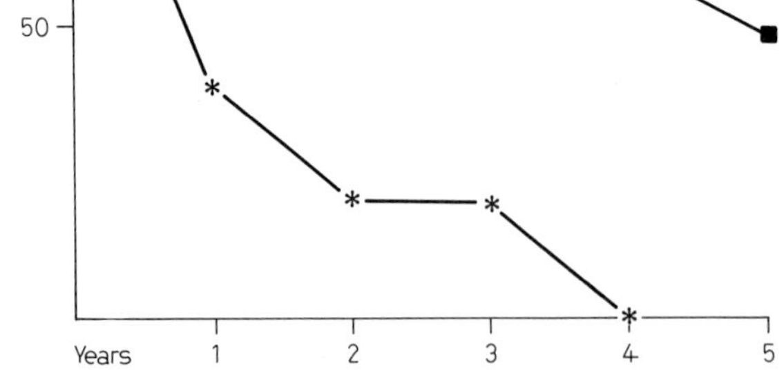

Fig. 3. Survival curves and operative risks: A (●); B (■); C (★).

Table V. Characteristics of patients with high survival rate

Absence of previous episodes of primary hepatic failure
Absence of ascites or easy management
BSP less than 12% at 45 min
Serum bilirubin < 2 mg/100 ml
Alkaline phosphatase less than double the normal value
Absence of seriously affected tests of cellular necrosis
Absence of serious hepatic fibrosis

Conclusions

Portal-systemic shunts are operations that we still carry out with conviction. Obviously, not all patients with esophageal varices can obtain benefit from this operation and patient selection is the main difficulty in this type of surgery. A strict selection slightly reduces the number of candidates for surgery, and in these cases alternative techniques, such as sclerotherapy or esophageal transection, although less defini-

If this policy is maintained and patients undergo frequent controls following surgery, portal-systemic shunt shows a low operative mortality rate and neurological complications and long-term survival rates are satisfactory.

Although selective SR shunt confirms its validity in the prevention of hemorrhagic relapse, it produced no striking improvement in clinical results, probably due to its temporary 'selectivity'. We find that its main advantages seem to be its capacity to protect the liver from post-

operative ischemia through the sudden loss of portal flow and the relatively less frequent, long-term chronic PSE. However, the definite confirmation of selective shunt validity requires a further period of evaluation.

Summary

The authors report their experience based on 584 portal-systemic shunts which, in their opinion, still represent the ideal solution in the management of portal hypertension. Of course, not all patients with esophageal varices obtain benefit from this shunt, patients selection being the main difficulty in this type of surgery. However, following this behavior and carrying out frequent follow-ups, the traditional shunt has a limited operative mortality rate, and the percentage of neurologic complications and the long-term survival rate are satisfactory.

As far as the selective shunt is concerned, although its validity is confirmed in the prevention of hemorrhagic relapse, it has produced no improvement in clinical results, probably due to its temporary 'selectivity'. Definitive confirmation of its validity requires a further period of evaluation.

References

1 Pezzuoli, G.; Spina, G.P.: La terapia di elezione dell'ipertensione portale Arch. Atti Soc. Ital. Chir., 76° Congr., 1974.

2 Pezzuoli, G.; Spina, G.P.; Galeotti, F.; Battaglia, G.; Baccaglini, U.; Opocher, E.: Risultati a breve e a lungo termine degli interventi di anastomosi porto-cava; in trattamento chirurgico dell'ipertensione portale. Atti della tavola rotonda (Piccin, Padova 1978).

3 Warren, W.D.; Zeppa, R.; Fomon, J.J.: Selective trans-splenic decompression of gastroesophageal varices by distal spleno-renal shunt. Ann. Surg. *166:* 437–448 (1967).

4 Pezzuoli, G.; Spina, G.P.; Galeotti, F.; Battaglia, G.; Rigotti, P.; Paruzzo, P.; Lion, R.; Sorato, R.; Opocher, E.: Clinical and experimental observations on post-shunt encephalopathy (PSE); in Orloff, Stipa, Ziparo, Medical and surgical problems of portal hypertension (Academic Press, London 1980).

5 Battaglia, G.; et al. Fattori prognostici nei pazienti operati di derivazione porto sistemica. Acta chir. ital. *37:*194–201 (1981).

6 Child, C.G.: The liver and portal hypertension; in major problems in clinical surgery, vol. I (Saunders, Philadelphia 1971).

7 Price, J.B.; et al. Operative hemodynamic studies in portal hypertension. Significance and limitations. Archs Surg. *95:*843–852 (1967).

8 Reynolds, T.B.: The role of hemodynamic measurements in portosystemic shunt surgery. Archs Surg. *108:*276–281 (1974).

9 Burchell, A.; et al. Hepatic artery flow improvement after porta-caval shunt. Ann. Surg. *184:*289–302 (1976).

10 Spina, G.P.; Galeotti, F.; Opocher, E.; Cornalba, G.P.; Rota, L.: Modificazioni emodinamiche dopo shunt spleno-renale selettivo. Atti del XXX Congr. nazionale. Incontri Mediterranei di Radiologia, 1982, p. 354.

11 Pezzuoli, G.; Spina, G.P.; Galeotti, F.; Opocher, E.: Faut-il encore pratiquer la dérivation sélective pour traiter l'hypertension portale? Chirurgie *108:*513–518 (1982).

12 Belghiti, J.; Grenier, P.; Nouel, O.; Nahum, H.; Fekete', F.: Long-term loss of Warren's shunt selectivity. Archs Surg. *116:*1121–1124 (1981).

13 Maillard, J.N.; Flamant, Y.; Hay, J.M.; et al. Selectivity of the distal spleno-renal shunt. Surgery, St Louis *86:*663–670 (1979).

14 Reichle, F.A.: Portal hemodynamics after distal spleno-renal (Warren) shunt. Ann. Surg. *176:*195–198 (1972).

15 Galeotti, F.; Opocher, E.; Maruotti, R.; Baccaglini, U.; Spina, G.P.: Risultati del

trattamento delle emorragie da varici esofagee con la sonda di Sengstaken-Blakemore. Urg. Chir. Comment. *4:*447–448 (1981).
16 Maillard, J.N.; et al. Les anastomoses de fortune dans les hypertension portale par thrombose du système porte. Ann. Chir. *32:*693–697 (1978).
17 Spina, G.P.; Battaglia, G.; Galeotti, F.; Rigotti, P.; Opocher, E.; Pezzuoli, G.: L'encefalopatia nel paziente operato di anastomosi portosistemica: incidenza e fattori prognostici. Chir. Gastroenterol. *15:*568–573 (1981).
18 Busuttil, R.W.; Brin, B.; Tompkins, R.K.: Matched control study of distal splenorenal and porta-caval shunts in the treatment of bleeding esophageal varices. Am. J. Surg. *138:*62–67 (1979).
19 Galambos, J.T.; Warren, W.D.; Rudman, D.; Smith, R.B.; Salam, A.A.: Selective and total shunts in the treatment of bleeding varices. New Engl. J. Med. *295:*1089–1095 (1976).
20 Langer, B.; et al. A prospective randomized trial of the selective distal spleno-renal shunt. Surgery Gynec. Obstet. *150:*45–48 (1980).
21 Rikkers, L.F.; et al. A randomized, controlled trial of the distal spleno-renal shunt. Ann. Surg. *188:*271–282 (1978).

Prof. G. Pezzuoli, Clinica Chirurgica II, Ospedale San Paolo,
Via A. Di Rudinì 8, I-20142 Milano (Italy)

Subject Index